CONFLICT IN THE COSMOS

CONFLICT IN THE COSMOS

FRED HOYLE'S LIFE IN SCIENCE

SIMON MITTON

Joseph Henry Press
Washington, D.C.

Joseph Henry Press • 500 Fifth Street, NW • Washington, DC 20001

The Joseph Henry Press, an imprint of the National Academies Press, was created with the goal of making books on science, technology, and health more widely available to professionals and the public. Joseph Henry was one of the founders of the National Academy of Sciences and a leader in early American science.

Library of Congress Cataloging-in-Publication Data

Mitton, Simon, 1946-
  Conflict in the cosmos : Fred Hoyle's life in science / Simon Mitton.
    p. cm.
  Includes bibliographical references and index.
  ISBN 0-309-09313-9
  1. Hoyle, Fred, Sir. 2. Astronomers—Great Britain—Biography. 3. Expanding universe. 4. Cosmology. I. Title.
  QB36.H75M58 2005
  520'.92—dc22

                2004030638

Printed in the United States of America.

For Jacqueline, Lavinia, and Veronica

Contents

Foreword

As Simon Mitton makes abundantly clear in the brilliantly illuminating biography that follows, Fred Hoyle was the quintessential outsider, entering Emmanuel College Cambridge from an impoverished family background and with a distinct Yorkshire accent, and leaving Cambridge in a misguided huff 39 years later. But in between he ascended into the highest ranks of British science, almost single-handedly returning Britain to the top echelons of international theoretical astrophysics and setting it on the path toward excellence in observational astronomy. It is a stirring Dickensian story of an inquisitive, rough-hewn lad making the grade in the tightly traditional world of Cantabrigian academia, yet with the depths of a Greek tragedy where the flawed hero finally becomes an outcast.

Fred Hoyle first came into my purview when I, as a young editorial assistant at *Sky and Telescope* magazine, read about Hoyle's *The Nature of the Universe* (1950) in Frank Edmondson's highly critical review. I learned there that Hoyle was a proponent of a dubious continuous creation cosmological theory, but I did not discover that he had just invented the term "big bang." I first met him at the 1961 Berkeley meeting of the International Astronomical Union—I have a slide of him and Geoff and Margaret Burbidge from there. After that I occasionally saw him in Cambridge, England, or at other international meetings. The last time I spoke with Fred was at the American Philosophical Society in Philadelphia in November 1981, when he had come to sign the roll book of this venerable American academy.

Some of Fred's prolific popular writing found its way to my bookshelves. His prodigious output at the height of his creative activity included a book on Stonehenge (1977) and one on Copernicus (1973). Both are idiosyncratic and rather forgettable, yet each contains a memorable idea. By providing an alternative to Gerry Hawkins' explanation of the 56 Aubrey holes at Stonehenge, Fred Hoyle inadvertently demonstrated that rational reconstructions of how megalithic observatories could have functioned were by no means unique and hence neither his nor Hawkins' was necessarily the right solution for the prehistoric use of the site. And with respect to Copernicus, he argued from relativity theory that it was an empty question to ask whether it was the earth or the sun that moved—except that in the earth's reference frame the equations would have been too difficult for sensible progress. More useful was his *Ten Faces of the Universe* (1977); in one of the essays, dealing with the problem of the earth's energy futures, Hoyle argued that if our destiny was the collapse of civilization through overpopulation, then the sooner the better to avoid the extinction of the human race. This essay became required reading in my introductory science course at Harvard.

In the fall of 1977 I worked at the Cambridge Observatories on the *Source Book in Astronomy and Astrophysics, 1900–1975* that I was editing with Ken Lang. Hoyle was one of a comparatively few authors (including Eddington, Russell, and Hubble) who had more than one paper in the final cut for the *Source Book*. Many articles had to be abridged, as was especially the case with the famous 108-page "B+FH" paper on nucleosynthesis (coauthored by the Burbidges and Willy Fowler), but at least we could represent what was Hoyle's most cited and surely most influential paper.

Unquestionably one aspect of the frequently negative reception of Hoyle's earlier *The Nature of the Universe* were his hostile remarks on traditional Christianity, seen as gratuitous by his critics. Yet his comment near the end of his account that "It strikes me as very curious that the Christians have so little to say about how they propose eternity should be spent" was fair and thought-provoking. Within a decade rumors flew that nothing had shaken Fred's atheism as much as his prediction and subsequent discovery of the resonance state of the carbon nucleus, which makes possible the substantial abundance of that element and consequently the possibility of carbon-based life in the universe, including Hoyle himself. In fact, Fred wrote in the November 1981 issue of the Cal Tech alumni magazine that

> Would you not say to yourself, "Some supercalculating intellect must have designed the properties of the carbon atom, otherwise the chance of my finding such an atom through the blind forces of nature would be utterly minuscule." Of course you would. . . . A common sense interpretation of the facts suggests that a superintellect has monkeyed with physics, as well as with chemistry and biology, and that there are no blind forces worth speaking about in nature. The numbers one calculates from the facts seem to me so overwhelming as to put this conclusion almost beyond question.

From time to time I had sat with Fred, generally discussing one historical issue or another, but I never asked him about the role of a superintelligence in the universe. Yet a decade later he expressed much the same view in closing a discussion on the origin of the universe and the origin of religion:

> The issue of whether the universe is purposive is an ultimate question that is at the back of everybody's mind. . . . And Dr. [Ruth Nanda] Ashen has now just raised exactly the same question as to whether the universe is a product of thought. And I have to say that that is also my personal opinion, but I can't back it up by too much of precise argument. There are very many aspects of the universe where you either have to say there have been monstrous coincidences, which there might have been, or, alternatively, there is a purposive scenario to which the universe conforms.

I wish I had had something like Simon Mitton's stimulating, sensitive, and sympathetic story much earlier, as I now realize that besides this question that are many other issues that I would have liked to have discussed with him. Alas, the opportunity is lost.

But we do now have this well-researched evaluation, *Conflict in the Cosmos*, which should convince most readers that Fred Hoyle was truly one of the creative giants of twentieth-century astronomy, that he deserved a share of the 1983 Nobel Prize, and that if he had not been so outspoken, he may well have received it.

Owen Gingerich
Cambridge, Massachusetts
December 2004

Prologue

I n the nineteenth century, savants in England had continuously improved the science of astronomy, bringing it to a high professional level by the end of Queen Victoria's reign. In January 1820, 14 gentlemen and scholars, one of them the future computer pioneer Charles Babbage, had founded the Royal Astronomical Society, which received its Royal Charter from King William IV in 1831. Sir William Herschel, the discoverer of Uranus, the builder of giant telescopes, and the most accomplished sidereal observer of his age, became the society's first president. In 1834, the British government provided the society with suitable premises free of charge, an arrangement that continued uninterrupted until 2004. The universities of Oxford and Cambridge had important observatories from 1794 and 1823, respectively, together with endowed professorships. At Greenwich, the Royal Observatory, one of the world's oldest scientific institutions, flourished in the age of Queen Victoria and was noted for its accurate observations of the positions of stars. In 1884, an international conference in Washington, D.C., convoked by President Chester Arthur of the United States, selected Greenwich as the world's prime meridian.

By the early twentieth century, British astronomy could hold its head high: A small community of professionals at the Royal Observatories and in the ancient universities conducted world-class research.

Furthermore, they encouraged the development of astronomy in the dominions of the British Empire, with the establishment of observatories in Australia, Canada, and South Africa, where the practitioners still looked to Greenwich for guidance.

After World War I, observational astronomers in the United States began to advance on their British colleagues. By 1917, the Americans had the world's largest telescope, the 100-inch reflector at Mount Wilson, as well as far superior observing conditions and generous funding from curious philanthropists. Britain, meanwhile, had suffered the dreadful slaughter of her young men in the war, followed by a deep and prolonged economic catastrophe, the slump. Although there were golden years for the physicists, particularly those in Cambridge, who won a string of Nobel Prizes by prising open the atom and its nucleus for their inner secrets, astronomy was in decline. British astronomers suffered from a failure to invest in new telescopes at home, as well as a distinct lack of enthusiasm for the long sea voyages needed to reach the cloudless skies of His Majesty's dominions. In 1944, with the untimely death of Sir Arthur Eddington, British theoretical astronomy lost the brightest astrophysicist of his generation.

Defense research during World War II drained the young talent from Britain's dozen or so better universities. Pure research in physics and mathematics became the pursuit of the older men who had survived the killing fields of the Somme in the previous conflict. A certain amount of astronomical research continued at bomb-strewn Greenwich because of the British Navy's requirement to maintain the expertise of astronomers for navigational and time-keeping purposes. When the university scientists were finally released from their secret war, they returned to laboratories and faculty buildings overflowing with students whose education had been suspended during the war. Furthermore, yet another economic crisis led to severe shortages of food, fuel, and equipment. By the middle of the twentieth century, astronomy in the United States was far ahead of that of Great Britain. The 200-inch telescope at Palomar had commenced observations in 1948, research on atomic weapons had produced highly trained teams of experimenters and theorists, the electronics industry was thriving, and the fledgling computer industry was rapidly growing.

Fred Hoyle completely transformed British astronomy in the quarter century beginning about 1950. By turns, Fred Hoyle startled and charmed his public with a copious stream of new ideas, implausible theories, and an innovative approach to research. While most of his professional colleagues specialized in a single area of research, such as comets, or the evolution of stars, or the nature of sunspots, Hoyle's approach had more in common with the leading intellectuals of the Enlightenment. He regarded the entire celestial realm—the universe and all its contents no less—as being within the compass of his enquiries. Naturally, this would bring him in conflict with those members of the academy who had a strong sense of ownership of their respective specialties.

By his example, he led a despondent research community away from a fading tradition, directing them instead toward the extraordinary richness and diversity of the new astrophysics that began to emerge in the 1960s. In the 1970s, thanks in good measure to his inspiration, as well as his considerable skill during the short period in which he directed, at the national level, the policy for astronomy research, Britain again became a world leader in the astronomical sciences. Many other distinguished astronomers played an equal, and some a superior, role to Hoyle's in advancing research expertise, thereby recovering Britain's international prestige. Such a claim is certainly true of Britain's radio astronomers, Britain's space research community, and those applied mathematicians who chose astrophysics and cosmology as their research areas. It is also important to recognize that the professional community as a whole, working through both the Royal Astronomical Society and the funding agencies, persuaded successive governments to increase the provision for jobs, new telescopes, and expensive space missions. Hoyle's personal contribution to the rebirth of British astronomy came from his outstanding ability to think outside the box, and his unfailing loyalty to international collaborations at a time when many British researchers regarded American astronomers as the competition rather than an opportunity.

An enduring feature of Hoyle's character was that in every sense he never let setbacks, rejections, or political maneuvers deflect him from his own research agenda. He always had a deep conviction that in his "search for the truth," which is how he expressed his life's mission, any opponent should be able to provide a counterargument from experiment or direct observation. He declined all opposition based on semantic arguments invoking the philosophy of science, or the deployment of a paradigm, or appeals to common sense. After all, as an undergraduate he had learned general relativity and quantum mechanics from two masters, Eddington and Paul Dirac. From both professors, he understood that accepting what is obvious could not enable him to discover the nature of the physical world.

After 1950, Fred Hoyle was a very public figure at home and abroad. In the 1960s, "according to Hoyle" became a catch phrase in discussions of the latest news from the cosmos. His broadcasts for the BBC in 1950 were just extraordinary and brought him immediate fame as a gifted expositor. With his gritty Yorkshire manner, his ability to be picturesque using words alone, and the universe itself as his topic, he transformed the BBC's approach to academic lectures, persuading them of the benefits of a less donnish style of presentation.

His lectures for radio audiences set the prelude for a brilliant parallel career as a popular science and science fiction writer. In the former genre, he followed his hero Eddington, soaring over the latter as a truly best-selling author. In science fiction, his first novel, *The Black Cloud*, remains his best, having now acquired cult status: In 2004, an opinion poll conducted by the British newspaper *The Guardian* to find the most accomplished science fiction writers placed Hoyle in third position!

Unfortunately, Hoyle's university career came to an undignified end in 1972 when a series of decisions by the University of Cambridge gave him the profound impression that envious colleagues had conspired behind his back to push him out. The publicity resulting from his resignation delivered a seismic shock to British astronomy, but fortunately, the professional community, by now large and diverse, quickly persuaded their political paymasters that all was well.

Fred Hoyle had a very considerable influence on my own career. I cannot claim, unlike many astronomers of my generation, to have been attracted into the subject by his radio broadcasts. My own trajectory started in high school with evening classes and then the opportunity to use the telescope of the Leicester Astronomical Society. As an undergraduate at the University of Oxford, I was strongly attracted to nuclear physics, where happenstance brought me into contact with Rudolph Peierls, who had taught Hoyle nuclear physics (but I did not know that at the time). Like Fred Hoyle, I chose astronomy rather than nuclear physics as a career. In 1968, the then-recent discovery of pulsars became the magnet that drew me to the Cavendish Laboratory in Cambridge, where my doctoral research involved daily contact with the radio astronomer Martin Ryle, Hoyle's archrival. After I completed my Ph.D., I had the distressing experience of a very sharp disagreement with Ryle, which led to my resignation from his group. Quite quickly, Fred Hoyle threw me a lifeline by offering a temporary position in his Institute of Theoretical Astronomy. This post brought me into contact with his associates and students. A few months after Hoyle's 1972 resignation, it was my fortune to be appointed to a management position in the new Institute of Astronomy, which enabled me to refresh and expand my network of Hoyle contacts. But for Fred Hoyle's initial appointment, I would not have been able to progress to such a satisfying career at Cambridge.

In researching and writing this biography, I privately compared Hoyle to Copernicus, Newton, and Einstein. These latter were achievers on a timescale that repeats only over the centuries rather than over a generation for the next decisive move forward. What is extraordinary about Fred Hoyle's science is that his impact derives equally from when he was right and when he was wrong! Generally within academia, an erroneous paper is quietly forgotten: It receives the silent treatment. Hoyle's contribution to the advancement of astronomy derived much of its impetus from the way in which his colleagues recoiled at his notions. His opponents deployed enormous resources to wrong-foot him. In the twentieth century, no other figure in astronomy had to withstand for such a long period the criticisms of both the invisible college of astronomers worldwide and the parochial college of Cambridge practitioners. Hoyle's scientific life was truly a conflict in the cosmos.

An End—and a Beginning

On August 19, 1972, Fred Hoyle sat in his office at the Institute of Astronomy in Cambridge for the last time. His summer had been busy. A record number of academic visitors had come to the institute to benefit from summer conferences, collaborations, lectures, and discussions. He had fretted to make sure the institute would be financed securely for the next 5 years. Just 3 weeks earlier, the Institute of Astronomy had been born through a merger of two astronomy departments, after the university had decided to join the historic Observatories established in 1823 with the pioneering Institute of Theoretical Astronomy founded by Hoyle in 1965. Hoyle had been the head of Theoretical Astronomy for 7 years, but now he had a new boss, because the university had not chosen him as the director of the combined institute.

On a sultry afternoon with a threat of thunder in the air, staff members who were in the old Observatories, including myself, made the short walk along the path through the parklike grounds to the building that had been the Institute of Theoretical Astronomy—IoTA for short—to take their afternoon tea in the library. This wonderful Cambridge tradition gave the researchers and their students an opportunity to exchange ideas, and maybe wish a departing visitor a safe trip back to California or India. But this afternoon, Hoyle would not be joining his colleagues for tea. He had spent the past two weeks clearing his vast office of personal documents, books, and drafts of scientific papers, the result of 36 years of scientific work, much of it carried out in Cambridge.

Affairs of state had eaten into a lot of his time for the past year, and we had seldom seen him in the institute. He juggled the duties of being both vice president of the Royal Society and president of the Royal Astronomical Society. At the February meeting of the Royal Astronomical Society he had presented the Gold Medals. One of them went to Fritz Zwicky, of the California Institute of Technology, where Hoyle himself had made astounding breakthroughs some 20 years earlier. After the presentations of the medals, Hoyle had given a lengthy lecture setting out his ideas on the origin of Earth and evolution of life on it. This was not a mere summary of accepted theory, but his own views on what might have happened. On the research front he had worked on new theories of gravity and had published two demanding techni- cal papers. He was still as productive as ever, working up novel ideas into papers, one after another.

Whereas most mature scientists would be content with two or three papers a year, Hoyle was still writing books at a furious pace. There had been two this year. One reviewed the scientific case for interpreting Stonehenge as an astronomical observatory and eclipse predictor, ideas that had brought him into sharp disagreement with archaeologists.[1] The following year, 1973, would be the 500th anniver- sary of the birth of Nicolaus Copernicus, who had taken the decisive step of abandoning the Earth-centered universe of ancient Greek philosophy. Copernicus had published a new, but flawed, theory of the universe with the Sun rather than Earth at the center of the solar system. Hoyle was hard at work for his London publisher on a book celebrating the achievements of Copernicus as one of the founders of modern science.[2]

His year started in turmoil and conflict about how the govern- ment's funding for scientific research should be handled. Two national committees of scientific gurus had reported to the government that a greater proportion of the money should go to projects of direct interest to the taxpayer and that the cash should be channeled through ministries other than the Department of Education. Hoyle smelled a rat. If the proposals were accepted, then future funding would be in the hands of politicized decision makers rather than the scientists and professionals in education. Hoyle set about vigorously lobbying the scientific community to resist "the setting up of more bureaucratic machines."[3]

The following month, February, Fred had flown to Australia to discuss plans for the new Anglo-Australian Telescope. He traveled with a colleague who was on the inside track of political machinations at Cambridge. This colleague had dropped a bombshell into the conversation: The university appeared to be considering the appointment of someone other than Fred as the director of the combined Institute of Astronomy. The news was a heavy blow—so heavy that, during a stopover in Los Angeles, Hoyle dashed off a letter of resignation to the vice chancellor. Hoyle was already feeling completely through with Cambridge by this time because of its dithering and (as he saw it) inept approach to making senior appointments.

Now, on his last day, he felt deeply that his beloved institute had been stripped of its international character. In 1961, he had developed the idea of an institute devoted solely to the theoretical side of astronomy. In only a few days' time, his world-class theory team would be united with astronomers who built instruments and looked through telescopes. He supported both communities but could not come to terms with the university's decision to disregard him when it came to appointing a director. He was certain he had made the correct decision.

For decades, Hoyle had been the best-known astrophysicist in Britain. His output of technical papers was prodigious, but he never confined himself to the ivory towers of academia. A gifted popularizer, he could make the most profound intellectual puzzles into entertaining radio talks and lucid television programs. Fred Hoyle's broadcasts and books influenced many of us who were drawn into astronomy. Most years, he wrote a book, sometimes two. The sweep of his accomplishments as a writer covered a spectrum from popular books to technical monographs. Imaginative ideas that were too speculative for journal papers and serious books were cleverly developed to be aired in the guise of science fiction.

Despite his fame and standing, matters in Cambridge had somehow unraveled in the past year so that, as Hoyle put it, "Now I really did want to be done with it."[4]

Even when the tea drinkers had drifted back to their offices, Hoyle still felt unable to make a break for it, not wishing to endure the embarrassment of further handshakes, eye contact, or best wishes. By early evening the institute building was finally empty. The time to depart had come. He would head straight for the main door and be done with the institute forever. He took a last look round the office and, as an afterthought, picked up the inky blotter on the desk as a memento. He seldom used ballpoint pens, always choosing to write confidently with a fountain pen and rarely revising manuscript drafts. Just as he left the office, which was at the end of a long corridor and some distance from the front door, he changed his mind about bolting for the exit. Instead, he took a nostalgic tour of the building, his pride and joy. Though founded by him, funded by his pleas for cash, and populated by his handpicked team of research astronomers, it welcomed his presence no longer.

Ray Lyttleton's office was nearby. Ray was Hoyle's earliest collaborator and together they had done important work on the origin of the solar system. In a long career, Lyttleton had made the strategic error of continuing to work in the same area, defending his early papers from attack. This approach was very different from Hoyle's, which was to keep moving into new areas before someone else came up with a better idea. I remember Lyttleton in those days as a sad figure, still the holder of a prestigious professorship, but not a scientist anyone listened to or read anymore. Taking his last look, Hoyle glimpsed Lyttleton's work on a spoof paper, trying to prove mathematically that it should be impossible to ride a bicycle.

Cyril Hazard's office was next. Cyril was making a huge effort to identify very distant objects now known as quasars. Most astronomers believed quasars were at immense distances from our galaxy, but Hoyle thought they could have been ejected from nearby galaxies. Hazard's objective was to get accurate distance measurements (a part of the project on which I myself collaborated with Hazard) and to nail the problem one way or the other. Hoyle looked at Cyril's office. It was a wasteland of scientific papers, piled so high and loosely against the window as to form a miniature landslide that extended halfway to the door. His desk was submerged under a mountain of photographs of candidate galaxies and quasars that Hazard worked on with collaborators in the United States. The blackboard was a mass of random prompts: names, ideas, galaxies to investigate. Here was a great friend and collaborator who, as a result of Hoyle's resignation, now faced an uncertain future without a job.

Further along the corridor lay the study of Sverre Aarseth, who had joined Hoyle years earlier as a graduate student and had then been appointed to the research staff in the foundation year. He and Fred had a strong interest in chess and were intently following the Spassky-Fischer world championship in Reykavik. On the last day of the Institute of Theoretical Astronomy, Aarseth had organized a boat trip up the river Cam to the cathedral city of Ely. At the end of the trip, when IoTA was no more, Aarseth presented the former director with the official visitors book, which had been kept since 1965. Such a volume normally would have been regarded as part of the university archives, but Aarseth knew it would give Fred great pleasure to have this personal reminder of all the visitors he had attracted.

Hoyle walked past the open-plan library, always the first place in Cambridge to receive from the United States the world's premier research publication for astrophysics, the *Astrophysical Journal*. Only the institute had an airmail subscription. On one table lay a large electronic calculator, purchased at colossal expense in 1968, with a display that used extremely complex neon lights to show each number. The airmailed journal and the huge calculator kept Hoyle's theorists a step ahead of the other astronomers in Cambridge.

In the lecture room, there was a wide expanse of blackboards, made from state-of-the-art ground glass and amply supplied with no-dust chalk. During the construction of the institute, Hoyle had squeezed the budget to afford blackboards of the highest quality because advances in theory require countless hours of argument at the blackboard. Now he looked at the jumble of words, mathematical symbols, and general squiggles. These were the remnants from the last seminar to take place under his leadership. He could make out some patterns of thought, but others he did not recognize: Even his towering intellect could not comprehend all the advances of modern astronomy.

Now he turned and swung the big heavy door to the outside world. It shut slowly behind him as he set off on the short walk home. To his right was a pasture where three horses grazed contentedly, oblivious to any cosmology more complex than a flat field of grass. To his left was a building housing the IBM 360/44 computer that Hoyle had purchased as bait to lure summer visitors from the United States, who used it to model the evolution of stars. Ahead, he could see the new buildings of the department of physics, which had recently moved out of central Cambridge. A turn to the left and he was walking past the entrance road to the Observatories.

I have made this walk countless times. Today the William (Bill) H. Gates Building for Computer Science blocks the view to the department of physics, but the horsefield is still there. Hoyle was retracing the footsteps of great astronomers who had worked at the Observatories. One of them was Arthur Eddington, an astronomer Hoyle had always particularly admired. Eddington had been one of Hoyle's predecessors as the Plumian Professor of Astronomy at Cambridge. Back in 1919, he had confirmed an important prediction of Einstein's general theory of relativity: the bending of the path of starlight by the gravity of the Sun. In the 1920s, Eddington's research on the structure of stars was groundbreaking and was an important launchpad for Hoyle's earliest researches.

By now, Hoyle was almost home, walking past the cornfield on Clarkson Road, to his home in Clarkson Close. Here, he could reflect on 35 years of achievements in Cambridge. He had produced more than 400 scientific papers, a couple of dozen monographs and textbooks, and several best-selling science fiction novels. He was the first professional astronomer to use radio, and later television, to bring modern astronomy to a vast public. Early in his career he had worked on the structure of stars and had investigated how a star's appearance evolves over its lifetime. The greatest of his early achievements was showing how stars make carbon, the element essential for life in the universe. Professional astronomers regarded his work on the origin of certain of the chemical elements as a soaring achievement, standing above everything else he had done. He had spent a quarter of a century arguing with his colleagues about the nature of the universe, always rejecting the popular notion that a universe produced itself out of nothingness in a "big bang." These clashes had made him very famous in the public eye.

Two days later, Fred Hoyle and his wife Barbara hitched their caravan to their car and set out on a long journey to a remote beach in Cornwall. They stopped to drink coffee from a flask on the Downs south of Swindon. As Fred sipped the hot liquid, the penny finally dropped: He was suddenly emotionally overwhelmed by the magnitude of the break he had just made.

Within 3 months, Fred and Barbara sold the house and left Cambridge for good.

Fred Hoyle was born on June 24, 1915, at the home of his parents, Ben and Mabel, in the countryside of west Yorkshire made famous by the Brontë sisters and elegantly described in the novel *Wuthering Heights*. His paternal grandfather, George Hoyle, came from the neighboring county of Lancashire and settled in Gilstead, a village founded in Saxon times. George's son Ben originally emigrated to the United States but returned to Gilstead to help his widowed mother.[5] He started working in the wool trade and married his first cousin, Mabel Pickard, who had considerable musical talent.

Today, that part of northeastern England is a popular tourist destination, which it certainly was not during the First World War. Fred's birthplace, Gilstead, is a village 1 mile from the town of Bingley and 6 miles from the city of Bradford. The latter was a great manufacturing center for the textile trade, with hundreds of chimneys belching forth sulfurous smoke, as they had done for a century. The urban industrial landscape of northern England is beautifully captured for us in the early works of the artist L. S. Lowry, although he painted the factory environments of Lancashire rather than Yorkshire. William Blake's poem *Jerusalem* immortalizes the "dark satanic mills" of the Industrial Revolution:

And did the Countenance Divine
Shine down upon those clouded hills;
And was Jerusalem builded here
Among those dark Satanic mills?
—William Blake (1804)[6]

Gilstead is in one of the valleys invaded by the textile industry. The local building material is sandstone, which takes up soot readily, and the smoke from factory chimneys has blackened all the stone within an area of thousands of square miles. The surrounding fields were, and still are, small and divided by drystone walls rather than hedges, walls that march in serried ranks up the steep sides of the valley to the bare moorland towering above Gilstead. The village today is scarcely larger than it was 90 years ago, when a simple house number (34 Gilstead) sufficed for a postal address,[7] and it is still surrounded on three sides by open farmland.

The village of Haworth, 5 miles away, to which the Brontës moved in 1820, was almost visible from the Hoyles' house. Also near to Fred's first home is a humpbacked bridge over the Leeds–Liverpool Canal. Its famous system of five locks at Bingley is now a major tourist attraction. Fred and his playmates had plenty of childhood distractions: the edge of the moor with its heather, a nearby stream that by turns could be a trickle or a raging torrent, a wood with numerous possibilities for bird nesting, deep snow in winter, the open fields, and the village high street. Fred never cared for the typical activity of little boys, robbing birds' nests: In later life he recalled his childhood excitement at finding a partridge nest with 15 eggs, which of course he left undisturbed. He and his mates had impromptu football matches, games of cricket, as well as fights and scraps.

Parents of that era permitted their children great freedom just to get on with playing, particularly during the long school holiday in summer. It is inconceivable today that a child of $2^1/_2$ would be allowed to wander freely around the roads and other parts of the village, but in those halcyon days, cars were a fantastic rarity. Road-building machinery, such as steam-powered rollers, arrived in the village in late 1918, causing small children to gawp all day at the noisy and smelly contraptions.

Ben did fairly well as a businessman engaged in the Bradford trade of woolens and worsteds, and his wife had been a schoolteacher before her marriage in 1911. In those days, when a woman working in the public service married, her job immediately came to an end. This all changed of course with the outbreak of the First World War. With Lord Kitchener of Khartoum signing up a volunteer army of tens of thousands of men a week, married women were able to return to careers such as teaching. But Fred's mother decided not to do so, partly because young Fred was considered to be of "frail disposition." In the days before a National Health Service was established in Britain, families took care to keep the doctor's bills down, and Mabel Hoyle felt she could not leave Fred with a minder during the day.

Kitchener raised his fighting force with an astounding propaganda campaign, launched on August 7, 1914. Recruitment was to be local, into Pals Battalions, the members of which came from the same town. At first, this was highly successful, with up to 30,000 being sworn in each day. Within a month, Kitchener had half a million volunteers to supplement the quarter of a million in the regular army. Ben Hoyle was too old to respond to this call to arms, so he did not sign on with the First Bradford Pals. Initially, the recruitment sergeants followed strict guidelines on minimum height, chest size, and age. By May 1915, however, many of the restrictions were relaxed as the carnage on the Western Front relentlessly destroyed the young volunteers. The maximum age was raised to 40 and Ben Hoyle could now join the war. However, he chose not to go into a Pals Brigade because of his dislike of "bull" (following pointless orders). Instead, he enlisted with the new Machine Gun Corps, a decision that would astound young Fred in the 1920s when he learned all about machine-gun warfare from his father.

At the outbreak of war in August 1914, the tactical use of machine guns was unappreciated by the British military. Consequently, the Army went to war with its infantry battalions and cavalry regiments each having a machine-gun section of only two guns each. These were supplemented in November by the formation of the Motor Machine Gun Service, administered by the Royal Artillery, consisting of motorcycle-mounted machine-gun batteries.

A year of warfare on the Western Front proved that, to be fully effective, machine guns must be used in larger units and crewed by specially trained men. To fulfil this need, the formation of the Machine Gun Corps was authorized in October 1915. By this time, the menfolk of entire villages were being wiped out in the Pals Brigades on the Somme. Voluntary recruitment was going sufficiently badly that the government had already passed a compulsory registration act as a prelude to enforced service from January 1916. Ben Hoyle, realizing that he could be conscripted into a unit he did not care to join, immediately enlisted with the Machine Gun Corps while he still had a choice in the matter. The Corps operated at the national level, and care was taken to train the men well. Between 1915 and 1918, the Corps drew 170,500 officers and men, of whom 12,500 were killed and 50,000 seriously wounded. The men were organized into teams of eight for each gun.

So how would Fred and his mother fare? As a mother married to a serving soldier, Mabel received a government allowance of 1 shilling a day for both of them to live on. On this pittance, it was absolutely impossible to maintain their lifestyle. Fortunately, Mabel had studied music at the Royal Academy, first as a singer, but later transferring to the piano. For the rest of her life, she played the piano for 2 or 3 hours each day. From 1916, she worked in the evenings at a cinema in Bingley as the piano accompanist for silent films. All her training was in the classical tradition, and it did not suit the local cinema management that she would embellish cops-and-robbers movies with the music of Beethoven. She was fired early in 1917, and the cinema attendance promptly plummeted. Within a week, she was back because, when the manager enquired around the town, he was told, "We didn't come to see your films but to hear Mrs. Hoyle play."[8]

Fred's appreciation of classical music began at an early age from listening to his mother performing at the piano. To the end of his life he enjoyed the works of Beethoven, which she played frequently.[9] He ascribed his own failure to learn any instrument to the fact that he knew so much music by heart before he went to school and subsequently could not bring himself to play what he regarded as boring and trite pieces for beginners.

Fred appears to have become adept at counting and numbers from an early age. By day, his mother had plenty of time on her hands, and she had taught him the numbers before he was 3 years old. He rapidly developed a plodding system for memorization that went something like this: "One and six make seven, two and six make eight, so three and six make nine, so four and six make ten, so five and six make eleven, so six and six make twelve." He would recite these ditties in his head after being put to bed at night. By setting little problems for himself, he found he could remember earlier results, and so, step by step, he constructed multiplication tables for himself.

As the war on the Western Front dragged on, Fred's mother lived in daily dread of receiving a post office telegram expressing the grief of the British government in informing her that her husband had been killed in action. Through censorship of the press, the government was able to keep from public gaze the scale of the catastrophes being ordered by the generals in London. Nevertheless, word spread that the life expectancy of a soldier on the front line of trenches was only a few months, maybe as little as 3 for fighting troops in the Machine Gun Corps. In a year of nonstop combat, only 1 among 20 would survive, apparently by luck.

Of course, there were lulls in the campaigns while forces regrouped or the weather intervened, but being in the front line with the Machine Gun Corps was exceedingly hazardous. In part, this was due to the design of the Vickers portable machine gun, which was water cooled and fed with belts holding 250 rounds. The gun had to be fired in short bursts, rat-a-tat-tat fashion. The gunner would put his finger on the trigger, shoot while saying "mam and dad" to himself, and then stop so the water could cool and the belt feeders straighten the webbing. Continuous firing in broad sweeps would have boiled the water, or the webbing would have jammed, or the gunners would have run out of ammunition too soon. The section leader was under instruction to fire random bursts across the battlefield every 10 minutes or so, a tactic that gave the enemy time to pinpoint the location and eliminate it with heavy artillery. Ben Hoyle's survival method was to remain hidden as long as possible and fire only at critical moments. This tactic directly contradicted the commands from London.

On March 21, 1918, it was misty in the Somme Valley. The Germans brought in reinforcements from the defunct Eastern Front. At 4:40 a.m., they launched the second Battle of the Somme. Exceptionally heavy German shellfire hit all areas of the British front occupied by the Fifth Army, most of the front of the Third Army, and some of the front of the First Army. The assault concentrated on British artillery and machine-gun positions. The German infantry used a new tactic: The front line pushed forward fast, while follow-up units besieged and engaged British posts, many of which only fell when entirely surrounded, outflanked, and reduced to a hopeless situation. Almost certainly, the mad rat-a-tat-tat system of firing enabled the German patrols to locate where the machine-gun-post foxholes were located, despite the swirling mist. Ben Hoyle's team did not open fire. The German lines swept by, leaving unscathed seven recruits of ages 20 or so under the command of the 40-year-old Hoyle.

By evening, Ben Hoyle realized that the Germans had achieved a complete breakthrough. He gave his men a choice: Stay in hiding or attempt to pass back through the advancing German lines to the British. Hoyle and one man decided to lie low, moving only small distances and concealing themselves in craters. On March 28, they rejoined the British forces at Arras. The other six crew members were never heard of again.

Ben Hoyle was demobilized in 1919 and returned home to Bingley. Although Fred did not realize it at the time—he was too young to have formed such an opinion—his father's retelling of his experiences on the Somme was to have a profound and lasting influence on the development of his character. Ben had acquired a deep and utter contempt for those in authority: He hated the stupidity of the allied high command and the gross incompetence of the government. The writer J. B. Priestley, born in Bradford, enlisted at the outbreak of war, and only two of his boyhood friends survived. He wrote: "They were killed by greed and muddle and monstrous cross-purposes, by old men gobbling and roaring in clubs, . . . by strong silent be-ribboned asses" (1933, p. 166).[10] In the Hoyle household of the early 1920s, similar views were loudly and frequently expressed as the country plunged into a horrendous economic slump.

Whereas returning officers were sent to university, the government did nothing whatsoever for the veteran foot soldiers and their families. The poor standard of living and health care made it very hard to rebuild family life. The Hoyles had very little money. If Fred wanted to go somewhere he walked. That was his only means of transport. Children's shoes were then in a terrible state. The poorest shuffled around in wooden clogs. Fred had one pair of clumpy leather boots, which he wore in all weather, summer and winter. The leather soles lasted 2 months or maybe 3 at best, then they had holes that let in water like a sieve. They went through a cycle of repair and deterioration until finally the uppers disintegrated. Clothes were as bad: Working-class people in Yorkshire were now dressed, literally, in rags, while Bradford was a world center for the manufacture and export of luxury woolen textiles. There was no item of outerwear equivalent to a modern water-proof anorak. On rainy days a coat quickly got saturated with water.

Young Fred blamed the winter for his first really serious illness. He walked four times a day to and from school, a total distance of 6 miles, without sufficient clothing, and so it was inevitable that he contracted many of the standard childhood illnesses. Though giving the killers whooping cough and scarlet fever a miss, he did get a horrible infec-tion of the middle ear, causing a week of unmitigated pain.

One of the things Fred's father did soon after returning to Gilstead was to fix an old grandfather clock, which stood in the corner of the main living room. The clock became a talking point between Ben and those who came to the house to help repair it. Fred became conscious of much talk about time. But what *was* time, he wondered? One sign of his precocious grasp of numbers was that he quickly taught himself how to read the time from a clock, a remarkable feat for a 4-year-old.

By now Fred could write out the multiplication tables up to 12 times 12. Multiplying by 12 was a very necessary skill until decimaliza-tion of the British currency in 1970, there being 12 pennies in a shilling. But it would be a further 3 years before he would read, at the age of 7. Eventually, Fred's reading began all at once by following the subtitles on the silent and jerky films at the cinema. "Within a week or two I was reading generally," he later wrote.[11]

Ben Hoyle's business ventures as an entrepreneur in the wool trade initially prospered from 1919, and the family of three was well off by the normal standards for most people in the north of England. In 1910, Mabel Hoyle had purchased two houses, 3 and 4 Milnerfield Villas, which were then among the grandest properties in Gilstead. They rented out number 3 and lived in number 4. Ninety years later, this house is still owned by the Hoyle family, being the home of Fred's sister Joan, born in 1921.

The woolen cloth industry had become firmly established in the area in the mid-nineteenth century, when Bradford became home to a considerable number of Jews who had escaped persecution in Poland and the Ukraine. Some of these immigrants set up as merchants, importing wool and exporting finished cloth on a world basis. In the early twentieth century, the numerous mills of Bradford, supplied with plentiful local coal and abundant soft water from the moors, were combing, spinning, weaving, dyeing, and finishing cloth for the London and international high-quality markets. Flawless fabrics were demanded, but, of course, the weaving process could result in minor defects that only an expert could see. These slightly inferior bales of cloth were "seconds," sold at a fraction of the price of the perfect article. Ben Hoyle kept on friendly terms with several mill managers and, on his daily rounds, built up his stock of seconds for sale to merchants in Bradford and tailors in Leeds. The most prosperous times for Bradford had been before the First World War, and after the war many of the big merchant houses disappeared. In 1920, however, Ben Hoyle was still riding the wave of a trade boom, with Britain's exports doing well and consumers at home making up for the exigencies of wartime. His export trade reached as far as China.

Young Fred learned something of the scientific method from his father. As a wool merchant, Ben showed his son how to distinguish a strand of real wool from an imitation. Fake wool burns to a trail of ash. Real wool shrivels up, with a little ball of soot accumulating at the burning end. Decades later, Fred Hoyle still remembered this test. He made good use of it in a 1982 paper, to describe ways in which carbon in space could have arisen from the degradation of suitable organic material.[12]

In June 1921, Fred had his sixth birthday, which meant he had to commence school in the autumn. He could have started a year earlier, but Fred had no interest at all in going to school and, because he was considered "delicate," his parents delayed until the latest time the law allowed. They decided to pay for him to go to a private school.

How did Fred's mother and father make their momentous decision about Fred's first school? Ben Hoyle had been forced to leave school at the age of 11 and go to work in the mill. His wife, by contrast, had completed further education and had been a schoolteacher herself. Thanks to Ben's energetic business deals, they had surplus income, so they could afford to send Fred to the local Dame school. Fred's mother knew some of the parents and children at the school, and she almost certainly influenced the choice.

Dame schools were founded in early Victorian times, as little more than child-minding operations by elderly women. At first, the quality of education was often low, with children learning only basics, such as counting and the alphabet. All the children were in the same room and were taught by the one teacher, or "dame," herself. Everything was informal, with children able to come and go at will. By the end of the First World War, however, there had been considerable improvement. By then, education was regulated by local authorities; registers of attendance were kept; and the schools taught reading, writing, and arithmetic. Nevertheless, it was still the case that all children up to age 11 were taught in one domestic room by one teacher. One outdoor lavatory served the whole school. The fees varied, from 9 pence a week for schools in working-class districts, up to maybe 2 or 3 shillings in the better districts. It is likely, then, that Fred took a florin (a 2-shilling coin) or half-crown (a coin worth 2 shillings and 6 pence) with him every week. He would have been given a slate and chalk for writing and may well have been expected to work and play on the floor. There certainly were no schoolroom desks and chairs in this type of establishment.

Fred started in July 1921, with some enthusiasm because school was somewhere different to go, with new friends to meet, and the possibility of some mischief making with the older boys. According to Fred, the old dame taught two things: reading and Roman numerals. After only a week or so, the school closed for the summer period. At this point, economic and family factors intervened in a drastic fashion.

The postwar trade boom had greatly benefited the woolen trade. Then, in 1921, the government took a momentous decision to put the pound sterling back onto the gold standard (so that paper banknotes could in theory be exchanged for gold sovereigns at the Bank of

England). Astoundingly, the government also announced that the exchange rate would revert to that pertaining before the war, despite the fact that Britain had suffered higher inflation than its trading partners. To sustain an exchange rate that was too high, the government followed a severe deflationary policy, with high interest rates. Exports completely collapsed, and the textile trades suffered twice over because competitors in America and the British Empire had built their own industries up while Britain was at war. A great slump was underway, and the first businesses to be destroyed would be those of middlemen like Ben Hoyle.

Compounding his business difficulties, Ben had to cope not only with the birth of a second child, Joan, in 1921 but also with the serious illness of his wife, possibly severe postnatal depression. The family decided to spend the summer as paying guests at a house in Essex, in southeastern England. At the same time, they rented out their Gilstead house to tenants.

After a summer of playing in the fields and lanes of Essex, the time came for Fred to enroll in the local village school. This was across common land with grazing horses and gorse bushes. Almost immediately, Fred worked out a truancy system with another boy and neither of them spent very much time at the school. Ben Hoyle's intention had been to sit it out in Essex (where the climate is better than in Yorkshire), until the business situation improved. He had no idea, of course, that the British economy would not significantly recover for a further 12 years.

In November 1921, a letter from Fred's maternal grandmother brought them all back to Gilstead in short order. The tenants had done a "moonlight flit"! In those days, this was not uncommon: A fraudster had rented a house and opened accounts with various businesses in the town in the name of the landlord, with no intention of paying any of the bills. The impressive address, 4 Milnerfield Villas, no doubt helped in setting up the scam. When the creditors started pressing for their money, the scheming tenant vanished into the night.

So, the returning Hoyles had a huge commotion to sort out. The tenant had paid them no rent. But it was worse than simply a lack of income. He had run up an enormous bill at the grocery, hired a Rolls

Royce and chauffeur, cleaned out the best dress shop in Bradford for his daughter, and enrolled the younger children at the fee-paying Bradford Grammar School. While the Hoyles were dealing with this unwelcome mess, Fred did not go back to school. He finally returned in January 1922, by which time he had escaped about 18 months of the education he should have had.

After only a few weeks he had an argument with the dame. Fred was proud that he could multiply, and the class was working on Roman numerals. So what about these Roman numerals, Fred wondered? "How did the Romans do multiplication?" This question brought an evasive reply from the teacher: You simply learned Roman numerals, and you did not need to multiply them.

With that unsatisfactory response, Fred was soon playing truant again. He persuaded his parents that he really was at school and got a message to his school friends that he was ill at home. The dame was not too concerned, except perhaps at the loss of fees, and did not make any enquiry of Mr. and Mrs. Hoyle. The subterfuge worked for about 4 weeks. During this time, and future spells of truancy, Fred obviously could not play in the woods and fields of Gilstead. Instead, he purposefully walked most days to Bingley, where there were very few motor cars or motor vans. Horse-drawn carts were used for shipping the finished goods, and the roads were plentifully supplied with piles of steaming dung. Fred wandered the streets, poking his head into the mills and factories. These were filled with thundering and clacking looms driven by steam engines. Perhaps the operatives mistook the small boy for the observant son of a manager.

Some days he watched the barges on the Leeds–Liverpool Canal, the longest canal in Britain, which has 92 locks to take the route up and over the Pennines, and he spent time at the Five-Rise Locks, opened in 1774, where river traffic is raised through 320 feet. It was still a busy waterway in Fred's school days. In later life, Fred Hoyle attributed an early interest in mechanics and hydraulics to the observations he made during his periods of truancy.

His 1922 truancy scheme unraveled in late April, when Fred's parents ran into a schoolmate who expressed concern at Fred's serious illness! The fact that Fred was not observing the requirement to go to school was worrying for Ben and Mabel. As they explained to their son, the "law" required him to go the school, and the "authorities" could punish the family with fines. They even threatened Fred with the specter of them all ending up in the workhouse as a result of serial fines. In later life, Fred claimed to have been deeply puzzled by all of this.

> How was it, I wondered, that the law could pursue so relentlessly a harmless boy like me while permitting the tenants to do a flit with all those debts unpaid? After worrying at this problem like a dog with a bone I concluded unhappily I'd been born into a world dominated by a rampaging monster called "law," that was both all-powerful and all-stupid, a view which has resurfaced from time to time ever since.

A family council of war decided that Fred could choose his next school and delay starting until September 1922, by which time he would have missed two years of education. He elected for an elementary school in Bingley, Mornington Street School, which then had a tough reputation. The school is still there, greatly changed no doubt, and renamed Priestthorpe First School.

For the first year all was well, but in 1923 he moved up one standard into a class run by a tyrannical woman. By Fred's account, she beat the living daylights out of the 8-year-olds, relentlessly thrashing them with the cane. Fred avoided attending for much of the autumn of 1923 by feigning illnesses. He also managed two months off when his tonsils were removed. In spring 1924, the boys and girls were all given the task of bringing in wildflowers for the nature table. A list of about 20 flowers was specified, and Fred had no difficulty finding them all since he spent so much time roaming the upland countryside.

The teacher asked how many petals a certain flower had. Fred looked at his bunch: "Six, miss." The teacher said five. Now here was a dilemma. The flower in his hand definitely had six petals. Fred puzzled about the numbers. Maybe this wretched woman couldn't count?

"Pay attention Hoyle!" shouted the teacher, simultaneously smashing her flat hand over his unprotected left ear. Fred had not seen it coming and had no time to duck. The pressure on the eardrum and middle ear must have been enormous. In his adult life, Hoyle became deaf in his left ear and attributed it to having his ears boxed.

The little boy, sobbing piteously, made a dash for the door, pleading the need to "go out" (use the lavatory), which was never refused. He never went back to the school, despite the head's offer of an apology. It was back to kingfisher nests in the sandy banks of the streams, the mills of Bingley, and the Five-Rise locks by way of education. However, after 3 weeks or so of nonappearance, the local authority again came down heavy-handedly on Fred's parents, insisting that he should attend school. His mother must have been at her wits' end.

Then she remembered a tiny school in the nearby village of Eldwick, run by a teacher for whom she had worked before Fred's birth. The school took children ages 5 to 14, who were taught in three sets. Fred was accepted into the oldest set, where one teacher took all 5 years (ages 9 to 14) simultaneously. A record, dated July 9, 1924, survives of a visit to Eldwick school by His Majesty's Inspector of Schools:

> The teaching conditions indicated in the last report still exist. It is impossible to expect that the children can be properly taught. The main room is awkward in several aspects, and indeed is suitable only for a single class.
> For the teachers there is no separate room, cloakroom or lavatory. The playground is encroached by ladders, broken iron pipes, and coke and coal dumps. Many of the desks are too small for the children to use them.[13]

Plainly, under these conditions it would be difficult to win a scholarship to the grammar school. But things immediately took a turn for the better.

The young woman teaching Fred drummed it into him that he must try for a scholarship. In this she had the support of his parents. Ben Hoyle had won such a scholarship to the grammar school but, because of his father's death, had been forced at the age of 11 to work in order to support his mother and younger brother. The school record shows that in January 1925, Fred was promoted to the senior standards and, month by month, moved up the class to be nearer the front, where the bright pupils were always placed, to catch the eye of the inspector. Fred now started to do the arithmetic problems for the girls, which made him popular. The school gave homework, which Fred would do by staying on at the school, often not leaving until 8:00 p.m.

The walk home along unlit roads took about half an hour. Sometimes his mother and sister Joan met him halfway.[14] The stars made an impression on Fred in a way that is impossible today. In the 1920s, the clear night sky in the Yorkshire countryside, free from the effects of outdoor lighting, would have been breathtaking, particularly on nights when the wind blew the billowing city smoke away from Eldwick and Gilstead. About a thousand stars would have been visible at any one time on moonless nights. As Fred recounted in *Encounter with the Future*, he had no idea at age 10 what these points of light in the sky were, but, after one frosty night, he wanted to find out more about them.[15] So Ben Hoyle took his son to see a man living some distance away who had a modest astronomical telescope. And then on Christmas Day, 1925, Fred tore the wrapping paper off a present from his parents to reveal a brass telescope purchased from Clarkson's Optical Stores in High Holborn, London.[16] This was one of many ways in which Fred's parents gave him every encouragement in scientific pursuits. Later, they visited another man who had a microscope, so Fred could see what small objects looked like under magnification.[17]

About this time, Fred Hoyle began the interest in chemistry that would stay with him throughout his life. There was a chemistry textbook at home belonging to his father, and he began to set up experiments in the small kitchen. So started a train of events that would bring him to the grammar school and ultimately to Cambridge. His father had collected some of the standard equipment for a school chemistry laboratory, such as a bunsen burner and retort stands, as well as a few jars of chemicals. His box of chemical weights and his chemical balance are now in St. John's College, Cambridge.[18] He made gunpowder and staged explosions as well as setting up standard experiments described in the textbook. He recalls how, still only 10 years old, he saved his pocket money and went to a wholesale chemist in Bradford for concentrated sulfuric acid and complex glassware.

His greatest triumph was the preparation of phosphine (also known as hydrogen phosphide). This colorless gas has an odor similar to garlic, rotting fish, or stale urine, so his experiments stank up the kitchen for a week at a time. Phosphine gas is potentially very dangerous: It catches fire spontaneously in moist air, which is why Fred needed special glassware so he could exclude air from the apparatus. It is also a toxic gas that can cause thermal burns, and modern chemistry texts classify it as an extremely hazardous material. Joan Hoyle still remembers the scary explosions he set off.[19]

Fred failed his first attempt at the grammar school scholarship. He had taken the examination organized by the local education authorities, consisting of two arithmetic papers, one English paper, and an essay to write. He showed the papers (which he kept for the rest of his life) to his father, who could see that Fred had gotten the right answers to the arithmetic papers. However, 2 months later he got the failure letter. Then there was a twist to the story: The results for the entire county showed that massive cheating had taken place in some areas. It was decided that the Bingley candidates would be reassessed on appeal, and Fred was told to attend an interview with the head of Bingley Grammar School, Alan Smailes. At the interview he spoke of his chemistry experiments and described articles on astronomy that he had read in an old encyclopedia.[20] The chemistry master was astonished at the phosphine experiment, saying "Well, you'll not be doing that here." A letter announcing the award of a scholarship duly came, and Fred Hoyle started at Bingley Grammar School in September 1926.

From then on he walked 8 miles (13 kilometers) a day, in two round-trips, because he had to return home for his midday lunch. The distance was actually a fraction under 2 miles, which meant the local authority did not have to provide the scholarship holder with money for the bus fares; 5 years into the slump, Fred's parents could not afford the bus fares for their son. Fred's route took him past the railway, where once a week he saw the boat train headed for Liverpool to connect with the transatlantic liners. He sometimes wondered if he would ever go the United States himself.

He approached the grammar school with a completely different attitude. From the outset he respected both the chemistry teacher and the head, who was a Cambridge mathematics graduate. By the end of his first year he had progressed from a middling performance to being top of the class. He also started to read a lot more widely, borrowing books from the public library, including Arthur Eddington's *Stars and Atoms* in 1927. Eddington was at that time one of the two most famous astronomers in Britain; the other was James Jeans.

Stars and Atoms was one of Eddington's several popular books, which immediately captivated an informed readership. In this book, Hoyle read about many of the problems of stellar astrophysics at that time. One of these was the source of the Sun's energy. In the nineteenth century, astronomers had assumed that the Sun was contracting little by little and getting its energy from gas falling in toward the center. This theory worked provided that the Sun was only a few million years old. However, by the end of the century, geologists were fairly sure that the Earth, and therefore the Sun, was hundreds of millions if not billions of years old. Eddington had plunged into the new science of what is now called nuclear physics, proposing that instead of gravity as the source of heat, the Sun must be extracting energy from atomic nuclei. He had suggested in 1920 that four hydrogen nuclei might somehow merge to form one helium nucleus. Should this be the case, a vast amount of energy would be released, because one helium nucleus is slightly less massive than four separate hydrogen nuclei. The difference can be accounted for as matter that has been transmuted into pure energy, just as Einstein had predicted it would be. Eddington was also feted as the astronomer who had proved that Einstein's general theory of relativity is correct. On rereading *Stars and Atoms* today, I am impressed at Fred Hoyle's level of understanding in his early teens: The book is technically more demanding than most of the popular science being published today. It introduced him to some of the greatest contemporary problems in astrophysics.

By now financial concerns at home were desperate. Ben Hoyle could hang on no longer as the slump continued. He decided to quit the cloth trade while he still had some money in the bank, but of course there were no jobs to be had because Bradford and Bingley had been totally dependent on the now-defunct woolen industry. The family would have to depend on government handouts for the unemployed. To make ends meet the family would have to cultivate the garden and grow fruits and vegetables for the table. Fred earned a few pennies by singing in the local church choir, but this source dried up when his voice broke. The complete absence of jobs meant that Fred might just as well stay on at the grammar school because there was no prospect of his being the breadwinner. When he was awarded a book prize worth £5 in a national essay-writing competition, he immediately sold the books in order to get cash in hand.

Back at school, Fred worked diligently at chemistry, coming in top of the class term after term, but in physics he was less successful initially, perhaps because the experiments were quite rigidly defined. However, he was putting on real spurts in the grades to such an extent that, at the end of the third year, he and three or four classmates skipped over the fourth year and were catapulted directly to the fifth year. He had thereby entered the class that worked for the matriculation certificate 1 year early. After just one term in the fifth, he was placed first in physics and chemistry and second in all other subjects. What now motivated him was the rate at which he could up his game, tackling progressively more difficult work with increasing ease. Some nights he used his telescope to observe the stars, which held a growing attraction.

The matriculation examination, taken at age 15 or 16, was a watershed for all pupils in grammar schools. Achieving the coveted award of matriculation would open the door to university admission at 18; about one in six boys made this top grade. Below matriculation was the School Leaving Certificate, which the majority of candidates could expect, and that opened the door to a job market with no jobs and millions of unemployed. A few candidates got nothing for their efforts. Fred matriculated and, at age 15, he returned to the school in September 1915 to a completely different regime.

Now he was in the class being groomed for university entrance two years hence, and known as the sixth form. Bingley Grammar had eight teachers for the 200 pupils in all years, so there was no possibility of intensive coaching in the special subjects Fred was studying. Laboratory space was given over to the younger pupils. Despite carrying a high administrative burden, Alan Smailes taught Fred advanced mathematics. In these circumstances the teaching amounted to a succession of research projects. Formal lessons were few, and instead the pupils were given tasks: chapters to read, problems to solve, and experiments to set up. The education authority made a grant of £15 a year to matriculated scholars who stayed on at school. Fred's parents allowed him to keep the full amount on condition that he bought his own clothes and shoes. For the first time in his life he had cash to spare, and so he started going to orchestral concerts in Bradford, given by the Hallé Orchestra and conducted by Tommy (later Sir Thomas) Beecham.

He also joined the local chess club, winning silver trophies in 1932 and 1933.[21]

In the sixth form, Fred developed his reading interests. In addition to Eddington, he tackled several astronomy and physics books by Jeans and also books on relativity and the new field of quantum mechanics. At that stage, I doubt that he fully understood the content of the books, but he began to grasp that cutting-edge physical science often concludes with a mass of equations. In the hands of physicists, mathematics was a tool to be used to crack nature's secrets. He started to realize that science cannot be done just with words, experiments, and hand-waving arguments. Again he was fortunate: The chemistry teacher was buying university-level textbooks, out of his own pocket, for the sixth form, and this introduced Fred to the fascinating world of university research.

Alan Smailes gave Fred and the other university candidates extra tuition in mathematics by teaching them at his home in the evening. This was marvellous for Fred; he was now studying in a tutorial group, just the way mathematics was taught at Cambridge. Fred had his sights on getting into Leeds University to read chemistry, for which he would need a local authority scholarship. However, Fred failed to win a scholarship to Leeds in 1932, and so Smailes, who had graduated from Emmanuel College, Cambridge, formulated a different plan: He wanted Fred and one other pupil to have a shot at the Emmanuel College entrance examination.

In December 1932, Fred went to Emmanuel for interviews and the examination papers. Smailes had given him only 3 months of private tuition—definitely too little. Fred was now in competition with boys from the best grammar schools and private schools in the country; the latter competed fiercely for scholarship places at Oxford and Cambridge. As he ate his dinner in hall on the first evening, he hoped the mathematics paper would have two or three problems he could tackle with some ease, but he feared most of them would be incomprehensible.

What Fred did not know that evening was the contents of a letter Alan Smailes wrote on December 3, 1932, addressed to the Master of Emmanuel: "The boy Hoyle has insight, energy, and originality. He is young, and his appearance is diffident and awkward. But," Smailes continued, "you'll see he has real ability in science." Smailes raised the question of how to pay for a Cambridge education. He explained that Hoyle came from a good home, with thrifty and intelligent parents. "His father has suffered very much in the slump and is now without situation. The boy will have to win scholarships and I am confident he will do so." On outside interests, and the potential to add to college life, Smailes notes that Fred has little interest in outdoor games but is good at chess.[22] But as to future potential, something college tutors always look for, he added, "I believe he will turn out to be a swan, no matter what sort of duckling or gosling he now appears."

In any event, when he took the papers in the great hall at St. John's College, the chemistry paper went well. Because Fred had learned chemistry from a first-year university text, he found the questions comfortable. Physics seemed quite good as well, and the mathematics paper was tolerable. The practical examinations were in the Cavendish Laboratory, an imposing Victorian-gothic building in the center of Cambridge. The examination was very different from experimentation at Bingley Grammar, and he had not received any training in laboratory practice, the keeping of notebooks, and so on. He struggled pitifully.

The Master wrote to Fred on December 20, charmingly starting the letter, "My dear Hoyle."[23] Fred had come within sight of exhibition class (worth £40 a year): His marks were chemistry 60% (good), physics 51% (just below standard), practical 44% (weak), and maths 28% (very weak). Smailes got a letter a week later, saying the college had "no emolument for Hoyle," but he should try again the following year. Smailes read this letter with dismay: If Fred could get a local authority scholarship, then he would be fully funded for board, lodging, books, and tuition. But the West Yorkshire authority would only make the full grant to those who reached Exhibition standard. At this point, Fred felt truly beaten. He was on the point of giving up.

Alan Smailes continued to have faith in Fred. In March 1933, he put him in for the examinations at Pembroke College. In the oral examination, Fred became flustered when the don, Philip Dee, tied him in knots, saying disdainfully: "Is there any physics you do know?"[24] The results from Pembroke disclosed that this time Fred had reached Exhibition standard, but Pembroke could not offer him a place because there were more candidates who met the standard than places available. But this achievement gave Smailes an important idea.

Fred's champion wasn't willing to give up the fight too soon. Fred stayed on at school and sailed through the Higher Certificate examinations in summer 1933, meeting the requirement for a scholarship by a fair margin. The costs of an education at Cambridge (or Oxford) were significantly higher than at the other universities. Smailes pleaded with the West Riding County education authority to top up Fred's grant should he be accepted at Cambridge. They added £100 to the potential award.

Smailes now implemented his important idea, to go back to Emmanuel, citing Fred's success in the March exams. On May 20, 1933, he wrote by hand to the Master of Emmanuel, saying, "I am anxious to secure a place for my pupil Hoyle," who had now satisfied the requirements for academic standards and finance. Dr. Giles replied that the application was very late, but the tutors would admit him if he could get the scholarships. On June 17, Giles formally wrote to offer a place to read natural science, subject to finance. By August 23, Smailes could confirm that the West Riding scholarship was in the bag, and so Fred was finally on the way to Cambridge.[25] He had been fortunate indeed to catch the eye of Alan Smailes, about whom he later wrote, "Without the encouragement and determination of Alan Smailes, I would never have reached Cambridge."[26]

CHAPTER 2

To the Frontiers
of the Universe:
Training for Cosmology

Early one morning in October 1933, Fred's mother and father, and his 12-year-old sister Joan, gave the young scholar a hearty send-off. He began the long journey to Cambridge at Bingley railway station. As the local train chugged along the valley, Hoyle looked back at the forest of mill chimneys clouding the air with smoke.[1] On the hillsides the trees were already showing their autumnal colors. His traveling companions included half a dozen of his classmates from the grammar school. At the first stop, Shipley, they hauled their bags and suitcases to another local train, which brought them to Leeds. Here they bought tickets for the express to Peterborough, a town about 35 miles north of Cambridge. From there a local train clattered across the bleak fenland of eastern England. Hoyle must have found this flat landscape strikingly different from the moors and dales of Yorkshire. From his window seat he could see farmers ploughing with horses, an extensive drainage system of ditches and dykes, and the fourteenth-century Octagon Tower of Ely Cathedral soaring over the fens. After some 8 hours of travel, Fred and a throng of students descended from the packed train at Cambridge, where they found themselves on the longest railway platform in the country.

The returning undergraduates and the freshers dispersed either to colleges or to lodging houses. Emmanuel College had assigned young Fred a room in a shared house about a mile from both the college and the railway station. Of course, he had no money to spare for a cab. After the long journey he had no choice but to hump his heavy bags to Mill Road, a working-class district of Cambridge, constructed in the mid-nineteenth century to provide homes for railway employees. No doubt a friendly landlady greeted him with sweet tea and some biscuits. Then she showed him to the small room that would be his home for the next 9 months. It probably had a bed, one chair, a cupboard or closet for his clothes, a gas ring for boiling a kettle, and a gas fire.

The following morning he might have had porridge, toast, marmalade, and tea for breakfast, served in his room by the landlady. She would have sternly told him the rules of the house: No girlfriends permitted and be back no later than 11:00 p.m. He would need to buy an undergraduate academic gown and cap, to be worn at all times when he was out and about in "the town." If he caused any trouble, she would report him to the college authorities.

After breakfast he strode along Mill Road, over the humpbacked railway bridge, and past half a mile of shops and houses. This street scene in the 1930s was similar to that in thousands of towns all over the country, with every kind of small shop, a horse-drawn cart delivering casks of ale to a pub, an occasional motor van, and a few cars. Unlike in Bingley, there were no factories at all, but there were hundreds of cyclists, taking advantage of the flat terrain of the small city. To his right he passed the former workhouse of Dickensian times, by then the city's maternity hospital, and a drill hall for army volunteers; the country was already moving onto a war footing again, following Hitler's assumption of the chancellorship earlier in the year. Fred cut across the diagonal of Parker's Piece, the largest public open space in Cambridge, created as a common pasture in 1613. By walking a further hundred yards, he crossed the threshold of Emmanuel College for the first time as an undergraduate. He had an appointment to keep with his tutor.

Cambridge, like Oxford, is a collegiate university with no well-defined campus. The 30 or so colleges control all undergraduate admissions. A college is self-governing and autonomous; members of its academic staff are known as fellows, and the heads of most colleges have the title of master.[2] The master and fellows are the collective owners and trustees of the college. Students look to the college to provide scholarships, living accommodations, social and recreational facilities, a library, a chapel, a dining hall, and close guidance on their studies. Every student has a personal tutor (a fellow) who is responsible for managing his educational progress. In the 1930s, tutors advised on what courses of lectures to attend and books to read.[3] They made the arrangements for the face-to-face instruction in small groups for which Oxford and Cambridge are still renowned. As far as undergraduates were concerned, the role of the university was to provide lectures, laboratories, examinations, the university library, and the statutes and ordinances, which governed all aspects of academic life. Surprisingly, many faculties then lacked any central facility: Their academic staff worked at home or in the colleges. Of course, an important faculty, such as divinity, had proper premises, and the experimental sciences had their laboratories, but many arts and humanities subjects, as well as mathematics, had no departmental buildings in the sense we understand today. Almost all university teaching officers could then expect to be elected a fellow of a college, an arrangement that bonds the university with its independent colleges.

In Fred's day, undergraduates spent only 1 year literally living in college; he endured 2 years in cramped lodgings in "the town." Colleges normally owned the freeholds of their lodging houses and controlled them tightly. In the 1930s, more students could be seen in the college courts in the mornings than is perhaps the case now, because so many university lectures were held there in the absence of faculty buildings.

Hoyle was 1 of 1,400 undergraduates who entered the university in 1933.[4] Emmanuel admitted 117, 7 to read mathematics and 22 for natural science.[5] Among this cohort, Fred must have been very much an outsider, with his pronounced northern accent, baggy tweed clothes, thick boots, shabby academic gown, jobless father, and inexperience of polite society. However, the mix at Emmanuel should have been very much to his liking. The elderly master, Dr. Peter Giles, strongly encouraged applications from grammar schools and from boys with big potential but little money. One-third of the undergraduates in Fred's group were from inner cities and grammar schools. At Emmanuel, at least, there were other undergraduates from northern England, with strong accents like his.

Fred had a very difficult first encounter with his tutor, Mr. P. W. Wood. He hesitantly explained that he did not wish to pursue natural sciences after all and would prefer to switch to mathematics. This must have shocked Wood: Emmanuel had admitted him on the strength of his physics and chemistry and certainly would not have let him in for mathematics. But one of the features of the Oxford and Cambridge colleges is that individual tutors have the power to let undergraduates change subjects. Hoyle explained to Wood that, when he had accepted the place at Emmanuel, he knew virtually nothing about the details of the courses on offer. Over the summer he had learned that at Cambridge he could do science, but receive very little advanced mathematical training, or do mathematics, but learn precious little about science. From what he already knew in science, he believed strongly that a rigorous training in mathematics would be needed for a scientific career and that he would be able to learn the science he needed to know rather quickly. He felt that the college had mistakenly assumed that his future career might lie in medicine or engineering, which would require him to start with the natural sciences. Fortunately for him, 2 days later, Wood gave him permission to switch to mathematics. This change would be of momentous importance.

Wood's plan was for Hoyle to switch back to natural sciences in the second year. Alan Smailes had sent Fred's Higher Certificate results to Wood, who would also have known about Fred's poor showing in the practical part of the entrance examination. Wood probably calculated that the first year of the natural sciences course, with heavy emphasis on experiments, would be beyond Hoyle's ability, given his lack of training in laboratory methods. But Hoyle chose mathematics for an altogether more positive reason: He was convinced that it was the gateway to all of the physical sciences. After all, plenty of great physicists had elected to study mathematics as their foundation course. James Clerk Maxwell, for example, graduated in mathematics from Trinity College in 1854 and then returned to Cambridge in 1871 as the professor of physics. His major scientific achievement was his theory of electricity and magnetism. William Thomson, later Lord Kelvin (1824–1907), graduated as the top mathematics student in his year. He made a huge contribution to the branch of physics now known as thermodynamics. J. J. Thomson, who discovered the electron in 1897, was another example, as was John Cockroft, who received the Nobel Prize in 1951 for his pioneering work in nuclear physics. Cockroft had worked in a laboratory only 3 minutes' walk from Emmanuel College. When Hoyle opted for mathematics he was certainly not opting out of physics.

The system of examinations at Cambridge is known as the tripos. As Hoyle found out in his first week, the university uses many terms that are unique to Cambridge: bedders, proctors, bulldogs, syndicates, gyp rooms, and so on. The word tripos, meaning an examination, is very old. From the medieval period to the late eighteenth century, examinations at Cambridge were entirely oral and consisted of disputations: The candidate locked horns with an official opponent who sat on a three-legged stool in front of the proctors, the university officials charged with enforcing regulations. This seated person became known colloquially as "Mr. Tripos," from the Greek noun for the stool. During the seventeenth century, proper examiners were introduced. The job of Mr. Tripos was downgraded to that of a jester: He made a witty speech. This requirement soon changed to the recital of a Latin verse, which was printed and published. From 1748, the reverse of this sheet had printed on it the candidates' names, in order of merit, and it came to be known as the tripos list. Mathematics was the original tripos, and in 1772 a written examination was introduced in addition to the disputation.[6]

So what had Hoyle let himself in for now?

The Mathematical Tripos at Cambridge was regarded as the pinnacle of academic learning within the university, and colleges competed fiercely to get the best students. It was (and still is) organized into a 4-year course of study, thus requiring an extra year compared to other undergraduate programs at Cambridge at that time. The full name of the course was Pure Mathematics and Natural Philosophy, the latter being interpreted as including any branch of applied mathematics, theoretical physics, theoretical chemistry, and theoretical astronomy. There were examinations at the end of each year. These were Part I, Preliminary to Part II, Part II, and Part III.

An average student would complete Parts I and II and then graduate with the B.A. degree after 3 years. Only students of outstanding ability could stay on for the heavy demands of Part III. The brightest students could skip Part I and take the preliminary at the end of their first year. These students had to do Part III because the Cambridge B.A. required a minimum of 3 years' study. The third option, which was the scheme to which P. W. Wood agreed with Fred, was to take Part I mathematics and then switch to natural sciences Part II for the second and third years.

In the 1930s and 1940s, no lecture room in Cambridge was capable of holding the 150 undergraduates admitted for the Mathematical Tripos, so all lectures were given twice, with the students divided into a slow stream and a fast stream. In the first year the fast stream was composed only of the dazzlingly bright students, mainly from private schools, and older candidates who had already completed a mathematics degree at another university. Hoyle came into neither of these categories.

In his first year he swam in the slow stream, attending lectures in analysis, algebra, mechanics, electromagnetism, statistics, and geometry. The latter was a double course, and the lecturer was none other than P. W. Wood. Hoyle regarded Wood as a breathtaking wizard at the blackboard. He would write down an ordinary-looking equation, then make a circuit of the lecture room, keys jangling in his pockets, arms waving in the air, all the while talking about the equation. As the circuit took him past the board a second time, he would make some subtle alterations to the notation and take off on another orbit, passing by with a further round of changes. Thus, in a rabbit-out-of-the-hat style, he would finally arrive at a solution. Twelve years later, Fred Hoyle himself was giving the geometry lectures to the slow stream, with an aged Wood, keys still jangling, taking the fast stream.

In May 1934, Hoyle took Part I of the Mathematical Tripos and, to his surprise, he was placed in the first class, for which Emmanuel College rewarded him with an Exhibition worth £40 a year (equivalent to about $3,000 today). The college also made an award of 2 guineas (worth about $150 today)[7] to him and 32 undergraduates in his cohort for their examination performance. Five of the college's eight candidates in mathematics gained firsts—a brilliant achievement.[8] Having caught up on the mathematics side, it was now time to think of returning to physics and chemistry for the second year, commencing in October.

During the summer of 1934, Hoyle did an enormous amount of walking, staying overnight in youth hostels, first in the wilder parts of the North Yorkshire Dales and then in the Lake District. These jaunts gave him plenty of time to mull things over, particularly his decision on whether to transfer from mathematics to natural sciences. He now saw the choice as follows: By continuing in mathematics he could become a theoretical physicist, whereas if he opted for natural sciences he could be an experimental physicist. Cambridge was immensely strong in both disciplines.

Two factors appear to have swayed the decision. First, Hoyle recalled the outstanding Cambridge physicists who began as mathematicians, including Arthur Eddington and Paul Dirac, at that time two of Cambridge's most famous scientists. These men started in mathematics and reached the highest levels in physics. Second, he was feeling pleased with himself, having hauled up his position from the bottom of the slow stream to first class in 8 months. If he could do that in the first year, why not join the fast stream in the second year and work up to first class in that?

When Hoyle got back to Cambridge, he ignored the advice of his tutor and decided to stay with mathematics, this time in the fast stream and with the ambition of being in the lead. This was an astonishing turn of events because the fast stream was, in effect, a year ahead of the slow stream, so Hoyle would be cramming 2 years' work into just 8 months. He had done no preparatory reading in the long vacation, was now in the company of the most intelligent math students in the country, and was doing the hardest math degree in the world. Furthermore, by rocketing straight to the Part II examinations, he was excluding the possibility of changing courses at the end of the second year: To graduate he would have to take Part III mathematics.

The examinations at the end of that second year were very tough and he was placed in the second class, not the first. In fact it was a middling performance, so he comforted himself, in June and July 1935, with the thought that, if he had got halfway up the class despite skipping a whole year, then he might further improve in the final year and graduate with high honors. The goal he set was to get in the top 10 in Part III, thus guaranteeing a grant to continue at Cambridge as a research student. He would concentrate on papers in theoretical physics.

The mid-1930s were a wonderful time for Hoyle to contemplate working in theoretical physics. At the beginning of the century, Lord Kelvin expressed his view that physics was becoming a closed book, with no new discoveries to be made. But, some three decades later, the subject had been transformed by two new theories: relativity and quantum mechanics. In Cambridge particularly, this was heady stuff because Arthur Eddington had confirmed a major prediction of relativity and Paul Dirac was taking quantum mechanics forward. In both areas, Cambridge was an intellectual powerhouse, and Fred Hoyle was in its midst.

In 1905, Einstein had published a new theory of mechanics, called special relativity. Looking at how the laws of physics behave when objects are traveling at a constant speed in a straight line, Einstein brought together Newton's theory of mechanics and James Clerk Maxwell's theory of electromagnetism. This amazing feat required assumptions, or "postulates." One stated that the laws of physics are the same in all frames of reference that are not accelerating and another that, whenever the speed of light through a vacuum is measured, it is always the same. To appreciate what these two postulates mean, imagine a futuristic space mission to Mars, traveling at very high speed. The astronauts on board can use exactly the same rules and equations to describe what happens inside the space capsule as they did back home: There will be no change in the way electronic circuits function, for example, or the temperature at which water boils under normal atmospheric pressure. Furthermore, if the astronauts measure the speed of the light streaming from the Sun, they will get exactly the same result as their colleagues on Earth, despite the fact that they are traveling rapidly away from the Earth and the Sun, in the same direction as the sunlight.

Once Einstein had brilliantly reconciled mechanical and electrical phenomena, he turned his attention to gravity, and this meant he must tackle acceleration. To see why acceleration complicates the issue, imagine being back on that space mission to Mars. Let us suppose the spacecraft does not have any windows for its passengers and is going at a uniform speed in a straight line, so you feel completely weightless. You then take a nap and, to your great surprise, when you wake up you are no longer weightless! A force is pulling you to the floor. There are only two explanations. Either the spaceship has landed on Mars and you are feeling martian gravity, or the pilot has decided to pick up speed and your body is sensing the acceleration. With no windows it is impossible to tell which is the correct explanation. This was Einstein's greatest insight: The effects on matter of a gravitational field or of a uniform acceleration are indistinguishable.

Einstein worked on problems to do with matter, gravity, and acceleration for several years. In 1905, he proposed that matter and energy are equivalent: In principle, matter could be completely destroyed, with the release of a matching amount of energy. He used geometry with devastating effect for handling the equations of physics in four dimensions of space and time. In late 1915, roughly when Ben Hoyle was leaving England for the Western Front, Einstein published his general theory of relativity, a new understanding of gravity, going beyond Newton's theory.

An immediate achievement of the general theory was its correct explanation of a small discrepancy in the motion of the planet Mercury, a departure from Newtonian mechanics that astronomers had long been at a loss to explain. It also correctly predicted the extent to which starlight could be deflected by gravity from a straight line onto a curved path. Using observations he made at a total eclipse of the Sun in 1919, Eddington showed that Einstein's prediction was correct and Newton's was in error by a factor of 2. The announcement of the Cambridge eclipse results caused a sensation, not only among scientists. It brought home to the public the transformation of physics, by Einstein and others, that was overturning established views of time, space, matter, and energy. Einstein became the world's symbol of the new physics. Reporters took a perverse delight in exaggerating the incomprehensibility of his theory, claiming that only a genius could understand it. More serious thinkers took the trouble to study the new concepts. Fred Hoyle already knew of the work being done at Cambridge by Eddington and Jeans.

The ideas of quantum mechanics were much newer than those of general relativity, and the curious public did not accord its founding fathers the same acclaim as Einstein. But once again, in learning this new theory, Hoyle was fortunate because the leading physicist in the field, Paul Dirac, taught him quantum mechanics in the Mathematical Tripos.

Quantum theory deals with physics on the scale of atoms and their particles. In the early twentieth century it was known that electrons orbited the nucleus of an atom. But if the atom was pictured as a miniature solar system, with the nucleus as a sun and the electrons as planets, then ordinary mechanics broke down: According to classical theory, the electrons would spiral into the nucleus in a fraction of a second, emitting a puff of energy. The way out of this predicament was to postulate that the electrons in an atom could not have arbitrary amounts of energy. Instead, just certain values were permitted. This device meant that the electrons could only occupy specific "energy levels" in the atom and move between them like stepping from one rung of a ladder to another. A spiral track down to the nucleus was impossible. When an electron jumps between energy levels in an atom, it emits or absorbs a package of energy, known as a "quantum." This is the origin of the expression "quantum mechanics."

The new quantum theory had other advantages over classical physics. The wave theory of light, first developed by Christiaan Huygens at the end of the seventeenth century, had been very successful in explaining many optical phenomena. But by the early twentieth century, certain experiments could be understood only if light behaves like a stream of particles, which were dubbed photons. It became clear that light has dual qualities and that photons are "quanta" or "packets" of energy with wavelike properties. In the 1920s, physicists showed experimentally that particles with mass, such as electrons, could also behave like waves.

Another aspect of the quantum theory that flew right in the face of nineteenth-century physics was the uncertainty principle. On the atomic scale, measurement is a delicate process. Suppose we want to know the position and speed of an atomic particle. How accurately can we measure them simultaneously? In the case of large objects, such measurements are not a problem. A speed camera on a highway can record the speed and position of a vehicle to the accuracy demanded by the legal authorities. But suppose we want to specify the speed and position of one atom using an atomic-scale speed camera. Could we do so? No. In 1926, Werner Heisenberg, then working in Copenhagen, deduced that it would be impossible to measure speed and position simultaneously with high accuracy. In part this is because atomic particles do not behave like hard billiard balls but more like exceedingly small clouds.

The two theories together, relativity and quantum mechanics, had set physics on an entirely new course, and Hoyle was determined to learn more about them.

For 3 months in the summer of 1935, Fred Hoyle read and studied. At school he had read Eddington's *Stars and Atoms*, and now he was taking lectures from the great man. In 1933, Cambridge University Press published *The Expanding Universe*, which was an enlarged and illustrated version of a lecture Eddington had given the previous year at a General Assembly of the International Astronomical Union. In 1929, Edwin Hubble, working at the Mount Wilson Observatory in California, had demonstrated that many galaxies are moving away from the Milky Way. He found that the more distant a galaxy is, the higher its speed of recession. Hubble had been given a head start by Eddington, who had written in 1923 that the galaxies had a marked tendency to be receding from our solar system. Eddington had no way of measuring their distances, however, and it was not even clear then that those "spiral nebulae" were indeed other galaxies beyond the Milky Way. Hubble solved this perplexing problem by measuring distances and speeds, using the world's largest telescope at the time. The relationship between velocity and distance was easily explained by postulating that the universe itself was in a state of expansion. In Cambridge, there was much excitement about the expanding universe, which Eddington sought to explain theoretically in the framework of Einstein's theory of relativity.

The Expanding Universe was a popular account of relativity, intended for the masses. What Fred Hoyle selected was Eddington's far more substantial work, *The Mathematical Theory of Relativity*, which Cambridge had published in 1923, when the general theory of relativity was still regarded as very hard physics. In Cambridge, Eddington's lectures were reserved for the end of the final year, by which stage it was felt the students should be prepared well enough. Eddington was by now world famous, not just for confirming general relativity but also for his work on the structure of stars. With *The Mathematical Theory*, Fred had in his hands an absolute masterpiece of modern physics, crafted by a superb expositor. The fundamental research papers on general relativity had been written mainly in German. In *The Mathematical Theory*, Eddington gave the substance of the original papers of Einstein and others, but departed from the original form in which they had been presented to give a continuous chain of deduction, including many contributions of his own. Einstein said in 1954 that he considered it to be the finest presentation of the subject in any language.

Hoyle also selected for vacation study Paul Dirac's *The Principles of Quantum Mechanics*, one of the first textbooks on quantum mechanics. Much of Dirac's impact on physics, and the impression he made on students at Cambridge, is a consequence of the success of this textbook. Throughout, it emphasizes abstract mathematical reasoning rather than experimentation and has neither illustrations nor an index. It is one measure of Hoyle's intensely powerful mathematical mind that he should tackle such an austere monograph without any access to a tutor. From 1926, Dirac lectured on quantum mechanics to final-year mathematicians, in a room at St. John's College. He continued to do so each Easter term for a further 42 years; the lectures I attended in 1969 were his swan song.

Fred combined his reading with vigorous activity, climbing hills and mountains. Together with an old school friend, then at Imperial College London, Hoyle did a good deal of walking in the Yorkshire Dales and the Lake District. On wet days he pored over his books. Evening conversations with other walkers over a glass of ale informed the pair of chums that the Scottish Highlands were far more daunting than the tame hills of northern England. They determined to try a new challenge.

Neither Fred nor his companion had a car, of course. For the first climb they took the steam train to Stirling and then a bus to Callander. A track about 6 miles long brought them to a flat upland called Dubh Choirein, a mile and a half from the summit of Ben Voirlich. This peak rises to 3,224 feet. As he stood at the base of Ben Voirlich, Fred thought how very much the mountain looked like an arrowhead. The two chums ascended to the peak, then went down the steep decline on the other side, whence they hiked a further 4 miles to a youth hostel, "feeling pretty tired on those last miles."

Much recovered the following morning, they set out for Stobinion, enshrouded in mist. They navigated by compass, reaching a youth hostel by evening after another long trek. In the following days they ranged widely, tackling one Munro (a mountain peak above 3,000 feet) after another. They bused and tramped across moors to the Grampians, winding up in Braemar on Royal Deeside. This is the most mountainous part of the British Isles—the parish of Braemar includes 24 mountains exceeding 3,000 feet—and is the venue for the annual Highland Games. From Braemar they walked through the Cairngorms to Aviemore, now a major resort for skiing and rock climbing. A grueling 40-mile hike took them to Fort Augustus, at the head of Loch Ness, in the heart of the highlands. Now they pushed westward, eventually reaching Loch Duich, on the west coast and within sight of the Isle of Skye. Here they put their tired feet up for a week. No doubt, Fred used the break to catch up on his serious reading. He was never one to while away his time aimlessly.[9]

By the end of the summer, Hoyle's reading had grounded him well in two of the most important theories to emerge in physics in the early twentieth century. He returned to Cambridge ahead in his work, rather than behind, and he was also physically very fit, as a result of 6 weeks of hiking in the Scottish highlands. From then on, all through his life, it was always important to him to be physically fit. He noticed that his academic work went much better after exercise.

For Hoyle's final year of the tripos, teaching was in much smaller groups. This was because many Part II students had taken their B.A.

degree and left Cambridge.[10] But there was another factor at work, namely, the huge choice on offer. In the examination, the candidates would take papers for between six and eight courses, chosen from a list of almost three dozen. Because there was no department of mathematics, the lecturers had nowhere to meet their pupils apart from rooms in the colleges. Nevill Mott, who shared the Nobel Prize in physics in 1977, describing the poor facilities in the 1930s noted that mathematics students had nowhere to sit, "except in the rather small and squalid library."[11] At the end of lectures, the students found themselves out on the street, in all weather.

Despite the demanding physical conditions, these lectures in small groups were a tremendous opportunity for the students because they were in the midst of a great intellectual explosion in physics and were learning from people who were, or who would become, Nobel laureates in physics. Even so, the standard of lecturing was very uneven. Hoyle regarded Dirac as "the cleverest expositor in Cambridge," whereas "Eddington was the worst—I do not recall Eddington ever finishing a sentence. The trouble was that he had very little correlation between what he was saying and what he was thinking, although he had excellent correlation between what he wrote and what he thought."[12]

All the lectures were also open to any graduate students, who were often in attendance to hear the masters firsthand. So Hoyle found himself rubbing shoulders with research students for the first time. Frank Westwater, a graduate student who was 1 year ahead of him at Emmanuel College, taught Hoyle in his third year. Frank had performed brilliantly in Part III of the tripos, winning the Tyson Medal as top student in applied mathematics. Frank and Fred became good friends, unaware that events would conspire to bring them together again, first in the Second World War, and subsequently at the Institute of Theoretical Astronomy.[13]

The one thing lacking in this third year was further instruction in the latest mathematical methods. Effectively, Hoyle's training was in nineteenth-century techniques, which was fine for most of the theoretical physics he was doing. But in reading Dirac's textbook, he had come across mysterious twentieth-century mathematics, which he simply had to pass over. Hoyle's knowledge of the mathematical toolbox was certainly incomplete when compared to a Cambridge graduate a generation later. His own development was going to depend on learning outside the classroom.

However, it was not all work and no play. At the grammar school and as an undergraduate, Fred Hoyle did not participate in team sports, but he followed cricket avidly. At Oxford and Cambridge the more important sports in the 1930s were athletics, boxing, rowing, hockey, rugby, cricket, Eton fives, and lawn tennis. In the Michaelmas and Lent terms, there were competitive encounters between the college teams. The annual competitions between the two ancient universities were much more important than intercollegiate rivalries. The annual Boat Race attracted a huge following and those men who were selected for the Oxford versus Cambridge events were said to have earned their "Blue."[14] This curious expression arose after the 1836 Boat Race, when Cambridge had fixed a ribbon of light blue to their boat. Oxford adopted dark blue.

Although Fred did not compete in college sports, he was a frequent spectator at college ball games, particularly cricket. He did, however, pursue his interest in chess to a high level. In his first year, the university selected him for the team to play against Oxford, for which he won a Half Blue, less prestigious than the award for sports, but still a fine achievement. The following year he was secretary of the college Chess Club, though during his tenure it collapsed because he did not find a successor. His main physical activity remained walking and rambling. In his professional life, walking and mountain climbing were to become of supreme importance, affording many relaxing opportunities to reflect on new ideas, sift intellectual puzzles, and just generally wonder about the nature of the universe.

One of the traditional summer pastimes for Cambridge undergraduates is punting: powering and steering a flat-bottomed boat with a pole on the River Cam. Most of the punts on the river today are "chauffeur punts," crammed with tourists, but in the 1930s, undergraduates dominated the river scene in central Cambridge. Excursions farther upstream to Grantchester Meadows and the Tea Rooms there could be taken on the weekend. There, better-off undergraduates, who showed off in fast sports cars, had cigarettes, picnic baskets, champagne no doubt, and attractive girlfriends, perhaps up for the weekend from the "season" in London. Fred neither drank nor smoked—he could not afford to do so. Clearly, an undergraduate from the north of England, with an unemployed father, could not compete on equal terms with these toffs, so he developed a different approach to enjoying his Sundays on the river.

Imagine the disruption that Hoyle and two research students inflicted on the tranquil Cam in the early summer of 1936. The three friends took out a long-term rental on a heavy wooden canoe, powered by one-sided paddles. At the front of the canoe, one man delivered strokes always on the left, the middle canoeist always on the right, and the man in the rear alternated. This needed a degree of coordination not much less than that required in a college racing boat. The sole aim of the trio was to pilot the canoe as fast as possible, certainly very much faster than a punt. Hoyle recalled: "For two consecutive summers we prided ourselves on being the fastest craft afloat on the Cam. I can't be *certain* we were, but I can say that nobody ever beat us in a challenge. Our time from Byron's Pool to the Garden House Hotel became incredibly short."[15]

As an undergraduate, Fred also developed a taste for scaling the ancient buildings in Cambridge, undertaking many difficult (and forbidden) climbs into and out of colleges. Night climbing seems to be a tradition peculiar to Cambridge. It dates back to the 1890s, when students would get into college after hours by climbing over the back wall.[16] Soon this stunt was deemed too tame, and the students turned to more challenging adventures. The 1920s and 1930s were the golden age for this unique form of student activity. Colleges such as King's, St. John's, and Trinity, with their crenellated walls, soaring chapel spires, fearsome gargoyles, and lead drain pipes, offered plenty of "climbs." Today, any student attempting such an escapade is immediately caught on security cameras. Hoyle practiced for college climbing by doing a circuit of the room in his lodging house without touching the floor at all, hanging by his fingertips from the picture rails. No doubt, it was also good practice for scaling the more difficult mountains of Scotland.[17]

Lectures for final-year students finished in about April, to allow time for revision and special coaching before the examinations. However, lectures for postgraduates continued, and it says something for Hoyle's enthusiasm for hard physics that he began to attend these advanced lectures even though his examinations loomed. Then, on a whim one Sunday, he strode off on a 40-mile hike; the aftereffects reduced him to hobbling to lectures in canvas shoes the following day. The punishing walk seems to have been a strange way to prepare for crucial examinations!

In the last 2 weeks before the finals, he spent the mornings working and the afternoons creating havoc on the river. The weather in the examination period in June 1936 was exceptionally hot, but Hoyle was so fit that the heat did not bother him unduly. Afterward, he felt he had done reasonably well on the papers.

In the run up to the examination period, Hoyle had made his decision about the future: He wanted to be a nuclear physicist. Although he could not know it at the time, the late 1930s were the culmination of a golden age of nuclear physics in Cambridge at the Cavendish Laboratory, where Joseph J. Thomson had discovered the electron in 1897. This was followed by James Chadwick's discovery of the neutron in 1932. In the years up to 1911, Charles T. R. Wilson developed the first cloud chamber, a device that enabled laboratory physicists to follow the tracks of particles emitted during the radioactive decay of uranium and radium. In 1919, Francis W. Aston discovered the phenomenon of isotopes, in which the same element has more than one nuclear form with different masses. He found that neon has two distinct atomic forms with different masses and shortly afterward found two similar variants of chlorine, chemically identical but differing in mass. The year before Hoyle went up to Cambridge, John D. Cockroft and Ernest T. S. Walton had perfected the first machine capable of accelerating atomic particles to very high speeds. They shot to world fame by "splitting the atom" as the newspapers put it, an accomplishment for which they shared the Nobel Prize in 1951. In fact, they were the first to smash an atomic nucleus apart by entirely artificial means, for although Ernest Rutherford, working in Manchester in 1917, was the first to split the atom, he used a natural source of radioactivity to do so.

The physics community in Cambridge and elsewhere was extremely keen on gaining a better understanding of the inner workings of the atomic nucleus. The previous three decades had seen remarkable progress in understanding how atoms work, largely thanks to the quantum theory. Could the nucleus, similarly, be made to yield its secrets to the theorists? Hoyle had temporarily dropped his childhood resolve to learn about the stars and the universe. Instead of reaching out to the vastness of the universe, he wanted to work on the world inside the atom.

Hoyle may also have been drawn to nuclear physics by the patriarchal figure of Ernest Rutherford, who had been elected Cavendish Professor of Physics in 1919. Rutherford had discovered the nature of the atomic nucleus in 1906, for which he won the Nobel Prize in chemistry in 1908. By the early 1930s, Rutherford had created for Cambridge the greatest nuclear physics laboratory in the world. He was the first to hint at the enormously large quantity of energy stored in an atomic nucleus, after he showed that the internal forces keeping a nucleus together must be millions of times more powerful than the bonds between an atomic nucleus and the surrounding cloud of electrons. This speculation led to another: Could the nucleus be the source of energy in the Sun and other stars?

One of Rutherford's charming habits was to walk around the laboratory on a regular basis, meeting the research students and questioning them rigorously on their experiments. Most of the apparatus in the Cavendish Laboratory was homemade, and so research students needed skills in glassblowing, metal work, carpentry, and electronic circuitry. Work at the laboratory bench never attracted Hoyle; he wanted to work on the mathematical theory of nuclear physics. In those days, the potential research student first had to identify a supervisor, who would be officially appointed by the university to guide the student in his research. Hoyle was going to need a theorist, rather than one of Rutherford's string of Nobel laureate experimenters.

Unfortunately, Cambridge's two brilliant theorists, Paul Dirac and Ralph Fowler, spent a great deal of their time overseas and so were unavailable to take research students. Undaunted, Hoyle identified Rudolf Peierls (1907–1995) as a possible supervisor. Peierls had been born in Germany, the son of a Jewish banker. He studied nuclear physics under quantum theory pioneer Werner Heisenberg. In 1931,

Peierls hastily married a student he had met at a scientific conference at Odessa in the Soviet Union. He managed to get her out of Stalin's Russia by train to the West. In 1933, they came to England to escape the anti-Jewish excesses of the Nazis and they initially settled in Cambridge.

With the examinations over, Hoyle went to see his potential supervisor in June 1936. Peierls, who had not met Hoyle before, hesitated. He said he would consider the matter. Being relatively new to university teaching at Cambridge, he was stalling while he made enquiries of his colleagues in the Cavendish about young Hoyle. They could not say anything about him, apart from asking "since he had failed to apply for a grant from the government,[18] how was he going to support himself?"

Fred came up with a novel solution to the finance problem! The West Riding of Yorkshire, which had awarded him a very generous local authority studentship, was prepared to extend studentships to a fourth year for exceptional reasons. Hoyle's Emmanuel tutor, P. W. Wood, was prepared to write to Yorkshire, citing exceptional circumstances, to get the funding for the first year of research toward a doctorate.

At Cambridge, the interval between the end of examinations and the graduation ceremony is very short. In the 1930s, as now, it was as brief as 3 weeks. Most undergraduates therefore stay in Cambridge between the end of examinations and the announcement of results and enjoy themselves. This is the season of the college May Balls— perversely always held in June—but Hoyle did not remotely have the financial means to participate, nor did he have a girlfriend. But that was of no consequence to him because he did not want to hang around in college every night with the aristocrats in tailcoats and their debutante partners in ball gowns.

One summer day, Fred and a group of friends hitchhiked to Devon and Cornwall, a distance of about 300 miles. On the first day they succeeded in flagging down a Rolls Royce, whose lady owner insisted on taking them to a stately home for a sandwich lunch. Out in the West Country they got lucky with the weather and spent their days swimming and sunbathing. Toward the end of their trip, a telegram came from Cambridge with the examination results. Fred Hoyle was placed in the top 10, gaining honors with distinction, so he had achieved his objective, and his future as a research student was now secure. Indeed, he exceeded his goal because the university awarded him the Mayhew Prize (jointly with another student) for coming in top in the applied mathematics papers. In effect he graduated as the top theoretical physicist in his year, an outstanding result. He returned to Cambridge in triumph a few days later to receive the B.A. degree. On this visit, Peierls confirmed that he would definitely take on Hoyle's supervision, with immediate effect.

Peierls and his wife had arrived at Cambridge in April 1933, just 6 months ahead of freshman Fred Hoyle. Scientists, particularly Jewish ones, were starting to leave Germany in considerable numbers. The Rockefeller Foundation had awarded a 1-year fellowship to Peierls, who spent the first half of it working with Enrico Fermi in Rome and the second 6 months with Paul Dirac in Cambridge.[19] At this time, Fermi was working on radioactive decay.

At the beginning of the century, Rutherford had undertaken pioneering work on radioactivity, showing that uranium and its products give out three distinct types of radiation. In early chapters of a 1904 book, he called them alpha, beta, and gamma radiation, terms still used today.[20] Alpha "radiation" is in fact a stream of particles—the nuclei of helium atoms consisting of two protons and two neutrons. Beta "radiation" also consists of particles, in this case electrons, produced when a neutron inside a nucleus changes to a proton. Gamma rays are the most powerful form of electromagnetic radiation, similar in character to light or x rays but more penetrating and potentially dangerous.

Fermi had started his career with stunning discoveries about how the electrons in an atom interact among themselves. But by the time of Peierls's visit, his attention had switched from the electron cloud swirling round the outer part of an atom to the nucleus itself and particularly to the phenomenon of beta decay. Fermi was asking how the nucleus could eject an electron: What was the underlying mechanism of disintegration that causes a neutron to transmute to a proton and an electron? Conversations with Fermi led Peierls to take an interest in the theory of beta decay.

Peierls's Rockefeller grant would last until October 1933. In April 1933, he turned down a firm offer of a good position in Hamburg, wanting to stay out of Hitler's Germany regardless of the cost.[21] The economic situation in Britain was still terrible, and the last thing British universities needed was a surge of job seekers from Germany. However, the refugee scientists were treated warmly and with compassion. Rutherford, for example, took a lead in organizing the Academic Assistance Council, to provide funds while the émigrés found their feet. At the national launch on October 3, 1933, Rutherford packed the Royal Albert Hall in London. By 1936, this program had run its course: Rutherford's initiative had found academic posts for 363 scholars, and a further 324 were being supported. A fund-raising group local to Manchester meanwhile gave Peierls a position for 2 years at Manchester University.

In 1935, by a curious twist of fate, Peierls was offered more secure funding back in Cambridge. The Cavendish Laboratory did far more than just atomic and nuclear physics. It was a world center for low-temperature physics and the physics of strong magnetic fields. The person in charge of the low-temperature physics laboratory was a Russian, Peter Kapitza. In October 1934, while on a home visit to Moscow, Kapitza was arrested and detained. On October 30, the Russian authorities decreed that all scientific manpower was needed in the mother country, and no further leave would be granted to work abroad. Kapitza's passport was canceled immediately. In 12 months of high drama in 1934–1935, Rutherford was deeply involved in exasperating negotiations with the Stalinist authorities over the future of Kapitza's valuable apparatus at Cambridge. It was eventually sold to the Russians at what was effectively cost price and shipped to Kapitza. Rutherford used the budget that had been set aside to fund Kapitza's salary for two research fellowships, one of which he awarded to Peierls.[22]

In 1935, the outward appearances of the Cavendish Laboratory belied its world leadership in physics. The Seventh Duke of Devonshire, whose family name was Cavendish, founded the laboratory in 1871. The building is in Free School Lane, a narrow street at the back of Corpus Christi College. The facade is in the style of the Victorian Gothic Revival, and the main entrance is a gatehouse, with a cobbled archway wide enough to take a horse and cart. Newcomers were generally quite shocked at the dingy passages and gloomy interior. Today, the Maxwell Lecture Theatre, opened in 1873 and hardly touched by the university since, seems a depressing environment in which to teach and learn. The fact is that the university was not generous in funding experimental physics research: Rutherford's reign is still talked of in Cambridge as the "sealing wax and string" era.

Today, research students go to the modern Cavendish Laboratory, on the western fringes of Cambridge, where they are allocated working space and access to computers. Importantly, they can interact with each other and with the academic staff generally, as an integral part of their training. Hoyle's postgraduate experience, by contrast, would hardly be recognized by a modern graduate student. For a start, he had to work in college rooms. Theoretical physics came under the jurisdiction of the Faculty of Mathematics, not the department of physics. So, in some sense he was an outsider, although he probably did not think about it very much at the time.

Famously, the Cavendish was thought to be hostile to theoretical physics in this period. Paul Dirac and Ralph Fowler were members of the mathematics faculty, for example. Rutherford himself placed a great emphasis on simplicity and had some impatience with complex theory: He preferred the "sealing wax and string" to "chalk and talk." There were no formal working relationships between the camp of theorists in mathematics and the tribe of physicists in the Cavendish. The former worked in colleges, the latter in the laboratory. Dirac and Fowler were difficult to find and often away. In any case, Dirac was always impossible as a conversationalist. One saving grace was that Ralph Fowler had married Rutherford's daughter, which gave the theorists a line to Rutherford through the son-in-law. Hoyle was now on the mathematics–physics divide at a time when the mathematicians and the physicists were beginning to come closer together as a direct result of an upsurge in theoretical nuclear physics.[23] And so, at the outset of his research career, he was aware of the suspicion that some of the physicists (though not Rutherford by this time) harbored toward the theorists. This is a theme that would recur throughout his Cambridge years.

Another important respect in which the social dynamic at the Cavendish differed from a modern university department was the absence of a place simply to meet and talk (a tea room, perhaps). The only departmental gathering was the weekly physics colloquium, and Hoyle paints a frightening picture of how this operated.[24] All the academic staff and their research students were expected to attend, and there appear to have been occasions that could open up serious and hurtful rifts. Outside the department, there were two private clubs for physicists, meeting in college rooms. Peter Kapitza had started the Kapitza Club, which mainly discussed experimental physics, while those with a mathematical bent had the Del-Squared V Club (Δ^2V). Both clubs followed an informal pattern: the presentation of a paper by a member or a visitor, followed by animated discussion. Membership was limited and by election only; as a new research student, Hoyle was not eligible to participate.

Doctoral students, particularly in their first year, are set research tasks on topics familiar to their supervisor. Peierls's research activity between 1933 and 1936 would directly influence Fred's first projects and papers, and he had an excellent project lined up for Fred. "You should improve the theory of beta decay of the nucleus," he was told. To which Fred replied, "I'll first have to learn what beta decay is all about."

The atomic nucleus is made of protons, which carry a positive electrical charge, and neutrons, which have no charge. More than 110 chemical elements exist, and what makes each one distinct is the electrical charge carried by the nuclei of its atoms. For example, carbon has six protons, nitrogen has seven, and oxygen has eight. With the exception of hydrogen, the nucleus also has neutrons. The stable version of carbon has six neutrons; because there are six neutrons and six protons, this version is named carbon-12, and it is the commonest form. Just over 1 percent of carbon in the natural environment has one additional neutron. This type of carbon is carbon-13, and it is also stable. But if we add another neutron, to make carbon-14 (six protons and eight neutrons), the nucleus is no longer stable. Carbon-14 is "radioactive." There is a tendency for one of its neutrons to convert to a proton and emit an electron, which turns the nucleus from carbon to nitrogen. This process is called beta decay. The three versions of carbon are termed the isotopes of carbon.

An atomic nucleus does not contain any electrons, just protons and neutrons. When the unstable form of carbon decays, one neutron changes into a proton plus an electron and a neutrino, an almost massless particle. The proton remains bound in the nucleus, but the electron and the neutrino are both ejected. It is important to understand that the emitted electron is not something that was always in the nucleus and then carelessly set free. Rather, it is created by the transition from neutron to proton and immediately fired out. Fermi, with whom Peierls had just spent 6 months, had, in 1934, published a simple theory of beta decay.[25] Fred's first task was to improve the match between what the theory said and what the experimenters were measuring.

In Hoyle's first term as a research student, the department decided that, as part of their training, each student would be allocated a section of a huge article on nuclear physics that had just appeared in the journal *Reviews of Modern Physics*.[26] They had to study one of its eight sections and give an oral summary of the main findings. Hoyle was the only first-year student to be assigned a section, the others going to more experienced second-year and third-year students. Hoyle was given the seventh section, on beta decay and nuclear forces, the topic of his doctoral research. Peierls offered to help Fred in the course of the term. Of course, there were no photocopiers, so the eight students had to share a single copy of the article, which must have been inconvenient for detailed study.

The weekly sessions started badly, Hoyle saying that "a great deal of blood was spilt," and the place became a torture chamber to taunt the theorists. No less a figure than Eddington was once crushed by the Cavendish cabal in this colloquium setting. Eddington was never good at taking questions from the floor and had failed to withdraw gracefully when an error he made was pointed out to him. At the first session on the nuclear physics review, the big shots pounded a third-year student, Maurice Pryce, producing a fracas that set the tone for the remainder of the term. Hoyle was in a sticky situation because the 26 pages in his assignment bristled with technical detail that he could not explain in simpler language. When his turn came at the bear baiting, he withered under the questioning. At the end of the session he had not got through all the material, and he was told to present the remainder the following week. This depressing demand turned out to be valuable advice—although he did not appreciate it at the time—because Peierls advised him to spend the whole week reading the paper very carefully and treating it as a reading project.

The full paper runs to 150 pages in double columns. It draws its material from 246 research papers and books and contains 253 equations and derivations. It is effectively a complete textbook, at postgraduate level, on everything that was known about nuclear physics in 1935–1936. Written in an accessible and direct style, it sets out with considerable clarity a number of the problems on which Hoyle would work in his early career. In particular, the paper draws attention to discrepancies between the energies of emitted electrons as actually measured experimentally and the predictions of theory. The paper hints at how these discrepancies might be resolved, without setting out a solution. The mathematical tools (matrix algebra, vectors, tensors) in the paper would have been meat and drink to a practitioner of Hoyle's intuition, but how was he going to explain them to the physicists at the colloquium?

For a week, Hoyle practiced his presentation, trying different approaches on the blackboard and rehearsing the arguments. He invented illustrations and diagrams, visualizing equations as pictures. This was truly the first time in his life that he had to think about not just absorbing new knowledge but also explaining the findings so that others with a much weaker mathematical background would understand them clearly.

He approached the second session boldly, and with northern bluntness pressed his advantage against the professors, who had not read the paper in detail. The second session went splendidly and launched a friendly relationship with Ralph Fowler. This was a great step toward being accepted as an accomplished theoretical physicist.

Near the end of 1937, Fred had to apply for a scholarship to replace the grant from Yorkshire. He decided to apply for a Goldsmiths' Exhibition. The Goldsmiths' Company, chartered in 1327, provided research studentships as part of its charitable remit. Fowler gave him solid support and provided a reference. A few weeks later the Goldsmiths' Company did indeed award an exhibition to Fred Hoyle. Now he was in clover: The Goldsmiths' award, at £350 a year, was way above the value of the government studentships. For the first time in his life, he was going to have money to spare, but for walking and climbing holidays rather than May Balls.

It is quite clear that Hoyle was an enthusiastic research student who deeply immersed himself in the new nuclear physics. He greatly admired Rudolph Peierls,[27] who in turn got on very well with his new protégé.[28] Hoyle's first scientific papers appeared, stimulated by Peierls. In 1937, he published an article, in the prestigious science journal *Nature*,[29] which considered the means whereby an atomic nucleus might capture an electron orbiting it. Another paper soon followed, extending Fermi's theory of beta decay.[30]

But then Peierls, needing to break out of his hand-to-mouth existence on short-term funding, accepted a lectureship at the University of Birmingham. This immediately posed a problem for Fred: Should he go to Birmingham, too, or stay in Cambridge and seek a new supervisor? For a few months he saw Peierls on a regular basis, hitchhiking or taking the train, when there was something to discuss on the beta decay project. Hoyle even stayed for a few weeks in boardinghouses but found the other guests no match for the research students at the Cavendish.

In the early spring of 1938, Hoyle was still undecided on Birmingham. He disliked large cities all through his life but, on the other hand, he had a great supervisor. In the end, Cambridge won the battle for his loyalty in a touching way. The great historic colleges line King's Parade, Trinity Street, and St. John's Street, which is in fact one road that merely changes its name for different sections. They stretch back, their buildings arranged in courtyards, toward the River Cam. The various colleges also own the land on the far side of the river. Between the Cam and the busy Queen's Road beyond, these grounds are planted as parkland and are open to the public. Locally, this open space behind the college buildings is known as "The Backs" and it is world famous for the vast numbers of crocuses that have naturalized there. There must be hundreds of thousands, if not millions. When Fred saw the yellow, white, and purple floral carpet in late February, that did it. He would not go to Birmingham, a huge city, grimy from the output of thousands of metal-working factories and motor car manufacturing.[31] It reminded him too much of Bradford and Bingley, from which he had escaped.

After Peierls had gone to Birmingham, Hoyle continued to work on beta decay, but an unfortunate misunderstanding led to an irreversible breakdown in the relationship between student and supervisor. On returning from a visit to the United States, Peierls contacted Fred with an intriguing thought: "Hans Bethe at Cornell University suspects that the departure between experimental facts and the theory of beta decay might be due to deficiencies with the experiments rather than shortcomings in Fermi's theory. Do you want to look into it?" Fred responded enthusiastically, "Oh sure, that's just the kind of puzzle I like, a disagreement between experiment and theory."

He was quickly on the case, talking to the experimentalists in the Cavendish Laboratory. He got help from Philip Dee, the don from Pembroke College who had given him a verbal pounding during his unsuccessful application to the college in March 1933. With his Cavendish colleagues, Hoyle found a way of getting the discordant results to agree with the original Fermi theory. An invitation to give a colloquium in the Cavendish followed, and afterward the physicists suggested he write the result up as a letter in *Nature*. This put him in a quandary: He had only done the work because of a tip-off from Bethe received via Peierls. Politeness demanded that he could not simply pinch another person's idea and rush into print. So, he acted tactfully and wrote to Peierls, "What do you advise?"

Soon Hoyle received an angry letter from Birmingham in which Peierls accused him of stealing Bethe's ideas. Evidently, Peierls had been in touch with the Cavendish, where they were still recommending publication, which may have needled Peierls even more. The stinging tone of the letter made Fred angry with Peierls's interference. As a compromise the paper eventually appeared in 1939 under the authorship of Bethe, Hoyle, and Peierls.[32] In it they showed that the experimental results had been wrongly interpreted, and they predicted circumstances under which gamma rays should be observed. The prediction was entirely correct, but Hoyle's relationship with Peierls (who I remember as a kindly person) never recovered from this spat.

To satisfy university regulations, Hoyle now required a new

research supervisor, one who would let him continue with the nuclear physics started under Peierls. Maurice Pryce, 2 years his senior and by now a fellow of Trinity College, agreed to take him, which suited Fred because they shared an interest in quantum theory. Pryce persuaded Fred not to bother with the official requirements for the Ph.D., then a relatively new degree established in 1919 to attract research students who otherwise would have gone to the United States.[33] He advised that it would be better for Fred to pursue a more traditional Cambridge route: Publish papers and get a research fellowship at a good college. Hoyle decided to take Pryce's advice.

Meanwhile, in the Cavendish Laboratory, matters were not going entirely smoothly. On October 19, 1937, Ernest Rutherford, aged 66 and at the height of his intellectual powers, died of a strangulated hernia while awaiting surgery. With his passing, nuclear physics research at the Cavendish Laboratory started to decline. Little by little, the key physicists took positions at other universities, and by 1939 experimental nuclear physics in Cambridge had essentially disintegrated. This decline prompted Pryce to move to Liverpool, so Fred now needed his third supervisor.

By great good fortune, Pryce suggested Paul Dirac as a supervisor, an inspired choice that would have profound implications for Hoyle's career. Dirac was normally quite reluctant to take research students but made a rare exception in Hoyle's case, perhaps impressed by the young theorist's dazzling mathematical ability. He was in any case bemused by the curious logic of being a supervisor who did not want a student, supervising a student who did not really want a supervisor (but had to have one to satisfy university regulations). The manner in which this came about illustrates once more Fred's luck in making connections, as well as the smallness of Cambridge circles before the Second World War.

During 1936–1937, Hoyle had twice been proposed for membership of the Del-Squared V Club but had not been elected. In 1938, Pryce, who was secretary of the club, proposed him for a third time, and this time he was swept in by a landslide, to the evident delight of the members. At the same meeting, Pryce became president, and he immediately nominated the newly elected member as secretary, which resulted in a second landslide. The most important task for the club's secretary was to line up speakers for meetings. Pryce suggested he try two fellows of St. John's: Dirac and Ray Lyttleton.

When Hoyle phoned Dirac, the great physicist made a surprising response: "I will put the telephone down for a minute and think, and then speak again." He did accept and, in March 1938, gave an outstanding talk on the theory of the electron.[34] And he agreed to be Fred's research supervisor.

About this time, Hoyle was making good progress with the quantum mechanical problem originally set by Peierls. Under Dirac, he was arguably working with the sharpest mind in the universe on quantum mechanics. In 1939, he published two huge papers, both of them formidably mathematical, on the new theory of quantum electrodynamics.[35] The first of these papers, on how an electron behaves in an electromagnetic field, is remarkable in view of what he was to achieve later in his career when the mysterious nature of force fields and elementary particles would be Hoyle's major research.

The other person he had approached for a Del-Squared V Club talk was Ray Lyttleton, who was working on the origin of the solar system.[36] Lyttleton's introduction to stellar astronomy had been through a course he attended in 1933, given by F. J. M. Stratton.[37] Hoyle called on Lyttleton in his rooms on Staircase I in the New Court at St. John's: "Please could you give a talk this term to the club?" Lyttleton's response was immediate and unhelpful: "I'm snowed under with teaching and research, and there are no circumstances in which I am going to be dragged away from it."[38] Throughout his life Lyttleton could be difficult about giving talks. By the 1960s, he had discovered a cunning ruse to avoid giving university lectures: He made them so narrow and technical that no student chose to continue beyond the first impenetrable lecture.[39] Disappointed by Lyttleton's refusal, Fred made his way back through the courtyard. Then, on the spur of the moment he went back to Lyttleton's room, popped his head round the door, and cheekily said, "I hope I can do as much for you another time." This impolite outburst earned him a laugh from Lyttleton, who was always good-natured. Soon the kettle was on, tea was made, and the two of them were arguing about the evolution of stars. Such disputations would continue in a cordial atmosphere for the next 30 years.

The social and political situation at Cambridge during Hoyle's research student years was influential on his career decisions. The young people who had experienced the aftermath of the First World War and the depression years developed an acute concern about the state of world affairs. Hoyle too was deeply affected.[40] He was disturbed by the steadily worsening political situation in Europe and felt a marked sense of frustration at the inability of the British government to curb the growing strength of Nazi Germany. He and his fellow students were naively unaware of how far Britain's standing in the world had fallen, with British imperial power by now an almost empty shell. In May 1939, Hoyle attended a lecture given in Trinity College by Bertrand Russell. The thrust was as follows: If Hitler is challenged, war could break out in a few weeks and all the major cities could be reduced to rubble by massive bombing; therefore, Hitler must be appeased.

At this point in his life, Hoyle nursed serious doubts about continuing research in nuclear physics. In March 1939, he read a chilling letter in *Nature*, by Frédéric and Irène Joliot-Curie, radiation physicists in Paris. Just 2 months previously, Otto Hahn, in Berlin, had startled everyone with the discovery that when uranium is bombarded by neutrons, the uranium nucleus breaks into lighter nuclei, and considerable energy is released. (Newspapers yet again hailed this as "splitting the atom.") The Joliot-Curies repeated Hahn's bombardment experiment. To their utter amazement they found that the fission (splitting) of the uranium nucleus also released neutrons.

On reading the *Nature* paper, Hoyle, along with many other physicists, realized the possibility of setting up a self-sustaining chain reaction, since the products of the first round of fission included the neutrons needed to induce the next round. Twenty-five years later, in *Encounter with the Future,* he wrote:[41]

> I instantly realised the consequences for weapons technology. I left the library in a thoughtful mood and walked towards the gates of the Cavendish, only to find a group already talking excitedly. They, too, had read the letter and realised what it meant. But whereas I had been wondering how the realisation of a nuclear chain reaction might be avoided, they were wondering how it might be done. The thought occurred to me that if a bomb could be made, these fools would do it. Even then I saw the road that was going to lead to Hiroshima, and influenced no doubt by my early

upbringing, and in particular by my father, I decided to quit nuclear physics. (1968, p. 75)

Within a few months of reading the *Nature* paper, Hoyle's Gold-smiths' Exhibition would run out. He once more had to search for funding to continue his research. The normal solution for a person of his age at Oxford or Cambridge was to win one of the research fellow-ships offered by the colleges. A year earlier he had tried his own college, Emmanuel, and Dirac's college, St. John's. Both rejected him, perhaps because he had been too honest in revealing that he was applying to more than one college. He needed to be bolder. In 1939, he applied only to St. John's, with Paul Dirac and John Cockroft as his referees. And he made a further application to a completely different source.

In 1851, a Royal Commission had established a Great Exhibition at Crystal Palace, in London, under the presidency of Prince Albert. It was the first international exhibition of manufactured products and was hugely successful. The profits of £170,000 from the exhibition were invested in land in the South Kensington area, close to the site of the Crystal Palace, and the income was to be used for the advancement of science and art education. The commission set up fellowships that con-tinue to the present day. The scheme of 1851 research fellowships gives a few young scientists of exceptional promise research funding for 2 years. Hoyle applied to the commission for one of their fellowships.

Astonishingly, in May 1939 he won both fellowships. The one at St. John's included a salary of £250 a year, plus free accommodation and meals at high table. The 1851 Senior Exhibition came in at the princely sum of £600 for 2 years, making his income £850, with hardly any living expenses! Average annual earnings for a British man were less than £250 in 1939, so he was now very well off indeed. He quickly bought his first car, a 1936 Rover, for £125. But there were clouds on the horizon. War with Germany seemed inevitable. Already by the sum-mer of 1939, scientists were moving from the Cavendish Laboratory into war work. But Fred did not follow his Cavendish colleagues, for a personal reason.

In May 1939, he had fleetingly met and been attracted to a young woman, Barbara Clark, who was in Cambridge for an interview at Girton College. He remembered which school she was from because she was being taught science by an undergraduate friend of Fred's. Sometime in early July, he drove to her northern school where he persuaded her to make a walking trip in late August, in the Lake District. On July 30, he wrote a rather formal letter to her, confirming the arrangements. Fred and Barbara did indeed make the trip to the Lakes, at the end of which Barbara visited friends in Cheshire, and Fred went home to Gilstead. At the very end of August, Fred was back in Cheshire with the Rover. He picked her up and they set off together for Barbara's home.

For months on end the news on the radio had been terrifying. In April 1939, Hitler issued a directive that the Nazi armed services should get ready to attack Poland by September 1. In May, two British warships moved a substantial amount of the Bank of England's gold to Canada for safekeeping. On August 31, Hitler ordered the attack on Poland. The imminence of war was uppermost in the minds of Fred and Barbara. On the car journey they talked of nothing else. How would it affect Fred's academic career? It would destroy his affluence and disperse his colleagues. The fellowships suddenly seemed pointless, futile even.

"It will swallow my best creative period, just as I am finding my feet in research," was his despondent summing up of the situation in which he found himself. Fred pulled the Rover over to the side of the road, and they got out to reflect.

"Barbara, this is all going to make a nonsense of years of courtship," he despaired. They decided then and there to marry forthwith. Their first proper conversation had been only 5 weeks earlier and they had spent less than 20 days in each other's company. They married on December 28, 1939, the start of a long, exceedingly close and happy marriage,[42] in which Barbara would give enormous support to Fred and his career.

Fred was staying at Barbara's house in Scunthorpe on September 3, 1939, when war broke out. On September 4, Fred's wake-up call was the terrifying sounds of air-raid sirens. Panic seized him. The whole family was awakened, gripped by the thought that Scunthorpe would soon be under heavy bombardment and they would all die.

CHAPTER 3

The Star Makers

E ven if there had not been a war looming, 1939 would have been a year of crisis, challenge, and change for Fred Hoyle. He lurched from one supervisor to another under circumstances that might have caused a weaker student to give up. He had to decide whether to continue in nuclear physics or switch to astronomy. And he and his wife hardly knew each other when they married. Thankfully, however, he had finished up with Dirac as his supervisor and had secured very generous funding from the 1851 Exhibition and St. John's. From now on, he could pick and choose what research to do.

For many scientists, the most creative years come after winning a postdoctoral fellowship, when they start to work with some degree of independence. To build a solid career it is important to choose the right place and the right people. Location and one's immediate colleagues are more important than choosing which intellectual puzzles to attack. Science has a social dimension and it is almost impossible to do research in isolation. Today, it would be very unusual for an ambitious scientist to do undergraduate study, postgraduate work, and postdoctoral research all at one university—even one as prestigious as Cambridge—as Hoyle had done.

Fred Hoyle chose to remain in Cambridge, when no doubt he could have gone to Birmingham or Manchester. Cambridge was a center of excellence in astronomy, and he was determined to add to the towering achievements of Eddington. Like any newcomer to independent academic research, Hoyle needed a mentor and a collaborator. Between finishing as a research student and having a program of his own, there was a transition to manage. Two people were crucially important to Hoyle at this time: Paul Dirac as role model and Ray Lyttleton as collaborator.

Dirac stressed to Hoyle the importance of mathematical fluency and encouraged him to adopt a rigorous approach to problems, using mathematics as his tool.[1] In 1931, Dirac had remarked: "The steady progress of physics requires for its theoretical formulation a mathematics that gets continually more advanced."[2] In the paper from which this quotation is taken, Dirac uses mathematical dexterity to predict the existence of a then-undetected particle, the antielectron (later renamed the positron). This feat impressed Hoyle enormously. Dirac also warned him of difficulties on the road ahead, with the following reflection: "A decade ago it was possible for people who were very good to solve important problems, but now people who are very good cannot find important problems to solve!"[3] So here was another reason to get out of mainstream physics and use his mathematical skills in an area where there were undoubtedly plenty of unsolved problems.

Election to the fellowship at St. John's brought Hoyle closer to Ray Lyttleton, also of St. John's. The organization of astronomy in the university was chaotic at this time. Theorists were associated with the Faculty of Mathematics, which consisted only of its academic staff and had no buildings of its own. A tiny group of theorists convoked with Eddington at the Observatories, situated 1 mile out of town. In the Cavendish Laboratory, the physicists showed no enthusiasm for plunging into nuclear astrophysics, a strange situation given that Eddington was convinced of the importance of nuclear physics to astrophysics. Eddington seems to have despaired of the dichotomy between physics and astronomy. With great foresight, he guessed that energy generation inside stars must have something to do with nuclear forces, but the Cavendish crowd said that the insides of stars were not hot enough to induce reactions between nuclei. Eddington famously retorted: "In that case the critic had better go and find a hotter place!"[4]

The broad preoccupations of astrophysicists 60 or 70 years ago were much the same as they are today. How did the solar system form, and how has it changed? How do stars form, evolve, and die? What is the story of the extragalactic universe? Are there components other than stars, gas, and dust in the universe?[5] In the 1930s, the prevailing theory of how a star produces energy and changes its appearance throughout its life was very unsatisfactory for many reasons. For example, although there was a general scenario for the lives of Sun-like stars, more massive stars did not fit into it, so Hoyle and Lyttleton found themselves addressing the serious business of making stars.

Hoyle and Lyttleton knew from Hans Bethe how four hydrogen nuclei, which are single protons, could go through a series of fusion reactions to make one helium nucleus consisting of two protons and two neutrons. In essence, Bethe had shown that Eddington was correct: The Sun and normal stars are hot enough to cook the protons into helium nuclei. (The Cavendish could now call off the search for Eddington's hotter place.) This reaction releases a stupendous amount of energy because the new helium nucleus is less massive than the four protons by 0.5 percent. This mass deficit is transformed into pure energy. By the end of 1938, all astrophysicists accepted the theory that nuclear energy powers the stars.

Although the details of the nuclear reactions in the interiors of stars were yet to be uncovered, it was clear that massive stars presented a special problem. The more massive a star, the brighter it shines, and very luminous stars consume their nuclear fuel at a prodigious rate. But the binary star system V Puppis posed a problem for Hoyle and Lyttleton. It has two components, each about 20 times more massive than our Sun. They knew its distance and, from its brightness as seen in our sky, could easily calculate the energy coming from each of the two member stars as 8,000 times more than our Sun gives out. But the V Puppis stars have only 20 times more material than the Sun. This meant that V Puppis would have enough fuel to last only a fraction of the lifetime of the Sun. In 1938, astronomers thought, wrongly as it turned out, that all the stars were more or less the same age, which implied that V Puppis had only "recently" become very massive.

To deal with stars over five times more massive than the Sun, Hoyle and Lyttleton proposed a revolutionary new idea: They envisaged a scheme in which stars are refueled by adding more matter to them. Superficially, this was just like filling up a vehicle's fuel tank. But the subtlety was that the fuel pump would also increase the engine's capacity and power rating with every refill, eventually turning stars like the Sun into blazing giants.

Another contemporary puzzle was the existence of binary star systems in which one of the members was a white dwarf, the collapsed remnant of a star at the very end of its life. In about 5 billion years, the Sun will have consumed all of the hydrogen in its central core, bringing energy generation to a close. As it cools down, it will collapse under the crush of its own gravity, becoming a white dwarf star, similar in size to the Earth but with the mass of the Sun. Astronomers had started to measure the individual masses of the stars in binary systems and invariably found the mass of the normal star to be greater than the mass of its white dwarf companion. Astrophysics in the 1930s could not make any sense of this: The white dwarf had reached the terminal stage *faster* than its companion, and so it should be *more* massive, not less.

Hoyle and Lyttleton stumbled into research on stellar evolution through an unlikely route. Their first meeting had revealed a mutual interest in the possibility that astronomical agents might be the trigger for the onset of ice ages. They felt that, if the solar system passed through a dark cloud of gas in space, this could trigger a drop in Earth's temperature. As they argued and scribbled feverishly at St. John's College, they realized that they had also hit on a clever way to explain how the masses of stars could increase. They speculated that stars were somehow scooping up gas and dust in interstellar space and so growing ever larger. Their solution involved a moving star tunneling through an interstellar cloud.

Their first joint paper is on ice ages and opens with these words: "There is direct evidence for the existence of diffuse clouds of matter in interstellar space. Any section of the Milky Way containing a large number of stars usually shows patches in which no stars appear, and the extent of these patches is often large compared with the average distances between the stars themselves."[6]

Interstellar matter is indeed quite easy to see without a telescope. On a moonless night and away from city lights, the Milky Way shows up as a misty arc across the sky, the combined light of myriad distant stars belonging to our galaxy. Dark patches and rifts across it are clearly visible. Nineteenth-century astronomers thought these might be holes where stars are absent, but by the twentieth century, astronomers accepted the likelihood that opaque clouds of cold gas and dust were blocking the starlight.

In their foundation paper, Hoyle and Lyttleton envisage what happens when the Sun passes through a cloud of matter in space. This is not an unlikely event. The Sun orbits round the center of the Milky Way galaxy, taking about 250 million years to make one circuit. At a guess, they said, the Sun's path would cut through an interstellar cloud every 10 million to 100 million years.

Now imagine, as they invited their readers to do, what happens as the cloud streams past the Sun. Let us forget all about gravity for a moment and pretend that the Sun is no more than a spherical obstacle in the flow. Its main effect is to bore a cylindrical hollow, like a tunnel, into the cloud. This simplistic picture is of course unrealistic. As a gas cloud streams by, it will be pulled toward the Sun by gravity. The deflected gas cannot fall directly onto the Sun because its streaming motion is too fast. Instead, it falls toward the center of the tunnel to create a concentrated tube of matter that eventually falls into the Sun. The beauty of this scheme in the eyes of its inventors was its simplicity and effectiveness. With the help of gravity, the Sun would be sweeping up gas and dust from over a vast area. Putting in some plausible numbers, Hoyle and Lyttleton suggested that a tube of captured material crashing to the solar surface would raise the Sun's power by anything from 0.1 to 1,000 percent.

At this stage, many researchers would have dashed off a quick note or letter to a journal to establish priority for their idea. Fred never really operated like that, although he was always a fast worker. He liked to consider the immediate implications of his discoveries, and this was a major one: a method of increasing the mass of the Sun. If this really had happened, what would the consequences have been?

In a spectacular leap across disciplines, Hoyle and Lyttleton decided they could explain ice ages by invoking collisions between the Sun and interstellar clouds. The Earth has endured several epochs, some lasting millions of years, when glaciers invaded entire continents, completely covering them in ice. The explanation offered by Hoyle and Lyttleton was that a collision with an interstellar cloud would raise the temperature of the Sun, as a result of matter pouring onto it. The warmer Sun would heat the Earth's atmosphere, which would then hold more water vapor. This extra water would, in turn, lead to excessive clouds, thus preventing the Sun's rays from reaching the ground. Snow accumulations in the winter would no longer entirely melt in the summer, and so ice would steadily increase. They felt that the Sun's last encounter with a cloud was very recent. "If the conclusions of this paper are correct," they wrote, "the solar system must have emerged from such a cloud within the last few thousand years since an ice age took place as recently as this."

Not content with a solution for the problem of ice ages alone, they moved on to the carboniferous era and coal formation. This warm and wet period they also explained by extra solar heating caused by the accretion of interstellar matter onto the Sun, but on a much larger scale, so the temperature of Earth's surface never fell to the freezing point. Their speculations were, of course, at a time when continental drift was not yet discovered. In fact, the Cambridge earth scientist Harold Jeffreys, also of St. John's College, had told Hoyle that landmasses do not move, and so an atmospheric explanation was needed for the geologic evidence of a warm carboniferous era in parts of the world that are now much cooler.[7]

Their whole paper is a mild invasion of the intellectual territory of earth and atmospheric scientists. Despite the fact that Jeffreys was the most distinguished earth scientist in the country at the time, Lyttleton never had any respect for him, feeling that he was an inferior mathematician. So it is not surprising that he did not bring his senior colleague at St. John's into the project. In concluding their paper, Hoyle and Lyttleton took a sideswipe at their earth science colleagues, with a provocative statement that their investigation "relates to a process that seems hitherto to have been overlooked." They accuse other researchers of proposing a variety of less satisfactory mechanisms to explain ice ages and assert that their own suggestion "has the merit of giving effects quantitatively in agreement with the requirements" of a successful explanation of ice ages. This rhetorical style of presentation would store up trouble for both of them in the future.

The impact of this early paper was to be enormous, although neither author can conceivably have imagined that would be the case, nor the reasons why. By 2003, it was Hoyle's tenth most highly cited publication, not bad for his first paper on astronomy.[8] But it was not until the 1970s that their insight into a hitherto unknown astrophysical phenomenon, the process of accretion, would find applications on a grand scale. With hindsight, of course, we know that their attempt to use the process to explain Earth's climate record over geologic timescales was misguided, however plausible it seemed at the time. Their enduring achievement was to realize the existence of a process that could be invoked correctly in a variety of cosmic situations.

The story of the paper's publication records the first big row between the young upstart Hoyle and the grandees at the Royal Astronomical Society (RAS). Lyttleton, as senior author and a fellow of the RAS, submitted their theory for publication by the RAS. A standard practice at any learned society is to send a new paper out for external "peer review." This process went badly, producing adverse criticism. The Council of the RAS then considered the matter: No paper was rejected unless two referees had so recommended. The Council listened to Robert Atkinson, the acting secretary of the society who handled papers. "Hoyle and Lyttleton are unwilling to make the changes demanded by the referees, and a long correspondence has ensued. The paper must be rejected as unworthy of publication."

Several members of the Council sided with the authors in the interminable discussions at the RAS, but in the end the president insisted that the Council could not override the advice of referees to decline publication. In disgust, Lyttleton withdrew the paper, thus shielding Hoyle from a formal rejection letter. The rebuff led to an unfortunate lack of confidence between Hoyle and Lyttleton on the one hand, and the RAS Council on the other, a lack of trust that continued for many years.

The two chums, thoroughly beaten by these establishment astronomers in London, many of them employed at the Royal Observatory, Greenwich, sought refuge in a parochial organization, the Cambridge Philosophical Society, which did publish their work. The published *Proceedings* of this society were mainly papers from the physicists in the Cavendish Laboratory and mathematicians in the colleges.

In June 1939, Hoyle walked round to the Cambridge Philosophical Society and dropped in a second paper,[9] applying the accretion theory for the Sun to stars in general. This offered an explanation of how a star evolves, or changes, throughout its lifetime. In the two decades following the end of the First World War, astrophysicists had made considerable progress in understanding the lives of the stars. Before 1926, Eddington had produced a dozen great papers on the structure of stars, and by 1939, astronomers had a reasonable notion of the relationship between the mass of a star and its luminosity, that is to say, the amount of energy it sends out. The connection between these two observable quantities, mass and luminosity, had been obtained without reference to the nuclear reactions firing the stars. In the first few months of 1939, Hoyle and Lyttleton worked on a description of the evolution of stars that likewise did not require detailed knowledge of the nuclear energy processes.

They explained the evolution of a star as a process in which new material would be added through accretion from interstellar clouds, thus increasing both the mass and the luminosity of the star. Their pioneering paper, on solar accretion and ice ages, was about to be attacked by a theorist at the Royal Observatory, Greenwich. Meanwhile, in their second paper, they owned up to some imperfections in their initial theory but pushed ahead nevertheless. "Our conclusions following from the accretion process are of such importance to theoretical astronomy, we have adopted the customary attitude of this science of not awaiting an accurate solution." They were so convinced by the simplicity of their accretion model that, to avoid any delay in publication, they dared to ignore their critics.

Their approach was a bold one: "If we consider the stars generally, then why not interstellar matter generally, rather than just dense clouds?" And the outcome was startling: Stars have voracious appetites, they said, gobbling up the interstellar gas as they sweep on their paths around the Milky Way, gaining in mass and becoming ever brighter. A star's evolution, and therefore its physical properties and appearance at any stage, was entirely in the hands of the interstellar gas, if the theory was correct.

They assumed that the interstellar matter must be atoms of hydrogen gas—12 years before the first direct detection of neutral hydrogen in interstellar space! In 1944, Hendrik van de Hulst, working in the Netherlands under Nazi occupation, correctly predicted that hydrogen would emit radio waves at a wavelength of 21 centimeters.[10] Later, in 1951, at Harvard University, Edward Purcell and his student Harold Ewen detected the 21-centimeter radio waves from hydrogen,[11] confirming both van de Hulst's prediction and Hoyle's speculation.

In the special cases of binary stars including white dwarfs, Hoyle came very close (but not quite close enough) to uncovering the true evolutionary mechanism. His solution to the problem of the normal star being more massive than its spent-out companion was accretion from interstellar space. Hoyle thought that the normal star continued to suck in hydrogen from space, to the extent that it became the dominant member of the pair. The modern picture, in which the normal star swells and the white dwarf accretes new matter *directly* from its companion star rather than from an interstellar cloud, unfortunately eluded him.

As a method of building and evolving stars, the Hoyle-Lyttleton accretion model received no support from other astrophysicists, for whom a nuclear physics approach seemed more promising. Independence from the details of the nuclear reactions, Hoyle thought, was one of the strengths of their model, but that was not how others saw it. Hoyle and Lyttleton were challenging the patrician figures of the day: Hans Bethe, Sir Arthur Eddington, Otto Struve, and the RAS Council.

Atkinson, in particular, was not going to let the matter rest with Lyttleton's withdrawal.[12] He had published groundbreaking papers on the generation of stellar energy through nuclear processes at very high temperatures deep inside stars. He wasted no time in recognizing that certain problems of stellar evolution could be greatly eased by allowing stars to sweep up fresh matter at a very high rate.[13] But he found flaws in the Hoyle-Lyttleton theory. In his first attack he dashed off a scathing criticism to the Cambridge Philosophical Society and followed it with a lengthy riposte at the RAS. Publication of the long rebuttal through the prestigious RAS was a shrewd move by a man who only published in top-ranking journals. In the RAS version, he claimed that Hoyle and Lyttleton did not contest the holes he had picked in their argument. He concluded that the energy requirements of bright stars cannot be solved by any accretion mechanism, unless some additional (and then-unknown) properties of the interstellar gas were brought into play. Nevertheless, Hoyle and Lyttleton refused to be deterred by Atkinson's bombshell. They continued to work in theoretical astronomy, even though most of England's scientists were now on defense work.

In 1939, much of the Cavendish Laboratory had been transferred to war work, mostly at the Royal Air Force radar establishment. Hoyle was overlooked on this round of drafting scientists into the war effort, probably because he registered his subject as mathematics rather than physics. The Cavendish physicists disappeared almost overnight, but it took the authorities longer to find roles for the mathematicians. Hoyle remained in Cambridge, almost a forgotten man, where he and Lyttleton were under instruction to stay put until called up. And so it happened that in the first year of the Second World War, Hoyle was able to continue with research into astronomy at St. John's College and to benefit from conversations with Eddington as well as Lyttleton. The following year, on February 6, 1940, Eddington, Lyttleton, and Maurice Pryce signed Hoyle's nomination to become a fellow of the RAS. He was proposed at the RAS Council meeting on March 8 and elected by the Council on May 10.

By November 1939, Hoyle and Lyttleton again tried to get new accretion results published by the RAS, which by now had decamped its administrative offices from London to the University Observatory, Oxford. Once more, the Papers Secretary was their enemy Atkinson, still working at Greenwich. He had the paper refereed and, when the Council considered the matter on November 10, it declined to publish.

One referee was again the Savilian Professor at Oxford, Harry Plaskett, who "has devoted a very considerable amount of his time and energies to you," according to a letter of explanation from Atkinson to Lyttleton. He continues, saying the paper should be stripped of its introduction and "its far reaching consequences."[14]

The consistent differences of opinion between Hoyle and Lyttleton and the RAS referees concerned the Council. As a device to deflect this new submission, Atkinson pointed to their recent papers in the *Proceedings of the Cambridge Philosophical Society:* "Their new paper overlaps markedly with what they have already published." The Council was in a quandry: Was it really being offered new work, or were Hoyle and Lyttleton trying to get publication twice over? At the conclusion of the RAS Council meeting, Atkinson wrote a firm letter of rebuke to Lyttleton, as a result of which the paper never appeared.

A major part of Atkinson's scientific criticism revolved around details of the accretion process. He asserted that collisions between particles would release too much heat. Atkinson calculated that the hydrogen cloud's temperature would soar to 10,000 degrees, reversing the infall of matter and blasting the cloud to infinity. Without a cooling mechanism to pump away the excess heat, the Hoyle-Lyttleton model was a dead duck.

Hoyle took this point to heart and produced the most complete treatment of the heating and cooling of interstellar gas to have been published at the time.[15] He needed a refrigeration mechanism and found one. With breathtaking foresight, he assumed that a tenth of the hydrogen in space, rather than being single atoms, must be in the form of molecules, consisting of two atoms joined together. The proof that this is indeed the situation lay 30 years in the future. The hydrogen molecules (to which, to be fair, Atkinson had alluded as a possible outlet) were the key.

Hoyle formulated his mechanism without reference to any laboratory data. Any value that he could not find in standard physics reference books he estimated on the spot. He wasted no time waiting for laboratory measurements, and, in any case, the competent physicists were now absent on war work. He piled assumption on guess, guess on arm waving, as he probed the mechanism of cooling. Astonishingly, history proved most of the assumptions he made to be correct. The versatility of Hoyle's intuitive approach to physics is wonderfully illustrated by this imaginative contribution to astrophysics.

With a cooling mechanism in place, Hoyle checked that his postulated hydrogen molecule could survive in the harsh environment of space. Material accreting onto massive stars would have to withstand powerful ultraviolet radiation streaming up from the star's hot surface. Ultraviolet light shatters molecules: It is powerful enough to strip a hydrogen molecule of its two electrons and break the molecule into two protons. To Hoyle's relief, molecular breakdown would not happen sufficiently quickly to arrest accretion.

Having discovered how useful molecular hydrogen would be for his purposes, Hoyle next puzzled over a mechanism for making molecules in deep space, without reaching any useful conclusion. In 1940, no observational evidence existed for interstellar molecules of any kind, although Eddington had speculated about them in 1937.[16] Unknown to British astronomers because of wartime restrictions, Walter Adams at Mount Wilson Observatory in California had commenced a detailed study of the spectra of starlight, using the world's largest telescope. In 1941, he announced the detection "beyond reasonable doubt" of the diatomic molecules CH and CN.[17] Molecular hydrogen was not discovered until 1970, when George Carruthers used a rocketborne experiment to detect its signature in the far ultraviolet spectrum of the star Xi Persei.[18] The ratio of atomic to molecular hydrogen in the star was quite close to the value at which Hoyle could only guess. More recently, astronomers have shown that interstellar molecules play a key role in the evolution of the galaxy: Every star and planetary system forms inside a molecular cloud. These are the only places where it is cold enough for gravity to overcome the tendency of warm gas to expand and disperse and for the dense "cores" that become stars to form.

Meanwhile, Hoyle and Lyttleton continued to stir the hornet's nest at the RAS. They kept up the pressure for publication of their continuing accretion research, which by now had extended to ideas on the formation of planets. In March 1940, Donald H. Sadler, a secretary of the RAS, wrote : "We cannot continue to discuss at length matters which it has already discussed," adding that "there is to be no further correspondence about accretion as a method of forming the solar system's planets other than a paper revised in accordance with the Council's suggestions."

Hoyle and Lyttleton battled on, but, by the end of the summer, the British Admiralty needed Hoyle for radar work, and so he and Barbara moved down to bomb-damaged Portsmouth. The Cambridge colleagues kept in touch by post, sending drafts of papers to and fro. Barbara typed these drafts, using Fred's 1930s Remington typewriter. Over 100 letters from Lyttleton to Hoyle (1939–1942) have survived, and they are a remarkable record of a feverish rate of research. From this archive it is possible to reconstruct some aspects of their approach. For example, on August 2, 1940, Lyttleton writes at length, giving a highly mathematical appraisal of a problem connected to Jupiter's reaction to the solar system passing through a dust cloud: The giant planet would suck in a huge amount of dust. This letter shows Lyttleton at his very best, reducing the problem to a system of solvable equations. The drafts hurtle back and forth, with Lyttleton issuing brisk instructions, such as: "get the draft back tomorrow, so I can work on it right away." Ray soon begins to find it hard to keep up with Fred's restless creativity. Several times, his letters, always to "Dear Hoyle," open with a short statement such as, "Thanks for your many communications," or "Thanks for your telegram," or even "Thank you for your bombard of letters."

By March 1941 the pair were ready with further refinements of the accretion theory, and this time the RAS finally accepted without a murmur that they should offer Hoyle and Lyttleton a right to reply to Atkinson.[19] The paper starts gently, reminding the reader of the formula at which they had previously arrived for the rate at which the mass of a star would increase as a result of accretion. But right in the first paragraph, they say the reason for writing is to deflect Atkinson's attacks: "Conclusions have been advanced by Atkinson that differ so widely from the views we have advocated that some statement of position in regard to this problem as it now stands seems desirable."

By invoking molecular hydrogen, they claimed to have dealt with the heating that Atkinson had predicted. By this stage, Lyttleton had learned of the successful detection of interstellar molecules, and this had strengthened their hand enormously. Their academic thrust was now to show how accretion could lead to great ages for the stars, up 50 billion years, and they advocated this as the strongest reason for believing in accretion. Their riposte to Atkinson is, by modern standards of academic debate, extremely nitpicking, although they own up to certain sweeping assumptions they were compelled to make. At the end, they boldly state: "We have developed . . . a consistent theory of stellar evolution."

And what did they feel they had achieved for astrophysics? "Our work," they write, "shows that a consistent theory can only be built if the parameters associated with certain factors take fairly definite values." What they are saying, in effect, is that the values they have guessed at, such as the amount of hydrogen in space, must in reality be correct because their theory, in which they have total confidence, falls apart with any other values. On reading the paper today, one can almost hear the ghost of Atkinson whispering "precisely"! This is a curious way to do science. Their conclusion is astonishing, with the vain assertion, "in regard to stellar evolution, the choice is between [our] consistent theory based on the idea of accretion and no theory at all." This is the first time in the scientific literature that Hoyle takes such a rigid public stance, refusing to believe that anyone else's ideas might have merit.

The RAS offered Hoyle and Lyttleton the chance to speak publicly about their accretion theory of stellar evolution, by inviting Lyttleton (as the senior author) to read a summary of the paper at the meeting of the society on Friday, May 9, 1941. Lyttleton at first accepted but, as a man who did very little public speaking, he must later have had regrets because he backed out at short notice. Then he evidently regretted that action because, 4 days before the meeting, he again wrote to the meetings secretary, D. H. Sadler, saying that Hoyle would read the paper. This put Sadler on the spot. He had already filled the gap in the program, but he nevertheless tried to squeeze Hoyle in at the end.

The society had moved to Burlington House, Piccadilly, in 1874, with a cramped lecture theater on the ground floor and a grand library above. Fred and Barbara arrived early for the meeting and went up to the library for afternoon tea, where a crowded scene awaited them, astronomers jammed into the library, wolfing down fruitcake, biscuits, and weak tea, a wartime treat indeed because no ration book coupons were needed for this bun fight. The high officers of the society were all there, along with the Astronomer Royal, the civil service astronomers from down the river at Greenwich, Thackeray from Oxford, and Eddington from Cambridge. The walls of the room were lined with what remained of the greatest astronomical library in the British Empire; most of the collections had been moved out of London to escape bomb damage. At night a fellow of the society camped out in the premises on fire watch, ready to leap into action with a hand pump and a bucket of water should an incendiary hit the RAS. Sadler kindly took Hoyle to one side and thanked him for coming, expressing the hope that his paper could be read.

At 4:30 p.m., everyone filed down the elegant staircase, lined with its portraits of past presidents, to the red leather benches of the lecture room. The leading astronomers of the day tended to sit at the back of the room, with the newcomers at the front. The president and the two secretaries took their places at a table on a raised dais, facing the meeting. Behind them hung the society's portrait of Sir Isaac Newton, and to one side, Sir William Herschel's portrait looked down on the throng. The president banged the gavel to bring everyone to order and called for a moment of silence in memory of Annie Jump Cannon, the society's only honorary member, who had died on April 13. An astronomer at Harvard College Observatory, she had classified the spectra of more than 225,000 stars and was the first woman to receive an honorary doctorate from the University of Oxford.

From Hoyle's point of view, the scientific proceedings started badly, by eating up time on a rerun of a paper that had not been fully discussed at the April meeting. Sir Arthur Eddington came to the podium next to present a very technical paper about the interior of a star. Barbara found it tedious in the extreme, with its complex math and Eddington's poor standard of presentation. At the conclusion of this paper, the president started a leisurely and unscheduled discussion about sunspots, followed by several questions to Eddington, most of which the great man declined to answer. Next on the agenda was a contribution from South Africa, read by a man from Greenwich. By now, Hoyle was getting very anxious, jammed in the bench without an inch to move, and nervously watching the time slip away. Finally, Eddington strode up to the podium once more, to read the second of his papers on the agenda. Once again, his ponderous and hesitating manner could not keep up with the sweep of the clock hands. By the time he was done, the clock stood at 6:30 p.m. The president, high officers, and the Greenwich men needed to scurry off to their London club, the Criterion, for a private after-meeting dinner, so the session could not overrun any further. Both Hoyle and another fellow were thereby denied their chance to showcase their work. Hoyle was fuming: As far as he could see, this was all a deliberate snub. He knew by now that Eddington did not believe in the accretion theory of stellar evolution. Hoyle felt that the officers of the society had secretly agreed, at their morning meeting, to string out the proceedings so that his paper would not be reached. Of course, he felt embarrassed too that his wife had wasted a day since she had not listened to his presentation. He was not invited to give the paper at the June meeting, unlike the other person denied an opportunity at the May meeting, who topped the agenda for June.

So far, the critics of Hoyle and Lyttleton had been based at the Royal Observatory, but now Oxford entered the fray, fielding Andrew D. Thackeray of the University Observatory, also a secretary of the RAS. Certain "establishment" figures had editorial control of a small independent magazine, *The Observatory*, which carried reports of RAS meetings, book reviews, short research papers, and news. Thackeray was one of its editors, and the issue of May 1941 carried a news note,[20] written by him, but unsigned, on interstellar molecules. Thackeray writes in a lively style that would not be out of place in a modern science news magazine. He launches off: "A discovery of first-class importance regarding the constitution of interstellar space has been announced by A. McKellar of the Dominion Astrophysical Observatory and W. S. Adams, director of the Mount Wilson Observatory." The existence of interstellar molecules "has been confirmed in a brilliant manner." The whole piece is good reportage, concluding with an unanswered question. "A great new problem is raised, that is, the explanation of how the molecules can survive in the almost perfect vacuum of space without dissociating [breaking up]."

Lyttleton hit the roof when he read this. Thackeray, as an officer of the RAS, would have been familiar with what the Hoyle-Lyttleton paper had to say about molecules in space, but here, in this newspiece, he gave no acknowledgment of their prediction. On July 11, 1941, Lyttleton dashed off a letter of protest to Thackeray, who showed it to his fellow editors of *The Observatory*. Within 2 weeks he got back to Lyttleton, saying his colleagues "are not in agreement with your views" and giving the reason why. Lyttleton rejected the rebuttal as untenable but then let 7 months go by, all the time festering on the rejection. Finally, he challenged Thackeray again on February 9, 1942. He pointed out that the final question in the news note had already been answered in the Hoyle-Lyttleton paper that preceded the announcement from the United States. Knowing that Thackeray had penned the unsigned news note, Lyttleton sarcastically added a calculated insult. "We feel we are now justified in asking you to forward our paper to your contributor [i.e., Thackeray] with a view to obtaining either a retraction of his expressed attitude or, alternatively, some indication of where our analysis fails, and in either event why it was not thought necessary to make any mention of a theoretical discussion of such great new problems." This nasty letter could only further widen the rift between the pair and the RAS officers.

Running battles with the RAS Council continued on other papers. For example, on April 9, 1941, an exasperated Lyttleton comments on a letter he has just received from the president of the RAS, which he says "amounts to a promise from him that the future conduct of the Council will not be so transparently corrupt." Lyttleton reflects on how they can make public the rough handling their papers had received: "We can eventually adopt [George Bernard] Shaw's plan of discussing unmentionable matters in the prefaces of our books, as a warning to another generation about how not to treat its prophets." At this time a Hoyle-Lyttleton paper on the internal structure of stars was under consideration. The July 1941 meeting of the RAS Council looked at the detailed statement of criticisms from the referee and again invited Hoyle and Lyttleton to revise their paper before the RAS would be willing to send it to a second referee.

Hoyle did the revisions and sent in a new version, but by now it was becoming very difficult for him to continue astronomical research with Lyttleton. In October 1942, the Admiralty moved Hoyle from Portsmouth to the new Admiralty Signal Establishment (ASE) based in Witley (Surrey), and this really was a watershed for their collaboration.

From 1939 to 1942, the Hoyle-Lyttleton collaboration had been characterized by a great richness in the range of problems they wanted to tackle using the accretion mechanism. Fred leapt in with all kinds of questions. What about sunspots? Could these be created by infalling matter? What about the Sun's outer atmosphere, particularly the corona, which is only visible to the naked eye during a total eclipse of the Sun? How could its temperature be as high as a million degrees? Could accretion heat the corona to this temperature? Yes it could. What caused the huge energy outbursts from the Sun, such as flares and prominences? These too yielded to accretion physics. Variable stars enter the picture too, and the nature of luminous red giant stars is explained. But this collaboration also shows the development of Hoyle's highly combative approach to critics. He reacted competitively, with a pugnacious spirit, feeling that he must defend not just his predictions but also his speculations against attack from any party. Because the British astronomical community was small and met monthly in London, the atmosphere became highly charged. Above all, Hoyle sought to "save the appearances" as the philosophers of ancient Greece had so often done, so that *his* models could be modified in response to new data. This trait would ultimately lead to his exclusion from the academy.

At Witley, Hoyle was made director of the theory group, code named Section XRC8, which was devoted to studying the propagation of radar beams. Hermann Bondi was the deputy director. Initially, XRC8 consisted of Hoyle, Bondi, and Cyril Domb, plus technical assistants. But soon the group had seven scientists, evidently carefully chosen because five of the group were eventually elected as fellows of the Royal Society: Bondi, Domb, Thomas Gold, Hoyle, and Richard Pumphrey (a physiologist).

Tommy Gold joined them in July 1943 and found a house to rent in the village of Dunsfold, about 6 miles from ASE. During the work-week, Hoyle house-shared with Bondi and Gold, returning on the weekends to Barbara and his infant son Geoffrey, who remained in West Sussex. At the end of the war, Bondi, Gold, and Hoyle were to make spectacular claims about the nature of the universe as a result of their musings. In wartime they had to content themselves with astronomy problems that could be solved with just a pad of paper and a pen. There had no access to the libraries, or to the literature, no research seminars, and no visits from eminent professors. Theirs was fireside research, done in armchairs or on the dining table, with outside contact being conducted through the postal service or by phone calls placed by manual exchange operators.

By day the trio worked on theoretical problems connected to the propagation of radar. Their evenings were full of a different kind of creative science, Fred always bubbling over with new ideas in astrophysics, on which the three of them would work together. Fred told Hermann about his new line in astrophysics: Space is not empty, and the stars are going through rapid evolution as a result of gobbling up gas and dust. Fred knew that the telescopes of the day could not investigate his accretion mechanism, and he confided to Hermann his disappointment that the papers he had authored with Lyttleton had failed to make any positive impact on the establishment astronomers.

Bondi found the accretion idea highly attractive and started further work. He pointed out to Hoyle and Lyttleton that "You do not go into sufficient mathematical detail, and perhaps that's why you are being ignored." Bondi's scientific experience at the time was on hydrodynamics, or fluid flow. His daytime war research centered on the theory of the magnetron, an electronic device that today is the source of the radio waves in microwave cookers, but back then was the guts of a radar transmitter. He also studied colliding streams of electrons. When Fred explained the accretion mechanism, Bondi realized that with his knowledge of fluid flow and colliding electron beams, he might be able to put a lot of flesh onto the bones of Fred's theory.[21] At this stage, Bondi was still registered as a Ph.D. student at Cambridge and, in 1942, had been unsuccessful in his bid for a college fellowship: His dissertation lacked the necessary bite or brilliant insight. Bondi worked hard on the Hoyle-Lyttleton accretion problem and submitted his results for the October 1943 competition. Clues from his work on the magnetron had enabled him to extend Fred's ideas, with the result that Trinity College elected him to a research fellowship in October 1943.

Excited by winning the research fellowship, Bondi worked the dissertation into a joint paper.[22] It made a confident start: By then, interstellar molecules had been detected, and it was known that the temperature of the interstellar gas is low enough for the Hoyle-Lyttleton mechanism to operate. Furthermore, as a result of the low temperature, the theorists could ignore any effects due to gas pressure, leading to an important simplification. Atkinson was completely ignored because most of his objections had by now turned out to be wrong.

Bondi's simplification enabled them both to model realistic situations. Importantly, drawing on his knowledge of hydrodynamics, Bondi calculated the drag a star experiences as it ploughs through an interstellar cloud. It transformed stellar dynamics at a stroke. Suddenly, instead of stars gliding serenely through the Milky Way, there was chaos, with stars slowing down and speeding up again by turns, as they encounter clouds and gravitational fields. The interstellar medium, far from being a minor add-on to the galaxy, became an all-embracing force, effectively controlling the movements of the stars and itself being churned and pummeled by stellar encounters. He cut the time taken for a star to gain appreciable mass to one-tenth of its former value, speeding everything up 10-fold. In a crucial twist, Bondi looks at the situation of a star steadily accreting matter from the ubiquitous interstellar medium rather than denser clouds. In the last quarter of the twentieth century this would turn out to have major applications for understanding the physics of collapsed stars immersed in interstellar gas.

Bondi and Hoyle hoped "the establishment" would sit up and take some notice of their lively paper. Its impact on the astronomy community over time has been most unusual. In astronomy, most papers that are destined to make an impact get cited during the first few years following publication and then become more and more neglected or overlooked by the rising generation of practitioners. Initially, the Bondi-Hoyle paper was greeted with complete silence.

In 1947, Bondi, Hoyle, and Lyttleton suggested that the Sun's

Fred Hoyle as an infant, with his mother Mabel. His father kept this photograph in his pocketbook as he fought in the trenches of World War I. (Hoyle archive, St. John's College, Cambridge)

Fred aged about ten (the height of his truanting phase) in the back yard of his Gilstead home. (Hoyle archive, St. John's College, Cambridge)

The front court of Emmanuel College, Cambridge, where Hoyle started as an under-graduate in 1933.

The Bridge of Sighs, St. John's College, Cambridge. (S. Mitton)

Fred Hoyle pictured about the time of his first radio broadcasts in 1949. (BBC)

Tommy Gold, Hermann Bondi and Fred Hoyle at the International Astronomical Union meeting, Rome, 1952. (Hoyle archive, St. John's College, Cambridge)

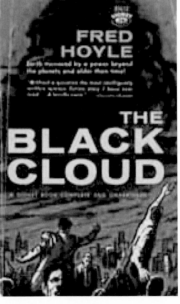

Fred Hoyle enjoys a joke with the physicist Richard Feynman in 1957 at Caltech. (California Institute of Technology)

Hoyle's first science fiction novel, The Black Cloud, is widely considered as a major classic in the genre.

Fred Hoyle standing between Sloan Laboratory and West Bridge Laboratory on the Caltech campus in February 1967. The dome of Throop Hall towers behind. Hoyle (age 51 here) had just invited Willy Fowler, Donald Clayton and Bob Wagoner to Cambridge UK for the opening of Hoyle's new Institute for Theoretical Astronomy. (Donald C. Clayton)

Fred's normal approach to research in the later part of his career: sitting at home in an armchair, which he always preferred to a desk in a faculty building. (Hoyle archive, St. John's College, Cambridge)

Fred Hoyle escorts Prince Charles at the dedication of the Anglo-Australian Telescope, 1974. (Australian Information Service)

Fred Hoyle opens the gate of his 200-year-old farmhouse on Cockley Moor above Lake Ullswater in August 1974. (Donald D. Clayton)

Fred Hoyle lecturing on solar neutrinos at Rice University in March 1975. Hoyle visited as Buchanan Visiting Professor of Astrophysics (Donald C. Clayton).

Fred (center) with his wife Barbara and daughter Elizabeth at the conference for his 60th birthday. Pictured with the Hoyles and the Burbidge (front left), Donald Clayton and Martin Rees (rear left) and Willy Fowler (front right)

Fred, seated here in his open top Austin Healey Sprite, liked powerful cars. (Hoyle archive, St. John's College, Cambridge)

Fred Hoyle in the New Cavendish Lecture Hall listening to a speaker. Behind him are Gustav Tammann and Martin Rees, who had replaced Hoyle as director of the Institute of Astronomy. The lack of focus on Rees seems to capture his emergence as Hoyle's successor. (Donald D. Clayton)

Sir Fred Hoyle, pictured in 2000, about the time of his 85th birthday (Cardiff University, UK).

The distinguished physicist Freeman J. Dyson, who is standing next to the bronze statue of Hoyle in the grounds of the Institute of Astronomy, Cambridge, shared many of Hoyle's interests in nuclear astrophysics and the origin of life. (Anna N. Zytkow)

corona, visible to the naked eye during total eclipses, is the result of the heating effects of accretion.[23] This was a theory susceptible to detailed mathematical study, and the work on it required extensions of the original accretion mechanism. The following year Hoyle used the Bondi-Hoyle-Lyttleton paper as the central plank of a monograph on solar physics he wrote for Cambridge University Press.[24]

One mathematician outside the group showed a spark of interest: William McCrea at Royal Holloway College, London. In the early 1950s, almost a decade after the publication of the Bondi-Hoyle paper, he used its results to model the evolution of hot giant stars. He showed that the giants had either grown by accretion or formed relatively recently, the latter explanation being the current view.[25] His student, K. N. Dodd, developed the work further by running calculations on the first digital computer at Manchester University.[26] Dodd appears to be the first researcher to refer to the mechanism as Bondi-Hoyle-Lyttleton (BHL) accretion. The alphabetical order is fair enough but hardly does credit to Hoyle, who had done most of the work. Leon Mestel, then a research fellow at the University of Leeds, took up the reins as well, by investigating how radiation from a star would affect the progress of accretion.[27] By now, cracks were opening in the original model: The stars would pass through the clouds too quickly for much material to accumulate, certainly not sufficient to affect stellar evolution. But Bondi's modifications had set up a picture of a star essentially at rest, accumulating matter from the interstellar gas, and this showed promise.

Ray Lyttleton also had made imaginative and independent use of accretion by developing a remarkable theory for the origin of comets. In 1942, he had written to Hoyle in a state of high excitement. "I have had a brainwave which I now put forward. The comets are due to meteoritic accretion when the Sun passes through a cloud of dust [all underlined]. This is so obviously the correct explanation that I cannot understand why we have not seen it before." Lyttleton had a lifelong aversion to any theory of the origin of comets postulating that new comets originated in a vast halo of would-be comets swarming on the edge of the solar system to far beyond Pluto. (As a matter of fact, this is our present understanding, and the distant swarm of icy worlds was finally discovered in the 1990s.) It took 10 years for his ideas to mature to the point where they were ready for publication, his completely different theory of comets seeing the light of day as a book, *The Comets and Their Origin*. He suggested that, when the Sun moved through an interstellar cloud of gas, ices, and dust, the accretion effect would concentrate the solid particles behind the moving Sun. Here, self-gravity and dynamic effects could assemble swept-up matter into something resembling a frozen pile of gravel, a few miles across.[28]

With his accretion model for the origin of comets, Lyttleton had set himself up for a new round of classic academic mudslinging between two conflicting theories—his gravel bank versus comets in hiding beyond Pluto. His model had merit in that it gave a crystal-clear and simple way of understanding how a comet forms. His opponents, the remote storage enthusiasts, were observers, not mathematicians, who never convincingly explained how their comet cloud had managed to end up so far from the Sun, conveniently out of reach of telescopic detection. Hoyle kept out of the ensuing squabble as it unfolded in the 1950s and 1960s, probably because problems in planetary science never really grabbed his interest. In 1941, he had ignored Lyttleton's explicit instruction to "think about accretion and comets." The planetary community never embraced Lyttleton's idea, essentially because it did not fit well with the observed behavior of comets as they blaze forth in the inner solar system. The observational astronomers saw comets as solid chunks of rock and ice, not a gravel-pile slurry. Yet Lyttleton's commitment to accretion remained one of fierce loyalty to his own ideas; by the time the Giotto spacecraft flew by the nucleus of Halley's comet in 1986, he cut a pathetic figure trying to defend a doomed theory.

Applications of BHL accretion had run into the sand by the mid-1950s, and between 1957 and 1971 the theory received just one acknowledgment in an independent research paper. But then a truly remarkable renaissance began. In 1967, Jocelyn Bell, working at the University of Cambridge, discovered pulsars—rapidly bleeping sources of radio waves. Theorists explained these as the long-sought neutron stars: the relict cores of massive dead stars that had collapsed until only a few kilometers in diameter and composed almost entirely of neutrons. Their density was about a thousand million million times higher than that of ordinary matter. In the 1970s, developments in "invisible" astronomy had reached the stage where x-ray and ultra-violet telescopes, operating in space on orbiting observatories, could detect the effects of neutron stars. Quite unexpectedly, the "invisible" astronomers really discovered places where accretion is happening.

For example, in 1973, Kris Davidson and Jeremiah Ostriker showed how to produce x rays in a binary star system that has a neutron star as one of its members. The neutron star captures material directly from the strong "wind" flowing from its massive partner.[29] Soon it was shown that neutron stars with companions having a mass similar to their own develop a thick disk of accreted material around their equators. In both cases, the infalling material produces very high temperatures where it lands on the neutron star, emitting x rays.

In the 1990s, the *Einstein X-ray Observatory*, a space observatory in Earth orbit, found for the first time faint x-ray sources located in globular clusters. Globular clusters are spherical clouds of stars and are among the oldest structures in the Milky Way. Astronomers think most globular clusters contain interstellar gas, the cumulative effects of material blasted from massive stars, though the gas has proved hard to detect. BHL accretion is one of several possible explanations for some of the faint x-ray sources. Isolated neutron stars may be accreting that gas, which releases energy in the form of x rays as it falls.[30] This is just one example of the remarkable revival of interest among modern theoretical astrophysicists in this early work of Hoyle and his collaborators.

Hoyle's first sustained research program, then, had a scientific impact that he could not possibly foresee at the time, although it is unlikely that this would have happened without the contribution of Hermann Bondi, whose improvements made the theory applicable to modern x-ray astronomy. What Hoyle achieved so early on came from working in a very different way than that of most of his contemporaries, who tended to develop a career in a particular area, such as solar physics or stellar evolution. The Greenwich and RAS establishment men had never seen an operator like Hoyle. Here was a person scarcely beyond the level of a research student who was firing away on all fronts, dismissing the work of others, and repeatedly pronouncing that his (incomplete) ideas must be correct, often just on the basis of the simplicity of the mathematics when very little direct observation under-pinned the work. No wonder the Council of the RAS was in disarray, with Hoyle apparently claiming to be an expert in virtually every area of astronomy.

Accretion provided the first landscape on which Hoyle clashed with a large number of people. He argued with the referees of his papers and with the RAS Council, as well as plunging into controversy with his interventions and questions at the regular meetings of the RAS. On most occasions when he spoke for a few minutes to state his views at open meetings, his remarks were followed by a fusillade of objections.[31] His manner of disagreement shocked the establishment, who were steeped in a tradition of polite behavior. Hoyle just spoke his mind and in a manner that little by little would store up trouble in the future.

CHAPTER 4

Hoyle's Secret War

Early in the war, Lyttleton and Hoyle had approached the Meteorological Office to see whether it had any use for the services of Cambridge mathematicians, but nothing resulted from this enquiry. They were concerned that compulsory military service might be introduced, though they figured they might be able to avoid it by supporting the war effort as civilian research scientists. Fred's call away from Cambridge eventually came from the Admiralty in Whitehall. A recruitment officer, Fred Brundrett, followed up a recommendation from Maurice Pryce, who had been hired by the Admiralty only a couple of months earlier. He set up an interview with Hoyle: The only offer on the table for him was a research position working on the development of radar. Brundett offered a salary worth about one-third of the emoluments from Hoyle's fellowships, which would be suspended. Fred accepted and agreed to be posted to the Signal School in Portsmouth.

The importance of radar technologies in combat had come to be appreciated soon after Hitler became a threat in 1933. The British government began to assess the possibility that Germany might launch an air assault on England, and in 1934 a large-scale air defense exercise was held to test the defenses of southern England. Mock raids were carried out on London. Even though their routes and targets were known in advance, well over half the bombers reached their targets without opposition. Prime Minister Stanley Baldwin's statement, "The bomber will always get through," seemed true.

To give time for their guns to prepare for the arrival of enemy bombers, the British Army experimented with detecting aircraft by their sound, using massive concrete acoustic mirrors with microphones at their focal points. In the first field test of this sonar, a wobbly milk cart rattled past the detectors, completely swamping the echo from the test flight. One of the experts observing these tests immediately wrote to the Air Ministry saying that new detection methods must be found, otherwise Britain would quickly lose in a major war. Whitehall responded by setting up the Committee for the Scientific Study of Air Defence, which turned to Robert Watson-Watt, then superintendent of the Radio Research station. At this time, "death rays" were all the rage in the popular imagination, and the committee asked Watson-Watt if a radio beam could be made sufficiently powerful to melt metal or kill a human being piloting a plane. The Ministry offered a prize of £1,000 to the first inventor who could kill a sheep at a range of 100 yards using a death ray.

Watson-Watt responded that a radio beam would have insufficient destructive power but that it should be possible to pick up a radio reflection from an incoming aircraft. He wrote a paper setting out the principles and asking for a grant of £10,000 to investigate this new method of detection. He demonstrated its feasibility with the reflection of a radio signal from a bomber flying over a regular BBC transmitter. Thus was radar (RAdio Detection And Ranging) introduced for military purposes, the British patent being granted in April 1935. The acronym "radar" was actually a later American creation; the British inventors called it radio direction finding.

The Air Ministry immediately started secret research into applying radar to the detection of aircraft. Rapid progress was made, and in December 1935 the Treasury authorized the construction of five radar stations for the defense of London. The Admiralty followed suit in 1936, with the controller of the navy directing the Signal School at Portsmouth to set up an experimental department to investigate the naval uses of radar at sea. Four years later Hoyle joined this department. By 1940, most battleships, aircraft carriers, and cruisers that had been constructed in British yards after 1936, as part of a massive rearmament program, were fitted with radar warning systems.

The importance of radar in naval operations was brought home to the Admiralty by the Battle of the River Plate, which took place in the South Atlantic off the coast of Uruguay, a neutral country, on December 13, 1939. None of the three British cruisers in this action (HMS *Exeter, Ajax,* and *Achilles*) had radar. The German pocket battleship KMS *Admiral Graf Spee* managed to inflict massive damage on HMS *Exeter* at a range of 9 miles with highly accurate fire, possibly assisted by radar range finding. However, the *Graf Spee* could not outrun the British light cruisers and ended up trapped in the neutral waters of the port of Montevideo by *Ajax* and *Achilles*. On December 17, the *Graf Spee*'s captain moved his ship from the port to the estuary and scuttled her. The British, posing as scrap metal salvagers, and using a Montevideo engineering company as a front, purchased the wreck from the German government for £14,000. On March 7, 1940, a British scientific officer managed to get on board, where he found all the equipment for gunnery radar—cathode ray tube, aerial array, electronic valves, and transmitter.[1]

The importance to the naval authorities of this discovery is to be found in a note written 2 years later by the Admiralty to the Treasury. When the bean counters at the Treasury complained that the £14,000 paid for the wreck "had gone down the drain finally,"[2] the director of the Signal Department made the following indignant reply:

> The examination of *Graf Spee* by our representative was most valuable in establishing the use of RDF [i.e., radar] by the enemy. It also provided sufficient technical detail in this matter to guide us in the search for enemy RDF in general and in the *devising and preparation of equipment for counter measures* [emphasis added].[3]

Hoyle would be a key player in the development of those countermeasures. The Admiralty now acted quickly to increase its research effort. The call went out to find suitable scientists and engineers. Before the outbreak of war, Fred Brundett had appointed about 100 men to the navy's research establishments. With the war on, he recruited at breakneck speed. Laboratory technicians, graduates, research fellows, and professors from universities, technical staff from industrial concerns such as Baird television, Philips, and Marconi, and experts from other government departments all poured in. Significantly, there were none from the Royal Greenwich Observatory or the Nautical Almanac Office, whose scientists were needed to run the Time Service and other vital navigational aids. Total staff numbers on Admiralty radar research rose rapidly to around the 1,000.[4]

As his train pulled into Portsmouth, to deliver him to his new job, Hoyle looked out at the gray city through thin rain. It was more burned and wrecked than any city he had seen so far. He took a taxi to the naval barracks, driving through dilapidated streets, and later recalled, "I didn't like what I saw."[5] In a few hours he had traveled physically and psychologically from the bright and cosy environment of Cambridge to the dismal, war-torn center of Portsmouth.

As a scientific civil servant, Hoyle settled into the standard office hours from Monday to Saturday, but spent his evenings and weekends on astronomy. He continued his research with Lyttleton on stellar evolution, particularly on the accretion model. But these were not peaceful times. Hardly a night passed when enemy bombers were not heard, their arrival being signaled by the opening up of the antiaircraft batteries. Hoyle said of this experience:

> Between the salvoes would come the steady patter of falling shrapnel. I must admit to being really scared by falling shrapnel. There was so much of it, and being killed by a couple of inches of jagged metal tearing into your skull was just as fatal as being blown to smithereens in the direct hit of a bomb.[6]

When the bombs fell, Hoyle, like many people, took a philosophical point of view: Short of a direct hit, he felt he was not in danger while sheltered from shrapnel and, if he did get a direct hit, he would know nothing of it. At this time, central London endured massive intensive bombing, its population cowering in basements, shelters, and underground stations by night. On September 21, 1940, the BBC Overseas Service made a live broadcast to the United States during a major air raid. Its reporters, out in Trafalgar Square and the shelters, talked with Londoners under fire.[7]

After the Battle of Britain in the late summer of 1940, Hitler switched from planning an immediate land invasion to the heavy bombing of cities such as London and Portsmouth. Nevertheless, the general public continued to fear that the German armies might try for an invasion. Shortly before Christmas 1940, Lyttleton wrote to Hoyle on scientific matters and concluded with a joke: "If the invasion should break out make in this direction and we can have a talk about things."[8]

Lyttleton was writing from Parkstone, Dorset, a rural location near Portsmouth. His letter looked forward to the ending of the war, even then thought to be at least 2 years in the future:

> I see that there will soon be nine chairs vacant at Cambridge. These will not be filled till 1942 at the earliest if then. Doubtless the same is going on at Oxford. There ought to be quite a few opportunities going when it is all over. If things are still the same Meave [Lyttleton's wife] thought we might take a bit of holiday somewhere in the extreme south-west this summer [1941]. . . . Still there is plenty of time for this yet. It would however give us a chance to talk things over and prepare a programme of work for next winter. I am looking forward to seeing your work on internal constitution [of the stars] and will work away at it as much as I can as usual without any rough stuff of course.

Much of this letter is an overview of the topics on which the two collaborators were working, all of which they intended to keep on the boil. Hoyle is thanked for reading and commenting on Lyttleton's paper about the origin of the solar system. Had Hoyle heard Eddington's latest 15-minute radio talk,[9] which Lyttleton had missed? Lyttleton also comments on a draft response to Atkinson's attack on the accretion papers. Next, he thanks Hoyle for recent comments on the motion of clouds through the Milky Way Galaxy. Abruptly the letter switches to the problem of transferring hydrogen to the center of a giant star to keep the nuclear furnace burning. In another sharp change of direction, the text brings in the idea that the redshifts[10] of galaxies could be due to their own massive gravitational fields. Without a pause (one can picture Lyttleton jabbing away at the typewriter keyboard with both index fingers), he is on to the sunspot cycle. Then there is a diatribe against Eddington and Atkinson, both of whom were in conflict with Hoyle and Lyttleton at this time:

I am just about as fed up with these two heels as I could be. Can you wonder there are wars when people like these are supposed to be our greatest thinkers and are looked up to and showered with honours, whereas blokes like [George Bernard] Shaw and Christ can hope for little more than ridicule.

From the outset of his naval career, Hoyle worked at a secret field station outside the village of Nutbourne, to the east of Chichester, rather than in the naval dockyard. The experimental station had been built well before the outbreak of war. It consisted of a large field with many aerials designed for naval radar, huts for the researchers, and machine shops for making prototype aerials. Secret equipment under test was hidden inside trailers normally used for moving horses or cattle. Security was high, with a 40-foot wooden watch tower and a high-perimeter fence patrolled by police and dogs. Those scientists who were based in Portsmouth took the train to Nutbourne Halt, where a lorry waited to shuttle them to the establishment. One mile to the north lay Funtington airfield, which could be used by test aircraft.[11]

Nothing that Hoyle had done in the previous 5 years qualified him to work in muddy fields tinkering with radar development. In his second undergraduate year, he had encountered electromagnetic theory, but that was it. He was now surrounded by technicians wielding soldering irons: It seemed that as long as you had a soldering iron (or an oily rag), you were seen to be doing your bit for the war effort. But Hoyle had no experience of instrumentation, so he decided that, if he could make any contribution at all, it would be through ideas, not hardware.

Fred settled down to read Admiralty reports on naval radar. Slowly it dawned on him just how vulnerable a ship was to torpedoes and dive-bombers. The navy had done everything in its power to protect ships from submarine attacks, but Hoyle was concerned about attacks from the air: How could a ship be better protected from them? British ships had early-warning radar and could call on fighter aircraft based on aircraft carriers to provide protection. However, the defending fighters knew only the distance of the enemy aircraft; they had no information on their altitudes. In June 1940, HMS *Illustrious* conducted flight trials off Bermuda, which brought home "the supreme importance" of height information for being able to deal with dive-bombers.

Hoyle saw that the difference between the attackers being repelled and ships being sunk lay in knowing the height of the incoming aircraft.

Hoyle studied the design of the Type 79 ship-to-aircraft early-warning radar. With dismay, he noticed that the designer had not fore-seen the need to measure altitude as well as range, possibly because the navy had initially expected to be able shoot down enemy aircraft with a barrage of antiaircraft fire from the actual ship being attacked. Real naval battles in the eastern Mediterranean, rather than pencil-and-paper exercises in Portsmouth, showed that this mode of defense was not effective, hence the need to use fighter aircraft. Between December 1940 and October 1941, Hoyle, ever the imaginative physicist, came up with a method to find both range and altitude using existing equip-ment. His first attempt at a solution did not work properly in battle conditions, but his improved method, introduced onto battleships and carriers from October 1941, greatly improved the success rate of defending fighter aircraft. It worked as follows:

Water reflects light and other electromagnetic radiation effectively. A lake with a dead-calm surface produces a mirror image of the sur-rounding landscape. Similarly, radar beams reflect off the sea. This means that radar emitted from a ship results in two beams, not one. One of them comes from the real transmitter and the other from the mirror image of it reflected in the sea. An aircraft flying toward a ship cuts through both of these beams. From the point of view of the ship's radar operator, the strength of the signal bouncing off the aircraft varies because the two beams interfere. It ranges from nothing, making the aircraft effectively invisible, to four times normal power.

Hoyle realized that the range at which an incoming aircraft enters the dead zone, in which it is invisible, depends strictly on its altitude. The radar operator could watch the signal slowly fade as the aircraft approached a position along its flight path where the destructive inter-ference between the two beams was total. The range at zero return signal could then be used to calculate the altitude. The radar officer could not possibly make the calculations on the spot, so Hoyle set to work producing a look-up graph relating altitude to range.

The range-altitude curve depended on many factors, such as trans-mitter power and the way the aircraft reflected radar beams. Hoyle, the beginner, was now learning from first principles. The data he needed on transmitters and receivers were not available at the Nutbourne field station, but in headquarters at Portsmouth Dockyard, where he was able to renew his friendship with Maurice Pryce. Having all the technical data at hand, Hoyle carried out an analysis of the errors involved. These depressed him profoundly. With the variations in performance from one radar set to another, it looked as if the inferred altitude would have an error of up to 50 percent, which was clearly useless to the pilot giving antiaircraft cover. In need of fresh air and time to think, Hoyle cycled to the downs for a ramble over the hills. Walking high up in the country air often cleared his mind and gave inspiration for the next move forward. By the time he returned to his bike, he had an answer to the problem.

Instead of having one curve for reading out the altitude, the radar officer should have a set of curves, calculated by Hoyle, for a range of circumstances. The operator would then run a trial with a plane of a type similar to those used by the enemy, flying at a known altitude. The test flight would then enable the officer to pick one curve from the set that most closely matched the actual performance of the particular radar set. The navy was already familiar with using test flights to calibrate its radar sets. Hoyle devised a simple graphical presentation, which soon became universal, in which the true height of an aircraft could be read off in seconds. His calculations embodied important corrections, such as one that allowed for the distorting effects of water vapor in the atmosphere. With Hoyle's graphs as a guide, each ship was recommended to construct its own individual version, either taken from worked examples in Hoyle's report[12] or calculated theoretically, and then to calibrate it experimentally during flight trials. Hoyle followed up his report by participating in several trials at sea.

Hoyle never received the credit he deserved for devising this method of measuring altitude. Radar work during wartime was cloaked in secrecy, for obvious reasons. His monograph had the classification MOST SECRET, so it could not have had wide circulation. It would not be released into the public domain until 40 years later. In the late 1980s, the UK's Naval Radar Trust arranged the production of an official history of naval radar in the Second World War. John Coales, who had received a major public honor (Order of the British Empire) in 1946 for his work on naval gunnery radar, took the lead in this history project. He wrote to many of the scientists who had worked in naval radar, asking for their personal recollections. Hermann Bondi had joined the naval researchers in 1942 and his response to Coales mentions Hoyle's solution to the altitude problem:

> The most important person is Fred Hoyle, who through making simple graphs, very early on enabled radar operators on [aircraft] carriers to determine the height of a plane from its fading pattern. I understand this was operationally very significant but he has never received proper recognition because it was a software not a hardware contribution.[13]

On another occasion, Bondi said: "It is of course typical of our regime that if Hoyle had made a clever black box or little gizmo of some kind there would have been a public reward for him once the war was over."[14]

During his first few months at Nutbourne, Hoyle had lodged in a large bright house with a family named Murray. As a consequence of the war, huge numbers of people were being displaced from their homes and needed temporary accommodation. Children were evacuated from London to the countryside; RAF airfields sprang up like mushrooms, bringing with them ground staff and pilots; there were city dwellers whose homes had been destroyed, and scientific researchers posted to secret establishments. Hoyle got a warm welcome at his billet: The Murrays wanted to "do their bit" for the war effort. On his first night he wrote to Barbara in Cambridge, saying that he would soon find a house to rent. In fact it took him many months to find a place, and though the small cottage had running water, there was no electricity. For cooking, the cottage had a coal-fired range, but the Hoyles hoarded their meager coal ration solely to heat one small bedroom. For 10 months they cooked every meal on a small primus camping stove.

For recreation, on evenings when he was not at work on astronomical problems, Fred could go to the Barleycorn pub in Nutbourne. In the public bar the locals played dominoes, crib, shovelhalfpenny, and darts. This was very much the locals' bar, with spittoons under the tables, and not at all attractive to the outsiders from the Signal School.

The saloon bar, with its padded seats, did attract the more refined types, but, even in here, privileged regulars had "their" chairs and woe betide any intruder who sat in them. In warm weather, Fred and his mates could use the garden bar, where there was a large aviary of canaries and budgerigars. The beer came in 36-gallon barrels, resting on trestles, and was always served warm. Beer and spirits were in short supply throughout the war. The Barleycorn got a delivery on Tuesdays, but the beer often ran out before the weekend. A shortage of glasses meant that jam jars would sometimes be pressed into service. The only bar food available was bags of plain potato crisps, each containing a twist of blue paper with salt inside.[15]

Out at Nutbourne, the field of antennas, with its sheep, huts, and scientists, was not a significant target for the German bombers. The same could not be said of the Signal School in Portsmouth: Bombing intensified during the autumn of 1940, bringing war to the dockyards and to the doors of the school. Back in September 1939, Churchill had urged a relocation of research and development away from Portsmouth.[16] The commander repeatedly tried to get his signalmen, telegraphers, and equipment moved out, but the top brass at the Admiralty in London instructed them to stay put until absolutely necessary—in other words, until bombed out. Finally, bombing massively damaged the school in January, March, and May 1941, and the staff then moved to standby accommodations. But this was most unsatisfactory because there were no air-raid shelters and the staff were crowded together in the basement of the Tactical School. One bomb could have killed them all.

By this stage of the war, the Luftwaffe was still much further ahead with the development of radar systems for air navigation than the British high command realized. On June 5, 1940, the evacuation of the British Expeditionary Force from Dunkirk had come to an end. The same day cipher breakers at Bletchley Park decoded a message from Luftwaffe headquarters, which indicated that the Germans must be using one or more radio navigation systems to guide bombers for attacks at night. Intelligence work soon showed that the system relied on two narrow radio beams, transmitted from different locations in Belgium and northeastern France, with both beams focused on the target. A bomber would fly along one beam and drop bombs where the second beam intersected its track.

Winston Churchill became prime minister on May 10, 1940, and 3 days later set up a small War Cabinet. The meeting of this War Cabinet on June 21, 1940, examined the possibility that the Luftwaffe did in fact have radio navigation systems. Churchill was extremely worried that such systems would allow the Germans to perform accurate attacks at night, overstraining Britain's air defenses. The senior RAF technicians summoned to the meeting proposed using countermeasures to jam the signals from the continent and throw the bombers off course. This was done, but, nevertheless, from mid-September the bombers were finding their targets at night. On November 14, 1940, the Luftwaffe spectacularly destroyed the city of Coventry, a vital center of weapons production in the English midlands. Immediately, it became a research question at the highest scientific level as to how far north these attacks could go, the limit being determined by the physics of radio beam propagation.

Maurice Pryce put Hoyle onto this problem. The standard method of calculating the bending of a radio beam by the Earth's atmosphere was cumbersome, involving lengthy calculations that could not be completed in time under conditions of extreme urgency. Pryce thought of a new solution requiring much less algebra and number crunching. He phoned Hoyle and set up a meeting with him the following day, saying he wanted an independent check of the mathematics and physics. Hoyle took the dozen sheets of paper and soon confirmed that what Pryce had done was correct. However, the formulas dealing with the intersecting radar beams could not be solved instantly using hand-cranked calculating machines. Hoyle must come up with a set of charts, as he had already done for aircraft altitude determination, for use under fire. A young mathematics graduate from Cambridge, Cyril Domb, had just joined them, and Hoyle sought his assistance. Between them, they cracked the problem in 6 weeks, producing a set of graphs from which the solution could be computed in about half an hour, a sufficiently brief period to allow countermeasures to be activated.

A few days later, Pryce was again in contact with Hoyle, with a new physics problem. The navy had put radars operating at the shorter wavelength of 10 centimeters (a frequency of 30 megahertz) on some of its ships, in a bid to improve the accuracy of range finding. The radar operators had made an astonishing discovery of great military significance. On some days, the radar detection range was much greater than expected, which meant the beam was traveling on a curved path far over the horizon! This did not happen every day, and the effect seemed to depend on weather conditions. Pryce suggested that water molecules in the layer of atmosphere above the sea might be responsible, and he asked Hoyle for an immediate report.

Hoyle's scientific investigations into how over-the-horizon bending of radio beams diminishes at longer wavelengths could have been invaluable in enemy hands. So, once again, national security compelled him to write up the science and the solution into a report classified MOST SECRET. He went to extreme lengths to keep it confidential, even arranging for the typing to be done under his direct supervision at Nutbourne. Despite the very limited circulation, the report reached Whitehall, where its reception would have important consequences for Hoyle's wartime career development. He would soon have his own theory group to manage.

On August 26, 1941, the experimental department of the Signal School was grandly renamed the Admiralty Signal Establishment (ASE). The change recognized the value of the contributions being made by its researchers, who would henceforth have their own "establishment," rather than being a small part of a large bureaucracy. Military strategists in London had by now realized that interservice rivalries could be hampering radar research, particularly the development of countermeasures. For example, scientists on war work for the Royal Air Force knew about German radars operating from Calais, but this intelligence was not passed to the Admiralty, whose ships in the English Channel were unwittingly exposed to accurate nighttime fire from German shore batteries on the French coast. To avoid further embarrassment and duplication of effort, Whitehall formed an interservice organization under Edward Appleton to coordinate signal intelligence efforts.

In September 1939, Appleton had held the chair of natural philosophy at Cambridge. The government had immediately appointed him secretary of the Department of Scientific and Industrial Research, the senior civil service post concerned with physical science. Appleton's own work on radio-wave propagation earned him a knighthood in 1941, and he was a member of the scientific advisory committee of the War Cabinet. Hoyle described him as "a small heavily built man in his mid-fifties, a scientific battleship on the London scene."[17] Appleton went on to win the Nobel Prize in 1947 for his discovery in 1924 of the reflecting layer in the atmosphere, now known as the ionosphere.

Appleton urgently needed to coordinate the research efforts on radar beam propagation. He set up a joint services panel to which various interested parties could make nominations. The Admiralty got two seats at the table and decided that one should go the ASE and the other to the naval hydrographer's staff. Considerations of seniority should have meant that Maurice Pryce would get the ASE place. However, he was too busy planning a major reorganization of the research department, so he nominated Fred instead. The hydrographer put forward Frank Westwater, the Cambridge mathematician who had been a year ahead of Hoyle at Emmanuel College and had become his friend and teacher. Westwater had already had an eventful time serving in the navy, having been sunk twice; he was the senior surviving officer of a ship that was lost while landing troops in Greece.

The joint services committee brought Hoyle and Westwater together. The committee had wide terms of reference and had to examine radio propagation at all wavelengths, long and short. However, most of the long-wavelength effects were already known, and so the bulk of the committee's work was to investigate short-wavelength aspects. Other mathematicians on the committee reworked the models of Pryce and Hoyle, finding them satisfactory. Soon, some trials would be set up to study the actual behavior of 10-centimeter wavelength radio waves over a distance of 100 miles.

Maurice Pryce continued to work at expanding and reorganizing the ASE. His younger brother, then at Trinity College, Cambridge, gave Maurice a tip-off: hire Hermann Bondi, a brilliant mathematician. Bondi thus started his war work on April 1, 1942, with a posting to Portsmouth as a temporary experimental officer. Bondi, a pencil-and-paper theorist, soon learned of the aerial research establishment at

Nutbourne, where "an interesting wild mathematician, Fred Hoyle" ran one of the departments.[18]

In the winter of 1941–1942, Britain was very much on the defensive. The Royal Air Force had fought and won the Battle of Britain in August and September 1940, with hindsight a major turning point in the war, since the defeat of Hitler's Luftwaffe deflected his ambition to mount an invasion. With the Luftwaffe fighters destroyed, the RAF switched to bombing raids as the only means of offensive action. However, Bomber Command faced an array of radars on the northern French coast and these posed a major obstacle in the way of launching surprise attacks.

The radar scientists needed a German radar set in full working order before they could invent a jamming technique. RAF reconnaissance discovered a radar dish on open ground outside a large farmhouse at the isolated village of Bruneval, France. French resistance fighters provided vital information: The farmhouse had 200 soldiers and radar operators and the nearby beach was not mined. On the night of February 27–28, 1942, twelve RAF bombers dropped three assault parties from the First Parachute Brigade. Most landed less than a mile from the radar setup. After regrouping, they launched an assault at about midnight. In the midst of an intense firefight, paratroops stormed the farmhouse and wrenched the equipment from its racks using crowbars. They captured a German radar operator and a large Würzburg[19] antenna. All this took only 15 minutes; then they headed for the beach with the prized equipment. At the beach, another firefight ensued and then a rescue by landing craft of the Royal Navy. The whole raid was brilliantly successful. British losses were two dead, six missing, and seven wounded. Back on the mainland, British intelligence examined the radar parts and interrogated the operator. From this intelligence the research scientists learned a great deal about the German radar system. At first the Würzburg looked impossible to jam, but eventually it was discovered that, if planes scattered a decoy of thin metal strips of varying lengths, it could effectively blind the antenna.

The Bruneval raid had an important consequence for Hoyle and the rest of the scientists at the ASE. So far, the London top brass had resisted all requests to move ASE out of the Portsmouth bomb zone, but now they realized the Germans might mount a counterraid to capture British radar sets on the coast.[20] The time had come to evacuate, the more so since the Experimental Division alone now had a staff approaching 1,000. ASE transferred its operation to a country estate, Lythe Hill House at Haslemere, and King Edward's School, Witley, at that time served by adjacent railway stations on the Portsmouth–London line. Other government departments had previously tried to requisition King Edward's School, but its trustees were a powerful body. The school had a Royal Charter and Queen Mary was its patron. The Admiralty referred the requisition request to Churchill, who obtained permission from King George VI to use the school only for as long as was absolutely essential to the war effort. The school reduced its roll from 280 to 120 boys, who moved out to the old workhouse at nearby Hambledon. Hoyle and the radar section continued at Nutbourne for a few weeks. They moved to Witley in the autumn of 1942, leaving Pryce in charge of the remnants at Nutbourne.

Hoyle now had important management responsibilities. In October 1942, he was appointed director of section XRC8 at ASE Witley, with Bondi as deputy director. Both held these positions until September 1945. As director, he ran a superb research group, with Bondi and Gold as key members. In the earlier part of the war, before joining Hoyle, the two men had been interned by the British authorities.

Hermann Bondi had been born in November 1919 in Vienna. He left his birthplace in 1937 to take up a place at Trinity College, Cambridge, to study for the mathematical tripos. In 1939, another Austrian émigré, Tommy Gold, arrived at Trinity to read engineering. Gold and Bondi first met, not at Cambridge, but on the concrete floor of an Army barracks 20 miles from Cambridge, on May 12, 1940. Both had been collared earlier that morning by the police and told to pack their bags for a few days. This was to be the start of internment. The British government, noting the speed with which the German forces had overrun Belgium and the Netherlands, had concluded (wrongly) that traitors or fifth columnists must have aided the Nazis. They therefore proceeded to round up the German and Austrian Jewish refugees who fled to England in the 1930s. Bondi and Gold were sent initially to the Isle of Man and from there were shipped to Canada, where they developed a firm friendship. Bondi in particular kept his academic work going in internment, despite having almost no books. By mid-1941, the government relented on the internment policy, so Bondi and Gold were finally released back to Cambridge in August 1941. Within 8 months, Bondi was applying his mathematics to radar, initially in Portsmouth and then a few months later with Fred at Witley.

Tommy Gold had graduated in engineering in the summer of 1942, unfortunately with poor results. Despite the fact that Gold had flunked his final examinations, Bondi remained extremely committed to getting his fellow Austrian Jew a good scientific position, and, supported by Maurice Pryce, he pressed Hoyle hard to recruit Gold. At first, this suggestion met with resistance from the Admiralty authorities because of Gold's poor degree. However, Hoyle placed great trust in Bondi's judgment; he pushed aside the bureaucratic objections and hired Gold as an experimental officer from November 1942.

A strong incentive for hiring Gold was his engineering background. Through the French resistance, ASE was receiving engineering blueprints from the German submarine base at La Rochelle. Within days of starting at ASE, Gold noticed on one blueprint an engineering drawing that looked to him like a snorkel. This device could be used to supply air to the submarine's diesel engine, which could then recharge its batteries, without the need to resurface. The snorkel, of course, had a much smaller radar signature than a surfacing submarine. To counter this, ASE would need to develop radar operating at shorter wavelengths that would have better resolution and so be capable of detecting smaller objects.[21]

Gold was delighted to join this small group of highly intelligent theorists. He recalls:

> I felt I could not possibly have done better for myself than join this group. The great enigma to start with was the director, Fred Hoyle. He seemed so strange: he appeared never to listen when people were talking to him, and his broad North Country accent seemed quite out of place. . . . [I]n fact he listened very carefully and had an extremely good memory.[22]

Their initial research area was the propagation of radio waves, particularly the effectiveness of short-wavelength radar at different distances. Shorter wavelength radars have the advantage of higher accuracy; the navy had equipped itself with 10-centimeter radars and had 3-centimeter sets in mind. As the push to shorter wavelengths developed, the Admiralty's major concern became the behavior of radio waves above the ocean, where variations in the amount of water vapor have remarkable effects; they can lead to long-range propagation under some circumstances or the absence of any such effect under others. For the navy, this was a curse, not a blessing, because an enemy well beyond the horizon could detect a ship using 10-centimeter radar.

By the autumn of 1942, Fred and Barbara Hoyle had a young son, Geoffrey, to look after, and at the time of the relocation to Witley, they had finally found a suitable house to rent not far from Nutbourne, at Funtington. Fred could not face house hunting all over again in Witley. So, for several months, he cycled 4 miles to the nearest train station and then took a 20-mile trip to Witley on a slow train; in wintertime the reverse trip was entirely in darkness. Hermann Bondi, meanwhile, stayed at his billet with the Palmer family. Tommy Gold, whose first contribution to the war effort had been working on a farm in the Lake District, also lodged with the Palmers initially. Tommy, however, was a strongly practical man, with an independent streak, and he did not want to camp out in a gracious house run by extremely polite people. Early in 1943, he found a little farmhouse with three bedrooms to rent. It was unoccupied because it was under the flight path of a nearby bomber station, where aircraft would sometimes crash on takeoff, blowing all the windows out of the cottage. Tommy and Hermann moved in and on weekdays Hoyle stayed overnight in the third bedroom. This was the start of a highly productive friendship.

About this time the trio organized themselves with private cars. Tommy got one first, an old Hillman, followed by Fred, who bought a 1928 Singer for £5. Thereafter, he motored over to Witley on Monday mornings, returning to Barbara on Saturdays. Gold got extra petrol ration coupons for them by arguing that they could not get to Witley before 9:25 a.m. using public transport. However, once they had secured the essential petrol ration allowance, in practice they rarely got started before 10:00 a.m! Food rationing was less easy to manipulate. A few special food items, such as chickens, stayed off rationing and, being in a rural area, they managed to get a chicken to roast most weekends.

Bondi recalls that the three of them spent all of their spare time at the farm cottage on scientific debates. They started private evening seminars on astronomy. In these sessions, Fred demonstrated great facility with "order-of-magnitude" calculations: Intuitively, he could figure out very roughly what a quantity was likely to be and would then go on to make further deductions on the basis of "ballpark" figures. According to Bondi:

> When Fred [stayed over] we spent all of our time discussing scientific problems. Fred's enormously stimulating mind, his deep physical intuition, his knowledge of the most interesting problems in astronomy, all combined to give me an outstanding scientific education in the few hours left after a hard day's work.

Bondi credits his wartime interactions with Fred Hoyle as being decisive for determining his postwar career.

> Fred Hoyle was always full of ideas in astrophysics, full of problems with which neither Tommy nor I were acquainted. When he stayed with us we talked late into the night about these questions.[23]

In winter they had insufficient coal to keep the drafty cottage warm. On chilly evenings they would keep their overcoats on and stamp up and down to keep warm, all the while talking science. What is particularly striking about this three-way collaboration is the manner in which Hoyle repeatedly made imaginative connections between areas of science that seemingly were not related at all.

A major aspect of their work at ASE was to assess the performance of existing equipment and forecast from the design the performance of equipment under development. The high-power 10-centimeter radar beams were produced in an electronic device known as a magnetron. The magnetron already had important industrial uses, but no theorist really understood why it worked. Hoyle gave Bondi the task of discovering how the multiple streams of electrons in the magnetron interacted.

During an evening discourse, Hoyle subsequently told Bondi of the work he and Lyttleton had done on accretion, but acknowledging that their theory had limitations. This was the moment when he engaged Bondi's interest in accretion, by drawing his attention to the common ground: Solutions to both the magnetron problem and the accretion problem would require an understanding of hydrodynamics (fluid flow).

Cosmology was evidently a significant feature of the evening discussions as well. The central dilemma in cosmology at that time was the current estimate of the age of the universe, which, at 1.8 billion years, was only half the age of the Earth! They talked a lot about the evolution of the universe and how that might affect the observations. Bondi felt that Hoyle was not such a good mathematician; he often found errors in Hoyle's solutions to equations, but Hoyle frequently got the "right" answer for the wrong reasons, such was the power of his intuition.

By late 1942, it became clear that Germany could not win the war, and the daily round of duties in Hoyle's theory group turned to giving advice to other sections. By now the different sections of ASE were working on improvements to the Royal Navy's equipment. Theoretical work continued apace on the propagation of radio waves above the ocean, with the emphasis on understanding the behavior of radio beams at a wavelength of 3 centimeters.

Hoyle and Bondi always found it difficult to arrange field trials with aircraft or submarines. The anomalous propagation of 3-centimeter radar beams was, in any case, unpredictable, the likelihood being that the effect would be undetectable on the day a test flight was allocated. To see the effect of changing weather conditions, Hoyle needed continuous observations rather than piecemeal data from intermittent flights. At this point, Snowdon, the highest mountain in Wales, beckoned.

The ASE had a station at Aberporth, on Cardigan Bay in south central Wales. Hoyle designed and implemented a physics experiment on a vast scale, involving a radar detector at Aberporth and a transmitter about 75 miles away as the crow flies, near the summit of Snowdon at an altitude of 3,560 feet. This setup for the radar beam gave a path over the sea for most of its length. A laboratory would be needed on Snowdon, and so the Admiralty requisitioned the cafe at the summit for Hoyle and his staff and surrounded it with barbed wire to prevent unauthorized access.[24] They also commandeered the Snowdon Railway, a rack-and-pinion line dating from the 1880s, for moving equipment and staff.[25] An old paraffin generator should have supplied electrical power, but it soon died, at which point Hoyle used his good standing with Edward Appleton to get two Lister diesel generators from the army. The huge radar set, designed for Arctic use, had an 11.5-meter dish as its aerial. Bondi ran the operation on the mountain, assisted in summer by an invalided-out former soldier and an elderly man and in winter by four sailors.

Fred sometimes visited Hermann Bondi on the mountain. At other times they met up at the military base on Cardigan Bay. If he needed to give Hermann a lift to the Snowdon train station, the timing was invariably misjudged. "I cannot remember a single occasion when Fred got us back to catch the train, and we made many night climbs in the naval car," Bondi recalls.[26]

Hoyle's original plan for the Snowdon radar station was the investigation of anomalous propagation. However, a new issue soon came to the fore. The Royal Air Force used radar to find enemy submarines operating in the Western Approaches. Bondi recalled the situation many years later:

> In that desperate struggle the one area that really worried Winston Churchill was the Battle of the Atlantic. To keep open the Atlantic lifeline was enormously difficult and the dangers [were] tremendous. Naval radar made a major contribution to this.[27]

The searching aircraft used radar to detect the submarines' periscopes, and the problem was scattering of the returning radar signals by ocean waves, particularly with the short wavelength (3 centimeters) needed for such low targets.[28] In 1943, the Americans had put a large radar into a DC3 aircraft and carried out successful field trials in Boston Bay. However, what would work in the calm waters of Boston Bay did not perform well in the rough Irish Sea. To learn more about wave "clutter," Hoyle had the brilliant idea of using the Snowdon radar set to study radar scattering from the rougher seas that occurred during the autumn and winter. This meant getting diesel fuel, and all the food that would be required, up to the summit of Snowdon before winter set in—a major undertaking. Bondi worked through the winter, until late January 1945, by which time he had made the earliest comprehensive measurements of the effects of wave clutter. Of course, by this late date, enemy submarine activity had much diminished after the liberation of France.

While Bondi worked on this experiment in late 1944, Hoyle had his first taste of science in America. This arose from his membership on Appleton's coordinating committee of radar research. When the committee assembled in autumn 1944, Appleton told the members that the U.S. Navy would be holding a scientific meeting on anomalous propagation at the Naval Research Laboratory. The U.S. Navy maintained a London Mission, which shared some secret information on radar technology, but matters had reached a point where a roundtable meeting was desirable. Appleton recommended that Fred Hoyle and Frank Westwater attend the meeting in Washington, D.C.

Westwater and Hoyle embarked from Greenock, on the Clyde, aboard the Cunard Line's RMS *Aquitania*, together with 7,800 American GIs who were returning from Europe for Christmas leave. *Aquitania* had seen service as a hospital ship in the First World War and, after a refit in 1920, enjoyed commercial success, providing a weekly express service to New York. In the Second World War, she carried about 300,000 Commonwealth and American troops to and from the theaters of war. A day or so out of Greenock, the twice-sunk Captain Westwater took Hoyle to the stern of the *Aquitania* and explained what to do if they were torpedoed: "When a ship is torpedoed, pandemonium reigns and nothing works as it should." The best plan, he explained, is to ignore the scramble of people rushing to the lifeboats and make your way instead to the stern. There, you should slide down one of the many ropes and wait in the water until the ship goes down, at which point an immense amount of jetsam comes up and you can find something to cling on to. Out in the mid-Atlantic in winter, with 30-foot waves, Hoyle took little comfort from this advice, or from his life jacket for that matter. As financial protection for Barbara, and his two children Geoffrey and Elizabeth, he had taken out as much personal life insurance as he could afford with Lloyd's of London. Life at sea included three sumptuous meals a day; the ship was resupplied from U.S. sources when in New York. Westwater's cabin was crowded, with eight men allocated to a space that would have slept two in the ship's prewar configuration. By day, they worked out the vessel's secret course, south of the Azores, by observing the altitude of the Sun at noon. And for relaxation in the evening, they took to dropping in on the radar officer's cabin for a drink.

Late on a short November afternoon, *Aquitania*, "the beautiful ship," majestically swept through the Verrazano Narrows (the present suspension bridge dates only from 1964) with its view of Brooklyn Bridge, past Ellis Island and the Statue of Liberty, and up the Hudson River to a berth at 42nd Street, close to the present site of the New York Port Authority Bus Terminal (constructed in the 1960s). For Hoyle, the sight of Manhattan at 6:00 p.m. on a November evening was absolutely amazing: After 5 years of blackouts in England, the dazzling lights of lower Manhattan seemed like fairyland, and the port area was a hive of activity.

A short cab ride took the two of them uptown to the Barbizon Plaza Hotel on East 63rd Street, then considered a stylish place, but they would enjoy its comforts for only one night. The following day they headed for Penn Station on 33rd Street, to take their seats in a Pullman car bound for Washington. Westwater sat resplendent in his gold-braided Royal Navy dress uniform, eagerly conversing with other passengers. The cocktails came in an unending stream once the train was out of New Jersey, then still a "dry" state.

Their first port of call in Washington was the British Embassy. Here, Hoyle received invitations to visit the radio propagation research teams at the Massachusetts Institute of Technology in Cambridge, Massachusetts, and the naval headquarters in San Diego, California. They pocketed $400 each (then an enormous sum) for "expenses," with travel tickets and hotel expenses on top. This was indeed VIP treatment.

It would be another 3 days before the radar conference that was the ostensible purpose of the visit would take place. Fred Hoyle used his time productively. He and Westwater took the train back north to Princeton Junction, where they stayed at the *faux historic* Nassau Tavern on Palmer Square.

Fred's Princeton agenda was to visit Henry Norris Russell, the director of the Princeton University Observatory and one of America's most distinguished astronomers. On June 25, 1913, Russell had announced, at a meeting of the British Astronomical Association,[29] his discovery that the total energy that stars radiate and their temperature are correlated, a relationship revealed by a form of graphical presentation long since known to astronomers as the Hertzsprung-Russell diagram. In the late 1920s, Russell had used the new quantum theory to determine the proportions of the different chemical elements in stars, and he found the relative amounts of 56 different elements in the Sun's atmosphere.[30] He also showed that the physical properties of a star at each stage of its evolution can be found solely from its mass, chemical makeup, and age.[31] Hoyle was already well versed in all the areas on which Russell worked: stellar composition and evolution, variable stars, dynamics, and the origin of the chemical elements. By 1944, Russell had already published almost 200 papers, many of them through the Royal Astronomical Society. Unknown to Hoyle, Russell had earlier heaped praise on one of the Hoyle-Lytteton accretion papers that the RAS sent him for peer review!

Russell spent six decades at Princeton, was revered for his encouragement and enthusiasm, and was an ebullient supporter of younger astronomers. He treated Hoyle and Westwater royally, engaging in expansive conversation. Indeed, the visitors found it hard to get a word in! The critical moment in this near-monologue came when Hoyle managed to mention that they would be going out West, to San Diego. Russell, who made lengthy annual visits to the Mount Wilson Observatory from 1921 onward, urged them to visit Walter Adams, the director of the observatory, at the offices on Santa Barbara Street, in Pasadena. The Mount Wilson Observatory was then the greatest astronomical observatory in the world, with its magnificent 100-inch (2.5-meter) telescope, through which Hubble had discovered the expansion of the universe. Russell immediately wrote a letter of introduction on their behalf to Walter Adams.

Back in Washington, they went to the conference organized by the Naval Research Laboratory, but this made no impression: Hoyle could remember nothing of it when he wrote his autobiography over 50 years later. He did remember the details of the flight west on November 23: departure at 7 p.m. from Washington National Airport in a DC3, stops in Knoxville, Tennessee; Little Rock, Arkansas; Amarillo, Texas; Albuquerque, New Mexico; Phoenix, Arizona; and finally arrival at San

Diego at noon. Through the long night they ate hamburgers or apple pie at the ground stops; the aircraft provided no food. Hoyle thought the flight "was marvellously quick."[32]

In the port of San Diego, the display of naval firepower in the harbor and the sight of an open-air production line building fighter aircraft amazed Westwater and Hoyle. They stayed at the lovely La Valencia Hotel in the heart of La Jolla, home to the University of California's San Diego campus.

A car journey lasting about 3 hours took them from San Diego to Los Angeles; there was then no freeway. In L.A., they paid a quarter each (real silver quarters) for the streetcar ride to the fabulous Huntington Hotel in Pasadena, where they stayed courtesy of the British Embassy back in D.C. The next morning they hiked the couple of miles up Lake Avenue, to meet Walter Adams on Santa Barbara Street. Adams was a dour New Englander who was already past the normal retirement age when Hoyle met him. He was not as friendly and outgoing as Russell had been the week before. Hoyle, in any case, probably focused the conversation around his own self-centered needs.[33] Adams did roll out the red carpet at the end of their conversation, however, by announcing that a car would be available within the hour to take Westwater and Hoyle to the Mount Wilson Observatory for the weekend, where they would stay in "The Monastery," dormitory accommodation of 15 bedrooms for astronomers using the telescopes. At the time, gasoline was severely rationed and Adams needed to get them onto the mountain by means of the daily shuttle service. Adams also said that they would be hosted by Walter Baade, who was observing that weekend.

On arrival at the observatory, which is at an altitude of 5,708 feet, the two English visitors were shown the solar observatory. Its 150-foot solar tower produced 17-inch images of the solar disk, showing dark spots in a wreathing cauldron of solar granulation, all deeply impressive to a man who would a few years later produce a major theoretical monograph on recent advances in solar physics.

The Monastery served dinner at 5 p.m. The gentlemen dressed formally in jacket, collar, and tie. A formal seating plan required Walter Baade as the observer on the 100-inch to preside at dinner, just like on

High Table at a Cambridge college. In a spirit of generosity, Westwater placed two bottles of wine on the table, a major *faux pas*: an iron rule of the observatory forbade alcohol on the mountain. The cook silently removed the offending bottles.

In those days the view from Mount Wilson was awesome, both by day and at night. The phenomenon of "smog" still lay about a decade in the future. In the day, the city of Los Angeles, sprawling below, looked close enough to touch. At night the city sparkled with light and, above the trees, the stars burned in a black sky.

After dinner, Hoyle and Westwater wanted to see the 100-inch in action. By the time they had strolled over, the huge dome was already open, the night assistant busy at a wooden console, his face dimly illuminated by a faint red light. The structure of the telescope rose high above them: 100 tons of steelwork forming the framework of an open "tube," embraced by a closed fork mounting. Most of the telescope's weight was borne by two large steel floats submerged in tanks of mercury, allowing the massive cosmic cannon to be targeted effortlessly on the starry vault beyond.

In the dome of the 100-inch, the great telescope was being readied for spectroscopic observations of stars and galaxies. The night assistant demonstrated the preparations required for making observations. With a series of push buttons, the main floor in the dome slowly moved up or down, the dome rotated to position the observing slit suitably, while the telescope itself swung around to point in the correct direction. All of this was accompanied by a cacophony of sounds: direct-current electrical relays snapping into action, huge fans sucking warm air out of the dome, the whirring of electric motors, and the trundling of the dome. This night at the 100-inch sent a tingle down Fred's spine. The thought began to form in his mind that henceforth he should perhaps look to the Californian astronomers for the best observations. Unlike the civil servant observers in Great Britain, they were independently funded as well as brilliantly equipped. And he would find, much later, that Pasadena too had theorists who deeply respected his knowledge of nuclear physics.

On the Monday morning, the pair set out to walk 7 miles down the mountain back to Pasadena. Mount Wilson is twice the height of

Snowdon, and Hoyle's wartime shoes had practically disintegrated by the time they reached Altadena, where by chance Baade recognized them and gave them a lift in his car. In his own account of this lift, Hoyle claims of the ride, "[Baade] never corrected the curve of a car as it went toward the edge of the road until almost at the disaster point. This penchant provoked trouble with the Pasadena police, with whom he had long-standing battles."[34] Fred tended to enhance little events and stories by making them sound dangerous or illegal! The police stopped Baade because he was an alien who had been granted special privileges, as we shall see.[35]

Although less well known outside astronomical circles than Edwin Hubble, Walter Baade (1893–1960) was arguably the most influential observational astronomer of the twentieth century and certainly the greatest observer Hoyle ever met. Baade's most important contribution to astrophysics was not, as is often contended, his revision of Hubble's distance and age scales for the universe. Rather, it was his discovery of two distinct stellar populations: old and young stars. This discovery opened up the study of stellar and galactic evolution. These research areas would be among the most fertile and exciting in all of astrophysics for decades to come, with Fred Hoyle making major contributions from 1945 onward.

Baade, a German, had gone to the United States in 1931 as a staff member of the Mount Wilson Observatory. He made the observations that gave rise to his concept of stellar populations during the wartime years, taking advantage of darker skies at Mount Wilson. A partial blackout was still enforced in Los Angeles and the San Gabriel Valley. Most Mount Wilson observers were working on weapons development programs, devoted to bringing Germany to its knees, while he, formally an enemy alien in their midst, was confined to Los Angeles County but had almost unlimited use of the most powerful telescope in the world. In Los Angeles, a military curfew operated from April 1942: All enemy aliens had to be in their homes between 8:00 p.m. and 6:00 a.m. Adams, as director of the observatory, appealed this rule to the highest possible level in Washington, and the army command issued an exemption allowing Baade out at night, strictly on the understanding that this was for professional purposes only.[36]

Back in his office, Baade was extremely helpful to Fred Hoyle, who had been cut off from the American astronomical literature for over 3 years. As a conversationalist, Baade was "almost the complete opposite of Adams," a fascinating talker "who never let a few facts get in the way of a good story."[37] Their exchanges turned first to Hoyle's current interest in research on novas—stars that briefly shine much brighter than normal as a result of an outburst. "If you are interested in such things," Baade said, "then why not look at supernovae, which are vastly more powerful?" So they talked about the implosion of massive dying stars and the nuclear explosions that are triggered by stellar collapse.

Their conversation moved on from supernova explosions. Baade had just made a very great discovery, which he now disclosed to Hoyle, who thereby became the first British astrophysicist to learn some dramatic news about stars.

Baade explained that he had recently turned his attention to M31, the great Andromeda galaxy about 2 million light-years from the Milky Way, and its two satellite companion galaxies, M32 and NGC205.[38] Baade handed Hoyle offprints of a couple of papers recently published by the *Astrophysical Journal* and took him through the main findings.[39] Among the local galaxies, M32, NGC205, and the central nucleus of the Andromeda galaxy stood out because no astronomer had succeeded in resolving the amorphous blur of light into individual stars. But the previous autumn (1943), in the most favorable months for observing Andromeda and its companions, Baade had used newly available photographic plates that were sensitive to red light to resolve individual red giant stars in these regions. This required an immense level of skill as an observer, and no other astronomer could have achieved this in late 1943. On examination, these precious photographs showed that the brightest stars were yellow giants. The real surprise was what the photographs did *not* record: red and blue supergiant stars that are so common in the spiral arms of the Milky Way.

Expressed in modern terms, Baade's great discovery showed beyond doubt that there are distinct "populations" of stars. In his paper, he described "Population I" as characterized by the "ordinary" stars in the solar neighborhood of our galaxy, the brightest examples of which are blue and red supergiants. "Population II" stars were the types found in elliptical galaxies, the central regions of spiral galaxies, and globular clusters. This excited Hoyle; he could immediately see that there must be profound implications for astrophysics. What physical mechanisms were at work to sift and sieve the stars into two such very different populations? Why did Population II not have any red giant stars? Why did the populations in large globular clusters and small elliptical galaxies look so similar?

As an indication of the profound importance of these discoveries, the editor of the *Astrophysical Journal*, Otto Struve, took the unprecedented step of accepting Baade's request that a high-quality photograph be reproduced with the paper. This required 700 prints to be made by hand, a task that took two assistants most of the summer. Baade loaded Hoyle with other offprints and articles, bringing him up to date on 3 years of progress in American astronomy.

The route back to England went via Boston and, for the third weekend in succession, gave Hoyle an encounter with a great astronomer, this time Harlow Shapley of Harvard College Observatory. Shapley's most significant contribution to science had been the determination of the dimensions of our galaxy and of the location of its center. Although Shapley's research career had peaked a decade earlier, his administrative skills were second to none, and he established an outstanding astronomy graduate school at Harvard, an achievement that stands today. Hoyle must have been slightly in awe of this grand elder statesman of American science.

From Boston, Fred went north by train to Montreal, ostensibly to visit radar research stations in Ottawa. In Montreal, he ran into Maurice Pryce, who had long since disappeared from ASE and was by now working in utmost secrecy on the British efforts to make nuclear bombs. The British had moved their atomic weapons research team to Montreal in 1942. Hoyle learned from Pryce that the British had made important advances in measuring the energy levels of atomic nuclei, and this classified information impressed Hoyle, who wondered if the new data there would have implications for astrophysics.

Fred got back to Barbara and the children, Geoffrey and Elizabeth, just before Christmas and, during the holiday season, turned over in his mind the momentous meetings in America. With Baade he had discussed supernova explosions, and with Pryce nuclear bombs. Could there be a connection? What if a supernova implosion and subsequent explosion was like the trigger and detonation of a nuclear bomb? There was no possibility of doing experiments. Could the conditions be calculated? How would one figure out the nuclear reaction chains inside the imploding star? It would be some months before he could explore these connections in a proper physics environment back at Cambridge.

By the spring of 1945, Hoyle's war service was drawing to a close. Many of the projects on which he had worked would go on to successful implementation after the war, drawing on the much more reliable electronic devices that became available from 1945 through the early 1960s. The Admiralty called him to a high-level meeting to discuss future plans for the Royal Navy. He claims[40] that, as the technical expert from ASE, he informed the assembled gold braids that "it might be possible to extend the range of naval gunnery to 25 miles by using aerial torpedoes with aerodynamic lift. These missiles could, in principle, be guided to their target." "What a horrible idea," boomed the admiral in the chair. The modern cruise missile is an implementation of Hoyle's idea.

Hoyle's last war duty was to visit Germany in late May 1945 to discover what the Nazis had done in the way of naval radar. This was mission impossible. He found that the destruction in the final stages of the war had been too great for him to be able to learn anything useful.

In July 1945, Fred climbed into his little Singer two-seater and left ASE for the final time. Back in Cambridge at last, he found that Lyttleton and Pryce, who were older than himself, had secured permanent appointments as lecturers in the Faculty of Mathematics. For the rest, temporary junior appointments were set up. Bondi and Hoyle each secured one of these, at the rather paltry salary of £200 a year. The St. John's fellowship had 18 months to run, at £250 a year. This, then, was Fred's reward for war service: a salary of £450 instead of £850 in 1939, decreasing to £200 in a year and a half. In Cambridge the prices of decent houses had doubled, the purchasing power of money had halved, and food rationing worse than anything due to Hitler was about to be introduced by Attlee's postwar government. It could have been easy to get depressed about the missed opportunities, the lack of a permanent position, and the low salary.

But Fred did not allow any of the trials of daily life to get him down. Intellectually, he had profited enormously from his wartime experience and the contacts he made. He knew a great deal about radar and the propagation of electromagnetic radiation. He had forged a wonderful working relationship with Tommy and Hermann, which would make up for the fact that Eddington had died in 1944. The research on accretion had gone tolerably well, and new lines opened up in terms of research on the structure and evolution of stars. The three of them had started to have ideas about the evolution of the universe, but nothing had come of this yet. Most importantly, he had met three of the most famous and productive astronomers in America: Russell, Shapley, and Baade—and the encounter with Baade had sparked new connections in Fred's mind. He felt a strong motivation to understand much more about the inner workings of a star and the role of supernova explosions.

CHAPTER 5

The Nature of the Universe

For about a quarter of a century, from 1948 to 1972, Fred Hoyle was one of the most famous astronomers in the world, renowned in the public eye not so much for his solid achievements in astrophysics but for his revolutionary ideas on the origin and nature of the universe. In Britain, Hoyle enjoyed two decades of being one of the best-known scientists in the country. His controversial ideas and the disputes they caused were often in the media. The series of events that would turn him into the world's most quoted cosmologist commenced during his war service.

When he decided to make a career in theoretical astronomy rather than nuclear physics, Hoyle originally gave cosmology a wide berth. He afterward claimed that, in the late 1930s, it seemed inconceivable to him that something as vast and complex as the universe could be understood on the basis of the sparse observations that had been made up to that time. By today's standards, astronomers then knew little about the universe. Edwin Hubble had shown that spiral galaxies contain stars and are very much like the Milky Way. He had discovered the relationship now known as Hubble's law: The more distant a galaxy is from the Milky Way, the faster it is moving away from us. Expansion of the universe offered a natural explanation for this link between the distances of galaxies and their speeds of recession.

Hubble and his collaborators found through subsequent observations that the universe at large looks the same from every vantage point and in every direction. If you were to be taken blindfolded from one place in the universe to another, on removal of the blindfold the galaxies around you would look broadly like those you had left behind, and you would be unable to distinguish from which direction you had traveled. In scientific terminology, the universe is said to be homogeneous and isotropic. These discoveries profoundly impressed astronomers at the time, who were astonished that the universe appeared so uniform.

The public and the professionals both held Hubble in awe, despite the fact that the age of the universe he obtained from his observations was in extreme conflict with the age of Earth as evaluated by geologists. Once Hubble became convinced by the expansion idea, he could estimate an age for the universe by calculating how much time had elapsed since all of the galaxies had been crowded together. The elapsed time came out at 1.8 billion years. But earth scientist Arthur Holmes had proposed ages of 3 billion to 4 billion years for the oldest rocks on Earth. How could the Earth be older than the universe? Could there be something wrong with the interpretation of the observations? No contemporary astronomer questioned Hubble's observations in this respect: He had never been seriously mistaken before. Furthermore, because he had unlimited access to the greatest telescope in the world, the Mount Wilson 100-inch, no skeptical observer could challenge his findings. Another two decades would go by before Hubble's successors could prove that his distance measurements were off by almost a factor of 10. By that stage, Fred Hoyle was very much a cosmologist himself, as we shall see.

The structure of the astronomy profession in the 1940s was very different from today. Theorists were thin on the ground and tended to be found in university mathematics departments. An observatory might have one or two theorists, but the bulk of the academic staff engaged in hands-on observational projects and instrument development. Theorists in the universities concentrated on tractable problems that could be solved without electronic computers (which did not exist). In practice, this restricted theoretical research to dynamical astronomy and astrophysics, with the emphasis on finding equations that could be solved mathematically rather than by "number crunching." Dynamical astronomy dealt with topics such as the orbits of minor planets, whereas in astrophysics the main activities centered on understanding the internal structure of stars and applying atomic theory.

During the Second World War, the Royal Astronomical Society's (RAS's) journal *Monthly Notices* published fewer than a handful of theory papers a year. Just four people contributed most of the output: Hoyle, Lyttleton, Eddington, and Harold Jeffreys, all at Cambridge. Small wonder then that the RAS Council had difficulty with the peer review process for theoretical submissions. Hoyle and Lyttleton were the most prolific fellows of the RAS, with scientific interests ranging far wider than any other British astronomers. But as far as cosmology is concerned, not a single paper appeared from this field during the war years.

Cosmology had nevertheless made some progress in the previous quarter of a century. Theorists began to construct mathematical descriptions or "models" of how the universe as a whole might behave over time, often making simplifications or assumptions so the equations could be solved and preconceptions taken into account. In 1917, Einstein's general theory of relativity became available in English. Einstein favored the notion of a universe that is neither expanding nor contracting on the large scale, but he faced a major problem with this preference. General relativity allowed the universe to be homogeneous and isotropic (as observed), but not stationary if it contained matter. The material universe had to be moving under its own self-gravity, a point already grasped by Newton two centuries earlier. Because he did not believe the universe to be collapsing, Einstein introduced into his equations an extra factor, which worked as if there were a repulsive force between the galaxies. In the static universe of Einstein, this arbitrary assumption canceled out gravitational attraction on the very large scale, without having any detectable effects on the scale of the solar system.

At about the same time, Willem de Sitter arrived at a different model of the universe, but in his simplified picture the universe contained no matter. However, he made the important step of demonstrating that more than one model of the universe is allowed by Einstein's general theory of relativity. With Einstein's extra repulsion added in, the space in de Sitter's model universe springs to life: It expands rapidly, but because there is no matter, nothing appears to change. It is an example of a steady-state universe, with an infinite past and an infinite future.

Between 1917 and about 1930, the only models on offer were Einstein's matter-without-motion universe and de Sitter's motion-without-matter universe. Very few observations had been made of distant galaxies, and there was nothing to help choose between them. In practice, the professionals gave de Sitter the cold shoulder and mostly believed in the static universe. With the collective mind-set closed, it is no surprise that the next innovator, Alexander Friedmann, initially shared de Sitter's fate of being ignored.

Friedmann was a superb Russian-born mathematician who made an enormous contribution to the development of the mathematical theory of the expanding universe. The Soviet Revolution of 1917, followed by the civil war and subsequent blockade of the Soviet Union, introduced a considerable delay before Soviet scientists became aware of general relativity. Friedmann was one of the first to appreciate the significance of the theory for cosmology.

In 1922, Friedmann wrote a classic paper on the curvature of space, published in German, in which he floated the idea of the universe changing its appearances over time. A further paper appeared in 1924. These two papers showed that the universe could in theory have both matter and motion without violating Einstein's equations and that the universe could be either expanding or contracting. Friedmann's work, coming from Soviet Russia, was ignored for at least a decade. He died of typhoid in 1925, before the significance of his work became appreciated.

In 1927, Georges Lemaître, a Belgian priest, advocated the idea of the universe having a beginning in the form of a "primeval atom," which he envisaged as a very high density initial state. He had earlier (1923–1924) spent a year at Cambridge, studying cosmology with Arthur Eddington. Lemaître independently discovered the same results as Friedmann, and so Lemaître's model universe starts with a primordial explosion, which is followed by rapid expansion, a coasting stage, and then a further phase of expansion. Initially, this model too received the silent treatment, but by good fortune Eddington became aware of Lemaître's work and started to promote it. Eddington saw the model's potential to resolve the conflict caused by the universe appearing to be younger than Earth. The solution to that difficulty was to add a repulsive force into Lemaître's model.

Eddington, the only serious cosmologist in Britain in the 1920s, triggered an important paradigm shift. He modified the "primeval atom" concept by proposing that the Lemaître universe be allowed an infinite period of time in the past to get started. Somehow, as a result of random disturbances, Eddington's universe awakes from its slumber and starts to expand. Galaxies form and it evolves to the state we see today. He popularized his theory in a small book, *The Expanding Universe*, which brought the idea of expansion to a much wider audience. Einstein gave this version of the expanding universe his blessing, and another Cambridge popularizer, James Jeans, made it newsworthy in a series of articles for the *Sunday Express*, a mass-market newspaper. Hoyle had become aware of these ideas while still at Bingley Grammar School.

A rigorous approach to cosmological theory finally emerged in the mid-1930s. Howard Robertson and Arthur Walker, both mathematicians, produced a geometric description for a universe filled with dust. Crucially, they showed how to construct a global four-dimensional grid system that encompassed both space and cosmic time, and their approach was sufficiently general to be useful in a variety of different models of the universe. In effect, they created a mathematical tool kit that allowed cosmologists to explore different initial conditions for the universe to see how it would evolve.

When Fred Hoyle started at the Admiralty in the autumn of 1940, certain ideas about the universe were more or less settled. There was agreement that it must be isotropic and homogeneous. The expanding-universe model gave a natural explanation for the recession of the galaxies. Finally, Robertson and Walker had shown how to handle geometry in a situation where space and time were inextricably linked.

Despite these advances, cosmology was in a state of turmoil. The numbers of professionals working in cosmology remained tiny. Not everyone accepted that expansion must be the cause of the redshifts observed in the light from galaxies. Even Hubble cautioned against interpreting redshifts as evidence for cosmological expansion. The age problem remained: Hubble stood by his 1.8 billion years, and because of his prestige, his methodology and conclusions remained unassailable. At the same time, some scientists were uneasy with a universe that appeared to have had a beginning in an explosion that came from nowhere. They felt it was philosophically unsatisfactory to have a creation event as the answer to the question "Where did the universe come from?"

In 1940, there was a wide range of cosmological theories from which to choose. For example, Paul Dirac and Arthur Eddington both worked on strange coincidences involving very large numbers in physics.[1] Dirac allowed the strength of gravity to vary with time. Robertson-Walker geometry permitted universes to expand or collapse, to be open (going on forever) or to be closed (having finite extent), to be created or to have existed for an infinite time. Eddington felt that cosmological studies would reveal entirely new physics but could not anticipate what that might be. Observational techniques produced data that left plenty of room for doubt. It was impossible to tell whether the apparent expansion was speeding up, constant, or slowing down. These circumstances combined to leave the field wide open to imaginative minds, and in the early 1940s Fred Hoyle was the most imaginative theorist in Britain.

Although Hoyle did not undertake research in cosmology during his war service, the topic was regularly featured in the evening seminars. At Witley, Fred was repeatedly goading Tommy Gold and Hermann Bondi into thinking about the nature of the universe. The nagging questions that never went away were, What did Hubble's data really mean and why did the age of the universe appear to be only half the age of Earth?

In their tripartite collaboration, each played a different role. Bondi, a brilliant mathematician, took up the technical aspects of the disputations, frequently solving the equations on the fly. Often Bondi sat crosslegged on the floor while Hoyle, behind him in an armchair, would prod him into activity. "Work this out" he would say, looking over Bondi's shoulder as he scribbled feverishly. "Now try this." Gold, meanwhile, played the bemused skeptic, the detached observer, there to question the assumptions and solutions. Hoyle, tough-minded and unconventional as ever, was dismissive of "establishment" ideas and became the instigator of difficult questions. According to Gold, "Hoyle was the driving force, always asking questions of us. He wanted to know how Hubble's observations could be understood. In those days he was full of Hubble trouble."[2]

Back in Cambridge, Barbara and Fred needed to find an affordable house for themselves and their two children. Fred resumed his fellowship at St. John's, where his income was too low to buy or rent in the city itself; prices and rents had skyrocketed during the war. After some weeks, they rented a house in the country village of Quendon, 25 miles from Cambridge, which the army had used until shortly before the Hoyles took possession. It was in a poor state of repair. Tommy and Hermann rolled up their sleeves and spent weekends assisting Barbara and Fred in making it fit to live in. The rent took more than half of Fred's heavily taxed salary, and they would soon slide into debt. During the week, Fred motored to and from Cambridge, where conditions for academic work were not good. The university and its colleges had admitted many more undergraduates when the war ended, and so, accommodation, whether for living, teaching, or research, was tight. Fred made his daytime headquarters in Bondi's rooms overlooking Great Court in Trinity College, where Tommy sometimes joined them.[3]

At Cambridge in the late 1940s, lecturers spent no more than 20 weeks a year on undergraduate teaching. Nevertheless, Hoyle shouldered a workload that would today be considered heavy, giving as many as six lectures a week in the Arts School, in Bene't Street. His annual quota was 112 lectures. In addition, St. John's College paid him to give six supervisions weekly for small groups of students, amounting to a further 150 hours of teaching per year.

The university admitted about 120 undergraduates for the Mathematical Tripos, but the lecture rooms held only half that number, so all courses were given in duplicate. Not every student attended these lectures, partly because the face-to-face supervisions in colleges were often considered more important.

The word quickly spread that Hoyle's lecturing style was excellent and he regularly packed the room. During 1945–1946 he presented 48 lectures on advanced statistical mechanics for Part III of the Mathematical Tripos. Roy Garstang, who took this course, recollected that "his statistical mechanics lectures were excellent and included work on degenerate matter and white dwarf stars."[4] The same year Hoyle gave 24 lectures on geometry, a subject for which he was initially ill prepared, but he again delivered outstanding lectures. In the following years, the faculty allocated to him standard lecturing in applied mathematics: statics and dynamics, electromagnetic theory, and thermodynamics. Meanwhile, Bondi was offering Part III courses in mathematical physics, general relativity, and cosmology: the observational background, Newtonian theory, and the relativistic universe. Garstang remembers Bondi as "a superb *tour de force*—he could lecture most of a term with rather few notes." Hoyle, as his lecturing confidence increased, did less preparation and would frequently erase or correct material on the blackboard, to the irritation of the note-taking students.

Poor time keeping tarnished Hoyle's growing reputation as an inspirational teacher. Undergraduates would arrive on time for supervisions to find a note on the door asking them to come back later or the following day. His research students did not know whether to hunt for the elusive Mr. Hoyle in the Arts School, in St. John's College, or in Bondi's apartments at Trinity College. One day in late 1946, Hermann went on the prowl to find him in St. John's. On this occasion, he met Christine Stockman, one of Hoyle's first research students in astrophysics, also looking for Hoyle at the top of the stairs outside Fred's rooms. The two conversed and agreed to go on a skiing trip together. They married the following year on November 1, 1947. After his marriage, Bondi had to move out of his bachelor's room in Great Court to a college apartment in Trinity Street. Here, Hermann joined Christine in her work on the structure of stars, while Tommy Gold and Fred Hoyle treated the flat as a walk-in headquarters. And so, their research discussions that had started in wartime Witley now continued in peacetime Cambridge, all day, every day, as lively as ever, and very fruitfully.[5]

According to Bondi, the areas in which they debated, researched, and wrote papers ranged very widely.[6] Cosmology certainly featured in the discussions, but it produced no papers before 1948. Fred has also remarked on the scope of his research over the same period:

> I was interested in all sorts of things. . . . If you look at my papers [for the late 1940s] you find they go in all sorts of directions. My head was exploding. Astronomy was at a stage that if you knew any physics at all, there were a hundred things you could do.[7]

The Council of the Royal Astronomical Society approached Bondi in 1947 to ask if he would write a critical review of all that was known—and unknown—in cosmology at that time. The following year the RAS published this paper in the *Monthly Notices*.[8] It is a meticulous work of scholarship, highlighting the gaps in knowledge rather than setting out a litany of the achievements at that time. Bondi covers the contributions of Hubble, Eddington, de Sitter, Lemaître, and contemporaries. He takes care to stress the importance of relativity to cosmology. The review brings in the new cosmological ideas of a growing community of cosmologists in England, with attention given to George McVittie of Queen Mary College, London; Edward Milne of Oxford University; and William McCrea of Royal Holloway College, London. Throughout, the review is informed by the progress that had been made in applying Einstein's general relativity to cosmology.

In his summary, Bondi foresaw a problem that would become an enormous challenge to, and source of argument among, cosmologists in the 1950s and 1960s: the consequences of the evolution of galaxies over time. He goes on to say, "Some models of the universe assume a catastrophic origin of the universe." This statement has its origin in the primeval atom concept of Lemaître and others. "Meanwhile, other theorists are more conservative, and do not embrace the concept of an explosive origin for the universe."

The strongest proponent of the idea of an explosive beginning to the universe was George Gamow, who was working at George Washington University on beta decay, stellar evolution, and the origin of the chemical elements. At a talk to the Washington Academy of Sciences in 1942, Gamow proposed that the observed distribution of the chemical elements could be the result of the breakup of a primeval superdense object composed of nuclear matter.[9] Four years later, in 1946, he advocated the seminal viewpoint that the chemical elements might have been forged in a chaotic process involving an explosion.[10] He envisaged the early universe as a highly compressed, hot, neutron gas. He thought nuclear reactions would rapidly build the array of elements that chemists organize into the Periodic Table. To prevent the neutron gas from fusing entirely to iron or nickel, he posited a sudden expansion of the gas, which would cool the material sufficiently for element building to cease.

His research student Ralph Alpher worked on the explosive synthesis of the chemical elements from 1946,[11] and by 1948 he and Gamow put a paper together on the origin of the elements.[12] Famously, the fun-loving Gamow, unable to resist the combination of names Alpher-Bethe-Gamow, added the name of Hans Bethe to this paper, which of course became the "$\alpha\beta\gamma$ (alpha-beta-gamma) paper" in talks about its results. Bethe made no contribution whatsoever and never even saw a proof! This short paper (it is less than one page), published on April 1, 1948, derives the relative proportions of the chemical elements in an expanding, hot universe with nuclear reactions truncated as the density of matter falls to a critical point. A few weeks later, Alpher and Robert Herman (*without* any inputs from Gamow) calculated that the heat left over from the hot primeval atom would make the temperature of the universe now about 5 degrees above absolute zero, a value remarkably close to the 2.73 degrees above absolute zero eventually measured to great accuracy by NASA's COBE (Cosmic Background Explorer) experiment in 1992.

Bondi prepared his review without mentioning Gamow's 1946 paper and without knowledge of the $\alpha\beta\gamma$ paper, then still in press. As he wrote the article, Bondi conversed with Hoyle, although in the published version he only thanks George McVittie and Tommy Gold. Hoyle later claimed it was he who steered Bondi into cosmology. "Bondi was not sure what he should do," Hoyle said.[13] "His expertise was in partial differential equations, fluid motions, things of that sort. So it was a big step when I suggested to him one day that maybe he should look at cosmology."

At around this time, Hoyle thought that cosmology had been in abeyance for a long time. In researching the article, Bondi had read Robertson's 1933 review of cosmology, an encyclopedic summary of progress in theory in the decade and a half since publication of Einstein's general theory.[14] Bondi showed this article to Hoyle, who went through it in some detail. For the first time, Hoyle felt a stirring of mathematical and technical thoughts about cosmology. He decided this was a masterly survey, but by then it was 15 years old.

In their discussions of this review, Bondi and Hoyle drilled down, searching for unseen implications. Had Robertson really thrown the net wide enough? Were there other possibilities in terms of modeling the universe using relativity? What had Robertson missed? Hoyle felt "a strong hunch that there must be something else, because there was no creation of matter in [Robertson's universe], and we felt creation had to be given expression at some point."[15] Matter creation had entered this discussion as a consequence of an amazing speculation of Tommy Gold, which arose serendipitously.

One evening the three of them went to a local movie theater that was showing the 1945 classic horror movie *Dead of Night*, starring Michael Redgrave. The action features a recurrent nightmare, and the movie has no beginning or end. Gold said of this movie: "It's completely circular. You can come in at any time and see it until the place you came in."[16]

Very soon after the evening out, Gold hit on a remarkable parallel between the movie plot and the universe. Hoyle yet again started puzzling over the implications of Hubble's observations. What did it mean, all the galaxies flying apart? Would space not soon become empty? Gold responded that the universe might be just like the *Dead of Night*, where you can come in at any time. Perhaps the universe had no beginning and will have no end!

Eddington's universe, with its long period in which nothing happened, had some characteristics of a universe without a beginning. Gold was now proposing something far more exciting: an ever-dynamic universe that had no beginning and would have no end. The implications were awesome: an expanding universe with no beginning? Hoyle and Bondi seized on Gold's idea. Could Hubble's expanding universe have been around forever, rather than having exploded from a primeval atom? The scenario would only work if the expanding universe created matter as it went along to fill the voids left by receding galaxies. The universe could exist in a steady state, always looking the same, if there were a process of continuous creation.

Hoyle dismissed Gold's proposal: "Ach, we shall disprove this before dinner," he chortled.[17] Bondi chimed in, too. "Tommy, we'll soon find the pitfalls to this one."[18] Dinner that night was very late.

The steady-state universe would be attractive to anyone familiar with Eddington and Dirac's ponderings on large-number coincidences. In an explosive universe, one would have to suppose that these large numbers were momentary coincidences, whereas they could be an enduring feature of a steady-state universe. "Isn't that funny?" Gold remarked.[19] None of the trio were shocked by the idea of continuous creation, which violated the laws of conservation of energy, because the explosive universe committed the same offense, although only for an instant in the moment of "creation."[20] In any case, ideas about matter creation were not new in Cambridge. Twenty years earlier, in a book that Hoyle knew well, James Jeans mentioned the possibility of matter creation:

> The type of conjecture which presents itself, somewhat insistently, is that the centres of the nebulae [galaxies] are of the nature of singular points, at which matter is poured into our universe from some other extraneous spatial dimension. [To an observer] they appear as points at which matter is being continually created.[21]

And Dirac, who was still Hoyle's official supervisor, had pointed out in 1937 that the continuous creation of matter can be related to wider issues in cosmology.[22]

Their long evening of discussions soon showed them that they might be on the verge of a totally new cosmological concept. They convinced themselves that the immediate objections would have solutions, and they set out to find them.

Bondi made quick progress with the mathematical ramifications and, within 3 weeks, reported that continuous creation worked nicely as a feature of the universe. Crucially, he found that the extent to which continuous creation violated the laws of physics was not as serious as they had first thought. While new atomic matter emerged spontaneously, the oldest galaxies receded, and their greatly redshifted light carried much less energy. The combined matter and energy content of the universe could remain constant despite new matter being created from nothing.

Hoyle meanwhile pursued another approach, trying to use field theory to explain continuous creation. A field theory describes physical reality, such as magnetic attraction or gravity, by means of the influence of a field on objects. The field equations of the general theory of relativity insist on the conservation of mass. Hoyle swept this requirement aside and recast the mathematics by adding in continuous creation, designated in the modified equations by the letter C: the C field would balance the density of the universe, its new matter replacing the receding galaxies.

Armed with the new C term, Hoyle found solutions to the field equations in which the average density of the universe is held constant by matter creation as the universe expands and the remote galaxies leave the scene. He felt that the created matter must be subatomic and suggested that the new particles would be neutrons. He was attracted by the radioactive decay of free neutrons. With a half-life of 10 minutes, a neutron changes into a proton and an electron, the two particles needed to make an atom of hydrogen, which is the most abundant element in the universe.

These discoveries excited him, and he proposed to Hermann and Tommy that they should work their bold cosmological ideas into a robust paper for publication in a major journal. However, the three were divided as to the thrust of the paper. Bondi and Gold wanted to concentrate strictly on the cosmological implications of the steady-state aspects, whereas Hoyle felt compelled to highlight continuous creation. The result was that Hoyle, ever headstrong in such matters, wrote a paper on the continuous creation of matter, and left Bondi and Gold high and dry. After some months, they produced a more philo-sophical paper on the steady-state theory of the universe.

Hoyle first submitted his paper to *Nature*, but it was rejected and then turned down a second time by the Physical Society. The loss of time caused by the two rebuffs enabled Bondi and Gold to win their little derby by a short head. They submitted their offering to the Royal Astronomical Society on July 14, 1948,[23] whereas Hoyle's did not arrive there until August 5, 1948.[24] The outcome was an interval of 2 months between the publication dates of the two papers.

The steady-state paper by Bondi and Gold is a masterpiece. At the outset, its authors advance a new fundamental axiom in physics, the *perfect cosmological principle,* which has become one of the enduring philosophical features of their paper. In considering the entire universe, they vault beyond the established cosmological principle that there is no preferred *place* in the universe by adding that there is no preferred *time* either: The universe will look the same from all places and at all times. The cosmological principle did not say that the universe must look *exactly* the same from all perspectives. Milne's point had been that, on the grand scale, our viewpoint must be presumed to be ordinary and every location in the universe must resemble ours in a broad sense. Bondi and Gold took the principle to the next logical stage: There is no preferred time to observe the universe. Explosive or evolving universes violate this new axiom because they imply that the universe was very different in the past.

This philosophy is intellectually attractive. It gives permanence to the abstract laws of physics, while regarding the present state of the universe as an outcome of the operation of such immutable laws. A bright light is now shining, extinguishing the long shadows cast by Eddington and Dirac, with their notions of the "constants" varying with time. Henceforth, physical laws cannot be assumed to be independent of the structure of the universe, and conversely, the structure of the universe depends upon the physical laws. They argue that the universe is in a stable, self-perpetuating state, but it cannot be static because Hubble's observations of the receding galaxies prove otherwise.

How rapid must the creation be? They calculate that it amounts to one new atomic particle per cubic meter every million years at most, a rate of spontaneous production then (and perhaps still) far below the sensitivity of laboratory experiments. As to the physics of creation, Bondi and Gold preferred the direct creation of hydrogen, which they, like Hoyle, envisaged as taking place in the vastness of intergalactic space.

In comparing the theory with observations, Bondi found it very striking that the theory agreed nicely with Hubble's results on the distribution of galaxies both nearby and at large distances. General relativity permits a wide range of models, and one could always be found to fit the facts because the number of free parameters is large. In the Bondi-Gold model, parameters are fixed by observations of nearby galaxies, not plucked from thin air. Their model, tightly constrained by observations, also meshed with counts of distant galaxies. This created a profound impression that the new theory had to be correct.

The final section of the steady-state paper begins the search for a field theory. Unlike Hoyle, they had no confidence in modifying general relativity, which they said could not be satisfactorily adapted to their theory. This sets the scene for a triumphant coda to their paper, a furious attack on Hoyle's proposition!

As a normal academic courtesy, Hoyle had shown Bondi and Gold drafts of his paper on continuous creation. His rivals said that Hoyle's theory had "a number of attractive features." They liked the way in which he made only slight modifications to the equations, since "this permits the use of the whole apparatus of general relativity," in which cosmologists had already invested an enormous effort. However, they protest strongly about the effects Hoyle's "creation field" will have on matter, which they describe as a "serious objection." They conclude: "In view of these objections we have no hesitation in rejecting Hoyle's theory, although it is the first . . . formulation of the hypothesis of continuous creation of matter." And they returned Hoyle's courtesy by showing him a draft of their paper before publication.[25]

Hoyle quickly got to grips with the objections Bondi and Gold had raised, which he considered to be rather technical and too driven by aesthetic considerations. At the meeting of the Royal Astronomical Society on January 14, 1949, he handed over a paper setting out his response to the Papers Secretary, and it was accepted for the March issue of *Monthly Notices*.[26] After dismissing mathematical technicalities, Hoyle locks horns with his close colleagues on the issue of the perfect cosmological principle: Their notion that neither place nor time is special would not do for Hoyle. "It is clear that the imposition, as a primary axiom, of the wide [i.e., perfect] cosmological principle is a radical new departure." Mischievously, Hoyle refuses to employ "perfect," substituting "wide" instead. He continued to feel that developing a model universe from a philosophical construct was putting the cart before the horse. He much preferred the approach of the physicist: Get the physical equations first and then derive the model to follow as a consequence of those equations.[27] In the end, this is a matter of individual preference.[28]

Both the steady-state theory paper and the continuous creation paper would have an enormous impact on cosmology in the two decades following their publication. They have received approximately equal attention.[29] Hoyle produced a total of more than 500 papers, articles, reviews, and books; this paper on continuous creation is his seventh most referenced work.[30]

Publication of the steady-state theory papers set the stage for the greatest clash of all time between rival theories of cosmology, a battle that would engage professionals and the wider public for 20 years. In 1948, cosmologists still had a variety of models and no easy means of distinguishing among them using the available observations. The steady-state theory burst on the scene as an upstart and in two variations: one philosophical, the other mathematical. The great Palomar 200-inch (now referred to as the 5-meter) telescope commenced observations in 1948. Bondi hoped it might make decisive observations,[31] but this was not to be. The merits of rival theories would have to be fought out in seminars, papers, and conferences by passionate argument, rhetoric, appeals to aesthetics, and an Olympian stance by Hoyle as to how research in physics should be conducted.

Bondi got his first chance to advocate the steady-state theory at the Seventh General Assembly of the International Astronomical Union (IAU), held in Zurich, Switzerland, from August 11 to 18, 1948. In the days before air travel became commonplace, international meetings were a fundamental necessity in astronomy. No single observatory can see the whole sky, and so astronomers would get together and plan worldwide observing campaigns to get a balanced view of the universe. Unlike most international scientific meetings, the Seventh General Assembly had very few papers: Most of the time was devoted to formal and informal meetings of groups of astronomers.

Christine and Hermann Bondi, as well as Fred Hoyle, were members of the Union, which had not met for 10 years because the world had been at war. They joined some 300 astronomers from 31 countries. When they arrived in Zurich, the contrast with postwar Britain was striking. In Britain, for example, food rationing was even harsher than it had ever been during the war, clothes were drab, and gardeners concentrated on growing food. Here in Switzerland's largest city, it seemed as if every window was bedecked with flowers. Meetings were held at the magnificent premises of the Eidgenössische Technische Hochschule (ETH, the Swiss Federal Institute of Technology), which boasted facilities far more modern than those available in Cambridge.[32]

At the start of the week, Bondi broached the new philosophy with his British colleague William McCrea, who would be one of the first to like the theory. He urged Bondi to describe the theory to Bertil Lindblad, who was president of the IAU and a renowned expert on galactic astronomy. Bondi chose his moment well to describe the new theory. The local committee had arranged a steamer excursion along Lake Zurich, a dazzling body of water ringed by jagged alpine peaks. The world's astronomers steamed toward Rapperswil with its castle and Polish museum, and, in these lovely and relaxed surroundings, Bondi conversed with his international colleagues. Lindblad and many others evidently took the theory seriously, although they generally did not like it.[33] In his many discussions with associates who preferred the evolutionary model propounded by George Gamow, Bondi kept throwing the same question at the skeptics: "If the universe has ever been in a very different state from what it is now, show me the some fossil remains of what it was like long ago."

Hoyle was like a hyperactive child at the Zurich meeting, bobbing into one room after another, speaking loudly and at length at every opportunity. He chimed in with calculations he had made on the chemical composition of stars. On the afternoon of Tuesday, August 17, he participated in a seminal meeting on the abundance and distribution of the chemical elements. In a meeting of the commission on extragalactic astronomy, he met Edwin Hubble for the first time, over tea and cakes during a break. Hubble spoke informally about the 200-inch telescope, which had been going through trials for 4 months. Hubble hoped to start observations of galaxies soon and perhaps obtain data that could discriminate between different theoretical models of the universe.[34] Fred found all of this truly exciting.

However, the week was not all positive for Fred. One incident in particular reflected the way he was regarded by senior members of the British contingent. The Americans normally threw a dinner for the entire Union, but the number of attendees at Zurich made it too expensive to continue this prewar tradition of inviting everyone. Instead they contacted each national delegation and invited them to send representatives. At this time the British astronomical establishment had been depleted by the untimely deaths of James Jeans and Arthur Eddington. Hoyle was himself becoming an establishment figure, though not of the traditional type. He had 28 papers to his credit, half of them produced during the war; the membership of the RAS elected him to its council in March 1948; and he made a point of repeatedly asking speakers difficult questions at the monthly RAS meetings. The British delegation decided that Mr. Hoyle was not a fit and proper person to represent British astronomy. This snub would be repeated 4 years later at the General Assembly in Rome, where Hoyle was a major plenary speaker.

Back in Cambridge, Hoyle busied himself on a raft of research projects in the Michaelmas term. Soon he was sending papers to the RAS every couple of months. He had a huge teaching load of lectures for the faculty and supervisions for St. John's College. Undergraduates and his research students were certainly finding him elusive. He attended an RAS meeting in Edinburgh in late October, which would have taken him away for at least 2 days early in the term. The meeting was low key. About 30 attended, the improvised catering arrangements being provided by astronomers who lived in the Royal Observatory, Edinburgh: Food rationing was still severe.

The Edinburgh RAS meeting gave Bondi, Gold, and Hoyle ample opportunity to promote their new theory.[35] The president let all three speak. The mood was intense, the outcome negative.[36] Gold went first, explaining that the steady-state theory implied a continuous process of creation. Bondi stepped up to the podium next to say that an observational test of the theory could be made by counting the number of faint galaxies: In the steady-state universe, there should be the same density of galaxies at great distances as locally.

Hoyle's confident presentation started with the observation that

Bondi and Gold "have shown that the astrophysical data . . . point *against* all models of the expanding universe that are *allowed* by Einstein's general theory of relativity." He went on: "I have attempted to build a new model by considering the possibility of a steady creation of matter." Turning to the blackboard, he chalked Einstein's equations in a form that the professional audience immediately recognized. Then with a flourish, he chalked a further tensor into the equations, to account for matter creation. He finished with a magnificent conclusion:

> Modern astrophysics appears to be inexorably forcing us away from a universe of finite space and time, in which the future holds nothing . . . towards a universe in which both space and time are infinite. The possibilities of physical evolution, and perhaps even of life, may well be without limit. These are the issues that stand today before the astronomer. Within a generation we hope that they can be settled with reasonable certainty.[37]

This was the first public occasion on which Hoyle risked mentioning that a universe infinite in both space and time would have profound implications for our understanding of the relationship between "life" and the "universe."

When Hoyle wrapped up, the president beamed at the audience and invited comments. Sir Edmund Whittaker, who had been Royal Astronomer for Ireland under Edward VII, immediately praised Hoyle, remarking that "the new term in the equation . . . proposed by Hoyle can hardly fail to have great influence on the future." He concluded by expressing warm admiration for the work as a whole. That is not how Edward Milne, ever a stern critic of Hoyle, saw matters. He accused Bondi and Gold of unnecessarily introducing new dogma. Sensing an increase in tension in the room, the president seized the moment to stop the temperature from rising any further. He turned to the physicist Max Born, who would win the Nobel Prize in 1954, and asked, "What do you physicists make of this?" Born responded with irony: "I am overawed by the whole character of the cosmologists! In cosmology we shall continue to discover new theories of world structure and evolution." Not persuaded by what he had just heard, an impassioned Born finished on a patronizing note: "I am filled with gratitude at hearing these papers, but I am unconvinced."[38] Afterward, Gold felt pleased with his presentation, which had gone well. His message had mesmerized the audience.

At this time, Hoyle was feeling stretched financially. Academic salaries were low, taxation was high, and the economy was in crisis. The government had launched Britain's welfare state the year before and this provided a handout of 8 shillings a week for their second child, Elizabeth. But he had the costs of motoring from the countryside to take into account, and, with purchase tax at 66 percent, good furnishings and household appliances were simply out of reach. The Hoyle's ran into debt over rent and motoring costs. To reduce their bank overdraft, they rented a house at Great Abington, 8 miles south of Cambridge and served by the public bus. That reduced their outgoings. Then, early in 1949, Hoyle had an exciting opportunity to enhance his income, by giving talks on the radio. The normal conduct of academic debate is through papers, reviews, seminars, and conferences. Hoyle now added the influential British Broadcasting Company (BBC) to the legitimate channels through which he would publicize his theories.

Hoyle's radio career started quietly. Sunspots gave rise to his first broadcasting opportunity.[39] The Sun's surface layers, where solar energy finally bursts forth as heat and light, show a great variety of action. Sunspots are one of the more obvious signs of solar activity. The number of sunspots that can be seen varies over a cycle that lasts about 11 years. At the peak of the cycle, 100–200 sunspots may appear in a year. The number then falls almost to zero, and a new cycle starts. Because 1948 was a year of solar maximum, with over 150 sunspots recorded, the BBC commissioned two radio talks on the subject from Fred. He had developed an interest in the solar atmosphere from his work on accretion[40] and had written a short monograph on solar research for Cambridge University Press in the summer of 1947.[41]

The recording for the first talk took place on New Year's Eve 1947. Fred received 15 guineas (nearly $1,000 in today's terms), a handsome reward for the work involved. We can well imagine his excitement as he traveled first class (at a cost of £1 3s 9d) on the steam train from Cambridge to Liverpool Street station in London. The broadcast was on January 8, 1948, in *Science Notebook*, a news feature for the Overseas Service, on the topic "The Atmosphere of the Sun."

The same day as the transmission, Dr. Archie Clow, a talks producer at the BBC Science Survey, wrote to Hoyle and asked if he could modify the talk for the Third Programme, the BBC's domestic radio service for high culture and academic talks. The Third Programme had a penchant for plucking erudite academics, diplomats, and military top brass from their desks to sit behind a microphone. At a time when the entire undergraduate population in Britain was no more than 50,000 people, the Third Programme brought university-style lectures of a high standard into every home that had a wireless set.

By return letter, Hoyle offered him *How Do Sunspots Affect Our Lives?*, promising a script within a couple of days. Clow asked if Hoyle could read the script live, at 10:30 p.m. on February 12, 1948. An excited Hoyle dashed off a reply by hand, saying that his teaching duties at St. John's made it impossible to do a live broadcast. As for the blurb about the program to appear in *Radio Times*, Hoyle demanded that he be described as a fellow of St. John's College: "I think it is unnecessary to refer to my post of university lecturer in mathematics." He clearly did not want the mathematics faculty to enjoy any prestigious publicity.

A few days later Hoyle relented on the issue of a live broadcast. "I have arranged my teaching on Friday Feb. 13," he wrote, "so that I can stay in London for the night of the 12th." This was a crafty move. The BBC put him up overnight at the Clifton Hotel at 47 Welbeck Street.[42] On Friday, February 13, instead of heading back to Cambridge and teaching, he took a 5-minute walk to the RAS apartments at Burlington House for the annual general meeting of fellows. The main business of the AGM was an address from the president, followed by reports from officers.[43] The meeting concluded with the announcement of the results of the elections for officers and councillors. Hoyle had stood for council and heard at the AGM of his success. In the forum of the RAS Council, he would be head to head with two of his adversaries: Thackeray was one of the secretaries, and Atkinson a councillor.

With Hoyle now on the council it would be difficult for the reactionary cabal to resume their blocking of Hoyle's papers. In the years 1945–1949, he submitted about 25 papers, which placed a heavy load on the society's officers as well as the small pool of referees. At one council meeting a member remarked that, if this flow continued, there would soon be no field left for other astronomers to explore. The council did spend considerable time discussing the papers and the referee reports because it was particularly concerned with the amount of speculative theory, based on "plausible" assumptions. Paper was still rationed at this time, and Hoyle's contributions required much work in the editorial office to get them into an acceptable form.[44]

A few months after his Third Programme broadcast, on July 27, 1948, the Turkish section of the BBC's Overseas Service wrote to Hoyle asking him to give written answers to questions sent in by listeners. These were to be translated into Turkish and read on his behalf. He responded with alacrity. The queries were just the kind of topics he'd always found intriguing. Is there life on the stars? When will the world come to an end? Can we use atomic power to reach the stars? Will telescopes search for life on Mars? For his efforts he earned 5 guineas ($300 now), then the weekly wage of a manual worker. He continued to offer spirited answers to listeners' hum-drum questions for some months, always at 1 guinea a time.

Early in 1949, he again heard from Dr. Archie Clow: Would he do a 20-minute talk on continuous creation, to go out on the Third Programme on March 28, 1949? The lecture should be "an account of the newest development in 20th century cosmology."[45]

This commission brought Hoyle into contact with Peter Laslett, then a research fellow in intellectual history at St. John's. The college did not require its research fellows to be resident in Cambridge, and they could have another job elsewhere. In Laslett's case, the second employment was as assistant producer with the Third Programme.[46] He signed off on Fred's requisition form: fee 30 guineas, plus first-class rail travel. The recording went on the air at 6:30 p.m. on March 28, 1949. This time the *Radio Times* gave his affiliation as university lecturer in mathematics.[47] In that year, he was one of the very few academics invited to speak on the Third Programme who was not already a full professor.[48]

Before the listeners heard Fred's gruff voice, a plummy-toned station announcer read the introduction. "This is the BBC Third Programme," he began. "In this talk Fred Hoyle gives his reasons for thinking that matter is being created all the time, so that the universe must have had an infinite past and will have an infinite future."

Hoyle cut to the chase immediately, launching off with, "I have reached the conclusion that the universe is in a state of continuous creation." He reviewed the state of observational cosmology, frequently using the rhetorical device of posing a question and then answering it, question and answer both of course being phrased suitably to suit his stance.

An obvious problem with a radio lecture is the absence of visual aids. Hoyle went to great lengths to get over technical points with word pictures. He explained the Doppler effect[49] by analogy with the fall in the pitch of the whistle of a receding train. Galaxies in the expanding universe he compared with dots on the surface of a balloon in the process of being inflated, the changing radius of the balloon being a measure of the flow of time. He had a lovely picture for the rate of creation of matter: "This means that in a volume equal to a one-pint milk bottle about one atom is created in a thousand million years."

Early in the talk he tackled rival theories of cosmology:

> We now come to the question of applying the observational tests to earlier theories. These theories were based on the hypothesis that all matter in the universe was created in one big bang at a particular time in the remote past. It now turns out that in some respect or other all such theories are in conflict with the observational requirements. And to a degree that can hardly be ignored.

Hoyle, the hill climber, positioned himself to insult his colleagues, using a mountaineering analogy:

> Investigators of this problem are like a party of mountaineers attempting an unclimbed peak. Previously it seemed as if the main difficulty was to decide between a number of routes, all of which seemed promising lines of ascent. But now we find that each of these routes peters out in seemingly hopeless precipices. A new way must be found.

At this point he included a last-minute insertion, jotted on the script:

> The new way I am *now* going to discuss involves the hypothesis that matter is created continually.

As to the method of creation, he invoked for his rapt radio audience, "groundwork that has already been prepared by H. Weyl, a German mathematician now resident in the United States." Hoyle then told his audience that it was not difficult for *him* to establish the consequences of the creation theory. The expansion of the universe receives a natural explanation as the receding galaxies move over the horizon (so to speak) while making room for the new matter. And then there is another down-to-earth analogy: "Although no individual person lives more than about 70 years, the human species replaces itself through the births of new individuals replacing the deaths of others." And so it appears with the universe.

Hoyle's talk had a startling end, one that would build up trouble for him in the future. He delivered a rebuke to those against the idea of continuous creation. His opponents claimed he was introducing an extra hypothesis into cosmology. Not so. Continuous creation is a new hypothesis, but it is not an *extra* piece of baggage because it *replaces* the hypothesis that all the matter was created "in one big bang at a particular time in the remote past." And "on scientific grounds this big-bang hypothesis is much the less palatable . . . an irrational process that cannot be described in scientific terms." Hoyle, the amateur philosopher this time, "cannot see any good reason for preferring the big bang idea." It is "distinctly weak."

To create a picture in the mind of the listener, Hoyle had likened the explosive theory of the universe's origin to a "big bang." Through this particular program he coined the term that would forever be attached to something he never supported. He repeated it in a radio broadcast in 1950. George Gamow promoted the expression as well, wrongly claiming that Hoyle had invented it as an insult during a radio debate between the two of them.[50] The following week, the BBC's magazine, *The Listener,* published the full text of Hoyle's talk, and the expression "big bang" appeared in print for the first time.[51]

To his opponents, the talk was vain, one-sided, insulting, not worthy of the BBC, and so on. One RAS fellow, E. G. Martin, was so incensed that he dashed off a letter of protest that very night:

5 Belmont Grove, London SE 13

Dear Sir

I listened to Dr. F. Hoyle's talk on Continuous Creation. His theories, put forward recently at the RAS, evoked strong criticism.[52] His manner is extremely disarming to the general public. In view of the dissent of a number of scientific men, a chance ought to be given to his critics to reply by a broadcast talk. Two suitable persons would be Prof. Dingle or Prof. McVittie. I trust the BBC will not allow only one side of the argument to be broadcast.

Martin spent his entire career at the Royal Observatory, producing fewer than two dozen papers on observational astronomy and the history of instruments. He made no contributions to cosmology. He worked with, among others, Atkinson, who of course continued his strong dislike of Hoyle's approach to scientific issues. I can only speculate that Atkinson may have put Martin up to writing the letter.

The BBC took Martin's letter very seriously. Laslett consulted both E. G. Martin and Hoyle. Fred felt that a straightforward talk from an opponent would be dull and technical. In any case he had not the slightest intention of engaging with Dingle, a spent force academically, whose field was the philosophy of science rather than astrophysics. Martin pressed the case strongly: Astronomers resented Hoyle's truculent manner. Hoyle responded by pointing out that none of the opposition had ever contradicted him in a scientific paper.

Laslett was now in a difficult position. He and Hoyle were of almost the same age and both fellows of St. John's College. He hesitated to overrule Fred's point of view. He asked for a ruling from his boss, R. A. Randall, the controller of talks. Randall referred the matter to an even higher level in the BBC, where it was felt that the Astronomer Royal should be consulted. The Third Programme tried to encourage academic debate, and the BBC did not wish to refuse the possibility of a reply, but it did not want to allow someone to broadcast who might discredit its reputation. This could have been a veiled reference to Dingle, a man whose already-fading career would end disastrously in 1968 with his ridiculous assertion that Einstein's special theory of relativity was wrong.[53]

In May 1949, Laslett visited the Astronomer Royal, Sir Harold Spencer Jones, by then ensconced in the splendor of Herstmonceux Castle in Sussex because the Royal Observatory was in the process of moving there from Greenwich in search of clearer skies. Spencer Jones had no personal stake in rival theories of cosmology. As treasurer of the RAS, he had been present at its Edinburgh meeting 7 months previously and knew about the ideas from the talks he had heard there.

After Laslett's visit, Spencer Jones wrote to the BBC. He noted the considerable diversity of opinion among astronomers, saying that "it is by no means certain that the cosmological views held by Mr. Hoyle are correct." Significantly, the Astronomer Royal concluded, "I do not find anything to object to in his broadcast." Hoyle had made it perfectly clear that these were his views and he did not claim that no other view was tenable. The BBC did not need to offer any "right to reply." The corporation heeded this advice.

Laslett fretted briefly over whether to commission any further talks from Hoyle. The correspondence file at the BBC had a note from Arthur Clow that read: "Do not use this man. His Yorkshire accent is too strong."[54] On the other hand, an experienced producer-director, Christopher Holme, thought Hoyle admirable: "His voice is clear and pleasant and he puts himself across well. I think there need be no hesitation in using him again."

The next commission came very soon. Laslett paired Hoyle with the distinguished geneticist Cyril Darlington, an arrogant scientist whose racist views became deeply unfashionable in the 1960s. Laslett took more care with the script this time, insisting that they go over it together on the occasion of a feast at St. John's College. The program—"Is There Life Elsewhere in the Universe?"—was broadcast on July 8, 1949, and caused little comment. The importance to Fred, however, was huge. He was now debating on air the question of life in the universe with one of the best-known biologists in Britain. It was a question to which he would return again and again.

He began his talk[55] with a bold assertion that "everyone is agreed that apart from Earth, Mars is the only planet in our solar system that could support life." Concerning the origin of the solar system, Hoyle speculated that it once consisted of two stars, that long ago the Sun's companion exploded, and the planets formed from a little of the debris. (This is now known to be incorrect.) He managed to fit in continuous creation, arguing that in an infinitely old and infinitely large universe, anything is possible. So, "life must not only be commonplace, but must occur in an infinite number of different places. Every event, no matter how impossible must be repeated at different places in the universe." To ram home his view that life is ubiquitous, he concluded thusly:

"Even now, in some distant spot, another fellow called Hoyle is broad-casting to an audience like you about an identical subject." Darlington failed to challenge Hoyle: After a ramble through what is now termed human molecular biology (he was speaking before the discovery of the structure of DNA), he backed Hoyle's position that life is common, not happenstance.

In late November 1949, Peter Laslett approached Fred Hoyle with a new broadcasting project. Laslett had been let down by the historian Herbert Butterfield (later to become vice chancellor of Cambridge). On the strength of Butterfield's radio talks earlier in 1949, Laslett had managed to secure five sessions during prime listening time at 8:00 p.m. on Saturdays, commencing at the end of January 1950.[56] Butterfield decided that he had no time to prepare the talks and asked to be excused. Laslett, acting on the instruction of Mary Somerville,[57] then assistant controller talks, offered the five slots to Hoyle. The fee of £300 amounted to a fair fraction of his annual salary from the university. Half a century later, astronomers who remembered listening to these broadcasts as teenagers often refer to them as Hoyle's Reith Lectures.[58] They are, however, mistaken: Hoyle was never Reith Lecturer, the most distinguished honor the BBC bestows on academics.

Throughout January and February 1950, Fred Hoyle was in a race against time to complete the 45-minute scripts, each running to 8,000 words. He had no access to a typist in Cambridge, and his method of working seems extraordinary today. Typically, Hoyle met Laslett in Cambridge a week before the broadcast and handed over a handwritten draft. This was posted to London, to be checked by the BBC and the official censor (another echo of wartime). Within 2 days, the script came back to Cambridge, where Laslett and Hoyle worked up the revised draft, which Fred then read live at 8:00 p.m. on Saturdays.

The BBC ran a high-profile marketing campaign in the *Radio Times* as well as in its literary magazine, *The Listener*. An unsigned trailer in the *Radio Times*, 1 week before the first broadcast, began with a mission statement saying that henceforth the corporation would offer its listeners something vital from the life of the universities and that it would seek out talent too good to be confined to the lecture hall or laboratory.[59] Hoyle would be the trailblazer.

Concerning the forthcoming series, the editorial said: "In two broadcasts he gave last year, he made it quite plain that he had an exceptional talent for lucid explanation of this complicated material. He succeeds by his blunt and vigorous way of putting things: He describes events in interstellar space as if commenting on a cricket match."

The Listener reproduced Hoyle's scripts in full, each taking nearly five pages.[60] An eye-catching full-page astronomical photograph, from the 200-inch Palomar telescope, graced the front page of the issue containing the first of the talks. It must have looked dramatic on the newsstands and was an immediate sellout, as were the next four issues.

On hearing of the demand for *The Listener*, the splendidly eccentric Oxford publisher, Basil Blackwell, snapped up the book rights and found he had a best-seller on his hands.[61] Blackwell produced the book in a matter of weeks, with publication in late April 1950. The first printing of 10,000 sold immediately, to be followed by a panic reprint in May. In Cambridge, Hoyle did a book signing session in Heffer's, the bookshop opposite St. John's College. Harper Brothers, in New York, promoted an edition for the American market. Hoyle claims[62] to have received £3,000 in royalties, equivalent to 5 years' salary: He was then on £600 a year.[63]

While working feverishly on his scripts, Hoyle kept his university lectures and college supervisions going, or just about. An undergraduate from 1950 recalls:

> In the Lent Term of 1950, Hoyle was just moving into the public eye with his broadcasts. We undergraduates at St. John's held him in awe. We really did sense that he was about to become very famous. Hoyle was rather difficult as a supervisor. He made no attempt to memorise our names. As often as not we would arrive for a supervision to find a note on his door "Sorry. Had to go to the BBC today. I will be back later." More often than not the delayed supervision did not take place. We tolerated all this because we felt so privileged to be taught by such a high-powered Fellow.[64]

The programs were immensely popular, with the final broadcast, which would lead to considerable trouble, airing on February 25, 1950. Fred recalls[65] that in writing the final script he went right to the wire, completing it only just in time. With the title "Man's Place in the Expanding Universe," this talk had majestic sweep, taking listeners far beyond the Milky Way.

It was all about other galaxies, and he estimated that there were 100 million of them, each containing a million planetary systems. According to Hoyle, life was ubiquitous and he wondered whether somewhere out there there was a cricket team that could beat the Australians.[66] He speculated on the origin of galaxies, which he said were forming all the time from background material. Next, he introduced the evidence that space is expanding, dismissed the explosive-origin theories, and propounded continuous creation, with due thanks to Bondi and Gold. The creation rate was "no more than one atom in the course of a year in a volume equal to St. Paul's Cathedral." Space was infinite, time was infinite, there was no biblical creation.

In his last 5 minutes, Hoyle (with the approval of the BBC and the censor) permitted himself a final philosophical outburst, on "how the New Cosmology affects me personally." He noted that, historically, cosmology had undergone a succession of intellectual revolutions. Would that happen as a result of the New Cosmology within, say, 500 years?

> I think that our [i.e., my] present picture will turn out to bear an appreciable resemblance to the cosmologies of the future. There are no new fields to be opened up by the telescopes of the future.

It was a breathtaking claim and, as it turned out, a ridiculous one.

In the printed version, the final remarks were headed "A Personal View," and he set himself the question, "What is man's place in the Universe?" He then launched into a major assault on religious and political beliefs:

> It seems to me that religion is but a blind attempt to find an escape from the truly dreadful state in which we find ourselves.

He attacked Marxists, out-and-out materialists, and, far more controversially, Christian belief. In the book version of the talk, he added an endnote saying, in effect, that if Christians imagine an afterlife without physical existence, then their brains must be endowed with a faculty not possessed by others. He claimed that Christian belief in an immortal soul existing without physical connections (and detectability) was nonsense. He referred to "several authorities on religious doctrine" who approached him after the broadcast.

The BBC Third Programme broadcast for only a few hours each evening, generally to an audience of about a hundred thousand listeners. The five talks went down extremely well with program controllers at the BBC, who were soon clamoring for repeats. The Overseas and European departments had both used Hoyle before, and they requested a series of ten 20-minute talks. Hoyle agreed.

Mary Somerville, recently promoted to controller talks (a powerful position at the BBC), wrote to Hoyle on March 17, 1950, inviting him to recraft the talks as a series of between six and eight 30-minute programs, this time for the Home Service. He responded by phone 10 days later, declining the opportunity, saying he had already agreed to do 20-minute versions for Overseas and felt he would go stale. However, she persuaded him to change his mind, promising that Laslett would be the producer, and so the bulk of the work could be done at St. John's College, where the recordings would be made in June after the university examinations. As a further incentive, she offered him a broadcast time of 9:15 p.m., directly after the national news, thus ensuring a large audience. Furthermore, the broadcast frequency of the Home Service had just changed, so that listeners in southern England could now enjoy transmissions free from interfering crosstalk from French radio stations. Fred took the bait.

Having gotten Fred's agreement, Somerville now turned to the tricky task of how to handle his personal religious views in the relaunch. In response to the first series, *The Listener* had already published more than a dozen letters.[67] Dr. Marie Stopes, an expert on fossilized plants, a very public figure and birth control pioneer, who also had clashes with religious authorities, wrote:

> He makes statements about geology that are just not true, thereafter using these mistaken ideas as foundations for some of his vast sidereal theories.

Several correspondents protested about Hoyle's antireligious point of view, including Fr. Frederick Copleston, S.J., an immensely accomplished historian of philosophy. A retired bishop asked, "Was anything so muddled and untrue ever written?"

The *Baptist Times* published an objection from a minister, and the *Church Times* reviewed the series with the nervous assessment that Hoyle's views were "forthright." Hoyle was emerging as a controversial figure in the broader public context, and the BBC became slightly alarmed. Somerville needed to ensure that the Home Service version would not produce another blizzard of protest letters. On March 31, 1950, a senior manager met Hoyle, who was accompanied by Lyttleton, to thrash things out. He reported back that Hoyle had been firmly told that the religious aspect must be toned down. Hoyle had been urged "to put the Christian case rather more fairly before proceeding to knock it down."

In May 1950, the need to control Hoyle became even more compelling. At the University of Manchester, the recipients of honorary degrees that year included Dr. Geoffrey Fisher, archbishop of Canterbury, the most prominent divine in Britain. For a little light reading on the train north, Fisher turned to *The Nature of the Universe*. The content of the final chapter profoundly shocked him. The dignitaries at the university honorary degree ceremony included the prominent Manchester industrialist Lord Simon of Wythenshawe, who had been a Liberal MP but had later joined the Labour Party. In 1947, he had been appointed chairman of the governors of the BBC, the ultimate pinnacle of power within the corporation. It was now Lord Simon's turn to be shocked: Archbishop Fisher, in his 15-minute acceptance speech, launched into a torrent of abuse against Hoyle, castigating his views on religious belief. Deeply troubled by this, Lord Simon turned to Sydney Goldstein, then professor of applied mathematics at Manchester, for an expert opinion. Goldstein, who had been a fellow of St. John's College, responded by saying it all depended what programming the corporation wanted. "If they want entertainment, the lectures are fine. If they want science they are not fine. The best astronomers would not agree with many of his conclusions. Hoyle has not the humility of a good scientist."[68]

Hoyle and Laslett did the new recordings in Cambridge in three sessions in mid-June. The original length of 28 minutes' running time had to be pared back to 24–26 minutes, in case the Nine O'clock News needed to overrun in order to broadcast government communiqués concerning the Korean War.

The *Radio Times* ran a trailer for the revised series, explaining that "listeners are reminded of the extent to which everything has had to be simplified. Mr. Hoyle presents in a way to make himself understood by any lay listener." Martin Johnson, an astronomer at the University of Birmingham, who had recently published a popular book on stars, signed the piece. As a prelude, a young radio astronomer, Bernard Lovell, spoke about radio astronomy some 3 weeks before Hoyle's broadcasts began.[69]

The series began on the Home Service on July 20; the eighth and final recording went on air on September 13. The 9 p.m. news received high ratings, and listeners stayed tuned for Hoyle. Hoyle's talk of September 13, 1950, had an audience of 5 percent nationally (2.5 million listeners), though it is listed as having an estimated audience of 7 percent in the West Region and 10 percent in Wales.[70]

Mary Somerville made sure that this time the BBC gave coverage to alternative points of view. She commissioned Herbert Dingle to consider "Does the New Cosmology Exist?"[71] He mounted a pithy attack on Hoyle's presentation, very much regretting that such a series of talks should be in such urgent need of corrections. Dingle accused Hoyle of presenting tentative speculations as established truths and entirely distorting cosmological thought. He concluded: "The picture which Mr. Hoyle has drawn with unusual skill and understanding is a false one."

After Dingle, Mary put on the novelist Dorothy L. Sayers, who by 1950 was much in demand as an essayist and lecturer on Anglican Christian theology. In her talk,[72] "The Theologian and the Scientist," Sayers accuses Hoyle of failing to understand the meaning of words Christians use. He is too literal about the meaning of "eternal life," for example.

The BBC, despite the best efforts of Somerville and her colleagues, now found itself scorched by a blazing academic argument. The problem was that, even though Hoyle had toned things down in his *talks* for the mass audience, the *book* of the unmodified first scripts was still selling like hotcakes: It reached the eighth impression during the currency of the Home Service broadcasts, and offended readers assumed its text to be identical to the radio scripts. The second series, of course, propelled the book sales to ever greater heights. Hoyle, a nonentity only 12 months previously, was now one of the most famous scientists in the country. His enemies regrouped. The upstart from Cambridge needed cutting down to size. He was making a fortune peddling ideas that the establishment rejected.

The BBC again published hostile correspondence on the Letters pages of *The Listener*.[73] Fred protested that his scientific work could not be shaken by polemical arguments, and he invited those who disagreed to publish research papers saying why. Dingle was dismissed as being "essentially trivial," but he fought back with the observation that Hoyle held a contemptuous view of all of British astronomy, save two or three individuals. He absolutely detested what he saw as Hoyle's vanity.[74] Lyttleton chipped in with letters criticizing Dingle's grasp of the history of astronomy. This was a nasty clash because Dingle worked professionally in history and Lyttleton did not. Toward the end of 1950, Lyttleton got personal:

> Dingle's arguments and personal abuse are of no more value than if Professor Dingle were to throw a dead cat into the room during one of Hoyle's lectures, apart from giving some indication of Dingle's state of mind.

One defense came from an unexpected source, Lord Brabazon of Tara, an aviation pioneer and a prominent person in public life.[75] He spoke up for Fred's manner: "Being presumptuous and cocksure is fine in a young man." The noble Lord felt that as a speaker Fred was so arresting, not because his listeners necessarily believed him but because *he* believed his own ideas. But by straying into theology he had put his foot into it.

Reviews of the American edition of the book were mainly positive. Only the *Christian Science Monitor* expressed hostility. It asserted that, in his efforts to simplify, Hoyle had failed to distinguish speculation and hypothesis from established fact. Readers would find it difficult to tell when Hoyle was speaking as Hoyle and when he was reporting generally accepted fact. The reviewer noted that "it is sufficient to say that the broadcasts have touched off a heated controversy."[76] The *New Yorker* was full of praise, hailing Hoyle as a worthy successor to Jeans and Eddington.[77] Writing in the *Herald Tribune*, the great Harvard astronomer Cecilia Payne-Gaposchkin felt it was a fascinating book that deserved to be widely read. "Rarely has so vast a panorama been presented in so small a compass," she wrote.[78] I am sure Fred would have liked this review. He had a soft spot for Cecilia, whom he may have met during his 1945 trip to Harvard. She was the first person to understand that the Sun could be mainly hydrogen and set out her views in her 1925 doctoral thesis.

What about the reception from professional astronomers outside Britain?[79] Frank Edmondson of Indiana University reviewed Fred's book for *Sky and Telescope*, an amateur astronomy magazine also widely read by professionals. Edmondson disliked Hoyle's provincialism: He had ignored the data and conclusions of recognized experts outside Britain. *The Nature of the Universe* was a philosophical work written by a skilled manipulator. "The New Cosmology is a large dose of philosophy spiced with a little science."[80]

Hoyle fared no better with professionals in Canada. In a talk for the Canadian Broadcasting Corporation, aired after Hoyle's programs had been transmitted by the Trans-Canada network, Ralph Williamson was specific about the irritation they had caused. Hoyle had concerned himself with a great diversity of astronomical subjects, without giving any inkling as to how discoveries are made through observation. He had no real experience of handling the large telescopes that make modern astronomy possible. Williamson expressed the view that theoretical deduction is a shaky business, likely to be much less trustworthy than observation. In an act of verbal savagery exceeding even Dingle's efforts, he ranked Hoyle as a pseudoscientist, almost as dangerous as the notorious demagogue Immanuel Velikovski.[81]

The Australian Broadcasting Corporation turned to Daniel O'Connell, S.J., recently appointed director of the Vatican Observatory and therefore an advisor to Pope Pius XII. O'Connell's view was that there were too few facts to make any sense of the universe at large. On the other hand, he favored Lemaître's explosive origin, "which so clearly implies a Creator." He dismissed the steady-state idea as hopelessly speculative and artificial. Forcefully and angrily he railed against Hoyle's cursory dismissal of Christian thought: "Hoyle sets out expressly to teach philosophers and theologians their business, but he makes no serious attempt to find out what they hold."[82]

A major effect of the broadcasts was to put Hoyle's continuous creation model in the spotlight, while sidelining the more philosophical

steady-state cosmology of Bondi and Gold. Many professional astronomers dismissed steady-state cosmology out of hand. Gamow and his
associates were silent on the steady-state theory until 1951. In his 1952
monograph,[83] Bondi gives Gamow the cold shoulder. Successive edi-
tions (there were five in all) of Einstein's *Meaning of Relativity* give no
mention to the steady state: Einstein remained confident that general
relativity was the only framework in which to do cosmology. Gamow's
popular book, *Creation of the Universe*,[84] published in 1952, dismissed
the steady-state theory out of hand, with no discussion or analysis, and
in another book, *Mr. Tompkins in Paperback*,[85] Gamow makes fun of
the Cambridge cosmologists with a satirical poem. Mr. Tompkins is
dreaming of being out in space. Hoyle "suddenly materialized from
nothing in the space between brightly shining galaxies" and burst into
song:

> The Universe, by heaven's decree,
> Was never formed in time gone by,
> But is, has been, shall ever be—
> For so say Bondi, Gold and I
> Stay, O Cosmos, stay the same!
> We the Steady State proclaim!
>
> The aging galaxies disperse,
> Burn out, and exit from the scene.
> But all the while, the universe
> Is, was, shall ever be, has been,
> Stay, O Cosmos, stay the same!
> We the Steady State proclaim!
>
> And still new galaxies condense
> From nothing, as they did before.
> (Lemaître and Gamow no offence!)
> All was, will be for evermore.
> Stay, O Cosmos, stay the same!
> We the Steady State proclaim!

Herbert Dingle served as president of the RAS during 1951–1953.
In his 1953 presidential address, he chose to mount a strong attack on
the new cosmology.[86] Bondi and Gold were in the audience, but Hoyle
stayed away. Fellows of the RAS crammed into the small lecture room
at Burlington House to hear Dingle. In the hot house he attacked the
views of the Cambridge mathematical cosmologists with high satire,

no doubt commanding the attention of those who agreed with him. By turns he renamed the new "cosmological principle" and the "perfect cosmological principle" as the "cosmological assumption" and the "cosmological presumption." He proclaimed: "It causes me considerable discomfort to use names which are clearly misleading." He continued: "We have, then, the strange position that in cosmology two impostors have usurped the throne of science, worn her crown and taken her name."

In Britain, William McCrea became the main supporter of the steady-state theory. It was he who had placed Bondi, Gold, and Hoyle on the speaker list for the RAS meeting in Edinburgh. McCrea's research style often led him to work at a fundamental level, with a philosophical approach. While at Cambridge he had attended Bertrand Russell's lectures. Philosophical analysis, much disliked by American astronomers, was something of a tradition in the 1930s and 1940s. McCrea shared Bondi and Gold's dissatisfaction with general relativity as a cosmological tool and started further development of the steady-state theory. McCrea had a deep insight into the physics of what Hoyle was proposing.

Overall, however, Bondi, Gold, and Hoyle were surely disappointed with the initial reception of their theories. Normally, in theoretical physics, the publication of a completely new paradigm stimulates many individuals and even groups to enter the new field and exploit it. This is what happened with the publication of superstring cosmology in the 1970s and 1980s, brane cosmology in 1998–2003, and the physics of extra-large dimensions in 2000. Without indulging in philosophical arguments on the reasonableness or otherwise, theorists plunged straight in to develop the theories further. Steady-state cosmology was largely ignored. Hoyle had been overwhelmingly successful in explaining his ideas to the public but had entirely failed to take his professional colleagues with him.

Lives of the Stars

F red Hoyle applied his creative genius to an extraordinary range of problems and worked in a variety of genres. His research covered topics from the solar system to the entire universe. Everything interested him: the Sun, stars, interstellar dust, galaxies, and cosmology. He even strayed into political and social science, climate change, archeology, paleontology, and molecular biology. He exploited every medium available—research papers, monographs, science fiction, children's books, and textbooks—to get his messages across. Any opportunity to make a radio or television broadcast delighted him, as did giving popular lectures, which attracted huge crowds. Most of the time he worked like a grasshopper in summer, jumping abruptly from one thing to another. In recounting his scientific career, it is inevitable that the narrative must by turns advance and backtrack. Take, for example, his work on accretion, on cosmology, and on astrophysics: These all overlap in time, but they are largely disconnected in intellectual terms. For that reason, I have generally chosen to cover one area of research at a time, considering together clusters of related papers and books. We now turn to his investigations on the evolution of stars, which began just before the war and remained an important theme in his work until the mid-1960s.

Most productive scientists today write their papers on a computer. Hoyle's working method was no different from that of scholars the world over before the age of computers: pen and paper. The manuscripts in St. John's College archives, as well as personal recollections from his surviving colleagues, tell us that his preferred way of working was to write on ruled paper with a fountain pen, using blue ink in the early years but later switching to black. Remarkably, his handwriting scarcely changed from the 1935 application form in Emmanuel College to the correspondence I had with him in 2000. His is a clear, confident hand, written with even pressure, always easy to read, and seldom revised. His signature hardly evolves. The early manuscripts are in a much better state than his first carbon-copy typescripts, now fading and disintegrating. From about 1939, he worked in hardback foolscap-sized notebooks, switching to loose sheets and then pads in the mid-1940s. Most of the surviving notebooks and pads are devoted to a single topic, so at any one time he would have a stack of these on the go. For conducting his own research, he preferred to be in a college room or at home. Even when he had his own Institute of Theoretical Astronomy between 1967 and 1972, the time Fred spent in the department was devoted to interacting with others rather than pursuing his own personal work. He never cared much for sitting at a desk, favoring a sturdy armchair. He often rested his papers on a slender hardback report of 1949, *History of the Cambridge Observatories.*

Without any question, Eddington stimulated Hoyle's interest and research in stellar evolution. At first, the popular books stirred Hoyle's enquiring mind, and then, as we have seen, he was deeply impressed by *The Internal Constitution of the Stars,* which set out the basic theory of stellar structure. He liked the content of Eddington's lectures, if not the halting delivery. On Tuesday, February 6, 1940, Fred had an afternoon appointment with Sir Arthur, who was foreign secretary of the Royal Astronomical Society (RAS). Fred wanted to be a fellow of the RAS, and to achieve that he needed three existing fellows to sign his application form and recommend him "as a proper person to become a Fellow thereof."[1] He wished to get Eddington's signature first, as the fellow recommending him "from personal knowledge."

As director, Eddington resided in the east wing of the university observatory. To call on him, Hoyle turned off Madingley Road and walked up the main avenue of the observatory grounds. The beauty of the observatory's neoclassical facade and the mellow appearance of the ashlar walls struck him. Completed in 1823, the building is described by the Historical Monuments Commission as "an example of the use of the revived Greek style [of architecture] for a structure intended for scientific purposes." Following restoration in 1991, it is today closer to the original plan than it was in 1940. The front facade faces due south, an essential requirement for a building that once housed a transit circle. Hoyle rang the bell by the imposing door leading to Eddington's apartment. The professor's sister, always called Miss Eddington, answered the door; she kept house for the great man. Hoyle was shown into the parlor and waited a few moments, admiring the spacious room. If you visit the observatory today, this part of the building retains many charming domestic features, despite being converted to offices in the 1970s. Eddington's sitting room was in the southeast corner (now part of the library), complete with a grand piano that had accompanied Einstein's playing of the violin when he collected his honorary doctorate in 1930.

Eddington signed Fred's form, and Hoyle turned to leave. Then unexpectedly, Eddington overcame his normal shyness and asked Hoyle what research he had done of late. Hoyle had recently become a frequent and energetic participant in the weekly seminar at the observatories, an event famed for the afternoon tea that preceded it.[2] Hoyle, at this date, could never contain his thoughts until the end of a lecture: The published records of meetings at the RAS show him jumping up and questioning speaker after speaker. Eddington must have noticed these earnest and voluminous contributions from the young man at the RAS meetings and the tea club seminars. Answering Eddington's question was easy: Fred was making calculations on the surface temperatures of stars.[3] This started a discussion on the source of energy inside red giant stars. Eddington's model for a red giant was not hot enough to permit the "burning"[4] of hydrogen to produce helium. Eddington responded that Gamow had suggested hydrogen-to-lithium burning instead, to which Hoyle countered that there was too little lithium around for this to be a viable possibility. And so on. Then a bell tinkled from the private quarters of the apartment. "My tea!" exclaimed Eddington, and he was gone. But from this moment on, Hoyle held Eddington in high regard and there was a rapport between the two, despite the difference in seniority. Hoyle was destined to become the next astronomer to hold Eddington's Plumian professorship.[5]

For several years, stellar structure and evolution featured prominently on Hoyle's agenda, and, in order to appreciate his enormous accomplishments, it is useful for us to review some of the fundamental things we understand about stars today and to relate them to the less complete state of knowledge when Hoyle moved into this fertile territory.

The chemical makeup of a newly formed star is almost three-quarters hydrogen, nearly one-quarter helium, plus some ballast, say one-fiftieth, comprising most of the other stable elements. For now, we only need to consider the hydrogen. What powers the Sun, and all "normal" stars, is the energy released in the deep interior by the conversion of hydrogen into helium. The commonest energy-producing cycle takes four protons and, in stages, transforms them into a helium nucleus, two positrons, two neutrinos, and energy in the form of gamma rays. A helium nucleus has less mass than four protons, and the lost mass—amounting to 0.7 percent—bursts out as pure energy.

Each proton carries a positive electric charge and like charges repel each other, so, for protons to merge, they must travel fast enough to overcome their mutual electric repulsion. What we call the temperature of a gas is in fact a measure of the speed of its particles. The proton cycle can only operate if the temperature exceeds 10 million K.[6] The energy released in the phenomenally hot central core of a star gradually makes its way to the surface. The main way this happens is by radiation, which is also the way the Sun's heat and light cross empty space. However, the photons are continually reabsorbed and scattered, which means that transport of energy by radiation is a slow process. But in certain layers, convection can take over as the chief way of moving heat. Convection involves the circulation of a gas or liquid, such as occurs when warm air rises in a room over a hot radiator, then flows down again after it has cooled. Any comprehensive theory of the physical processes in a star must take into account both radiation and convection.

Astrophysical theory can be applied to calculate what a star is like below its visible surface and how it will change over time. Stellar structure calculations involve generating "models," each describing a set of possible interior properties for a star. The predictions the models make about how stars will appear can then be compared with observations. If I create a model for a star consisting of one solar mass of material with the same chemical makeup as the Sun, then my model star must match the radius, temperature, and luminosity of the real Sun. By putting more than a solar mass into my model, I should be able to see how stars more massive than the Sun behave—that they are hotter, brighter, and have shorter lives. If I apply my model to a range of stellar masses, I should be able to reproduce the temperature and luminosity of an assorted population of normal hydrogen-burning stars. Stellar evolution calculations are designed to show how the appearance of a star will change during its lifetime. This again can be compared to observations of real stars. Although we cannot follow the evolution of an individual star, a sample of many will include stars at various different stages of a stellar life.

Theorists such as Eddington, Hoyle, and Lyttleton liked to work on stars because they can be treated as symmetrical spheres for many calculations. Inside a star, properties such as temperature, density, and chemical composition depend *only* on the distance from the center, not on the direction away from the center. This means that stars were easier subjects than many other celestial bodies in the days before electronic computers. Eddington's book, *The Internal Constitution of the Stars*, showed astrophysicists how to analyze the structure and evolution of stars using pencil, paper, and a first-class mathematical mind. In 1940, when Hoyle entered this field, the first digital computer at Cambridge lay 12 years in the future.

When Eddington and Hoyle met for the first time as equal colleagues, in February 1940, researchers could all agree that Eddington's basic theoretical treatment of stellar structure was correct. The source of the central energy was no longer mysterious because, a year earlier, Hans Bethe had shown how hydrogen-to-helium fusion could power the Sun and stars. Astrophysicists also understood that the rate of energy production would depend strongly on the central temperature of the star: the higher the temperature, the greater the energy that would be released. By now the surface temperatures of stars could be assessed from their spectra with considerable accuracy. Back in 1924, Eddington had shown (using a little sleight of hand) how to calculate a star's energy output, or luminosity, in terms of two observable properties: mass and surface temperature. He compiled a list of stars for which luminosities, surface temperatures, and masses were known and demonstrated that his result applied to all of them, despite a millionfold range in their luminosities. This correlation is now called the mass–luminosity relation: the energy output of a normal hydrogen-burning star depends almost entirely on its mass.[7]

One more point on which Eddington and Hoyle could agree concerned the chemical composition of the Sun. In common with most astrophysicists of the time, both believed that the *interior* of the Sun was made of about two parts iron to one part hydrogen.[8] The spectrum of sunlight is crossed by copious narrow dark gaps, or absorption lines, the majority of which are the signature of iron. For half a century, this made it natural for astronomers to believe in an iron Sun, and the idea that iron constituted the bulk of its interior persisted even though it had been demonstrated in the late 1920s that the Sun's outermost layers—its atmosphere—consist primarily of hydrogen. Six years later Hoyle would show that this assumption about the chemical composition of the Sun's interior was entirely wrong.

From 1939 to 1942, Fred and Ray Lyttleton worked together on stellar structure more or less in parallel with their work on accretion, as well as on research concerned with the origin of the solar system. Lyttleton's letters to Hoyle speak of the high priority both of them gave to stellar evolution. In one undated letter, Lyttleton confesses to having little war work to do and says that writing up "the solar system stuff is boring compared with our real struggles [stellar structure, accretion]."

The immense advantage Hoyle brought to the partnership was his understanding of nuclear fusion reactions, a legacy of his research directed by Peierls and Dirac. From the surviving correspondence, it appears to me that Lyttleton had a firm grasp of how to conduct a demanding line of research, and he may have been the superior mathematician. Lyttleton seems to have had the major hand in drafting their 1942 paper, "On the Internal Constitution of the Stars."[9]

The paper considers the structure of stars more massive than the Sun, such as Sirius, the brightest star in the night sky. In massive stars, the conversion of four protons into one helium nucleus takes place not through the proton cycle that operates in the Sun but in a different series of reactions, known as the CNO cycle, which uses carbon as a catalyst. The amount of energy released is heavily dependent on the star's central temperature. Hoyle and Lyttleton showed that, in massive stars, convection is strong in the layers immediately surrounding the central core but not close to the surface. This contrasts with the situation inside the Sun, where the reverse situation applies and a radiation zone surrounds the core, convection taking over nearer the surface.

This is a remarkable paper. It immediately moved the research frontier far ahead of Eddington's achievements. Hoyle is probably the first astrophysicist to use the nuclear reaction rates of Bethe and Gamow and the first to incorporate a rate of energy generation that is very strongly dependent on temperature. Hoyle and Lyttleton calculated the luminosity, radius, and surface temperature of a normal star directly, given only a mass and the initial chemical composition. Eddington's models had used surface temperatures as input data. Hoyle and Lyttleton dispensed with this approach by deriving the surface temperature. They discovered that Eddington had gotten away with hair-raising approximations. They did, however, cling to a chemical composition that is only one-third hydrogen.

Even in wartime, the council of the RAS discussed every paper submitted, and the referee reports received about it, before deciding whether to accept it. They received Lyttleton's final draft in June 1941. The officers conducted the review process in a leisurely fashion. After 8 months, they produced an outcome that inflamed the authors and rekindled the abrasive dispute between the two young men and the astronomers of the old school. The council first considered the paper and one referee report at its meeting in February 1942. After discussion, the council decided to send the referee's comments to Lyttleton, who duly responded, conceding next to nothing by way of amendment or revision; his response was a rebuttal of the referee's "points."

The council was in a quandary at the March meeting. The president read out the main criticisms made by the referee, followed by Lyttleton's letter. After discussion, the council determined to consult the referee again, with a view to reaching a favorable conclusion at the April meeting. They also decided to seek a second opinion from another referee. Meanwhile, Lyttleton, in a conciliatory gesture, agreed to work on shortening the paper, in view of the wartime rationing situation. After the April meeting, A. D. Thackeray immediately wrote to Hoyle and Lyttleton with a list of changes to be made by them so that the paper could be accepted. His letter[10] is a model of English politeness and offers a compromise. "Council welcomed your offer . . . to shorten the paper. . . . I think it might be possible to obtain Council's approval of the paper . . . if it were shortened to $^2/_3$rds the present length (instead of half, as supported by the referee)."

Thackeray's concession electrified Lyttleton. He made an immediate decision to accept the offer. Without delay, he copied Thackeray's letter (in handwriting and in pencil) to Hoyle, and for a cover letter dashed off, "I have begun rewriting the paper." He added: "Some of the remaining criticism is justified, but everything can now be answered." Buoyed by Thackeray's suggestion, he asks Hoyle to respond by phone, to make arrangements to meet in London, and to finalize the text. To achieve the desired shortening, Lyttleton resorted to the device of extracting an entire section of the paper, on red giant stars and cheekily resubmitted it as a separate paper, which was subsequently accepted as well.[11]

It is not hard to conjure up Lyttleton's excitement when he received proofs in July 1942. But that was not the only proof paper he had to attend to, because he also received a communication from the United States, one that touched a raw nerve. In March 1942, Lyttleton had sent a short note to the *Astrophysical Journal*, making a mathematical correction to a paper from a distinguished observational astronomer. Otto Struve, the editor of *Astrophysical Journal*, attached a letter to the proof of the technical note. He said: "I am sending a proof of your paper to Luyten in order to comply with my usual practice of informing in advance astronomers whose work is subjected to criticism in a controversial paper." Lyttleton boiled over with rage and immediately dashed off a 1,000-word despairing letter to Hoyle.

The letter enclosed the corrected proof of their joint paper for Fred to look over once more before it was returned to the journal. On telling Fred of Struve's letter, I can almost see Lyttleton wringing his hands. He writes: "I am so sick of this old trick of introducing a new rule as a general principle in order to gain an unfair advantage, that it nearly makes me vomit every time they do it." He worries that if he protests the journal editors will make things more difficult for him in the future. Lyttleton is spooked by what, in a torrent of indignation, he calls "all the lies, frauds, calumnies, slanders, insinuations, and shifty intrigues," and he suggests they both stop taking any notice of such things, as long as publication of their work is not held up.

At this date, Lyttleton still had not learned the fate of a joint paper on red giant stars. His anxiety is palpable because he knew that the RAS council would make the decision a few days later. Lyttleton is convinced that the papers secretary "will not have the courage to pass it on his own bat" and writes pessimistically: "It is very likely they will hold it up as the summer recess is upon us, and they have not failed to use this in the past to delay us." He fretted unnecessarily: Council nodded the paper through and published it in the autumn.

A red giant is one outcome of stellar old age. When our Sun becomes one, in about 5 billion years, it will swell until its outer layers swamp Earth and stretch as far as Mars. The paper by Hoyle and Lyttleton on red giants is the first attempt in the UK or the United States to understand the physical processes causing the huge expansion of a star's atmosphere when it becomes a red giant.[12] They accounted for this phenomenon by proposing that the atmosphere of an aged star has a composition different from that of its interior.

Lyttleton always felt that he and Hoyle were given an unduly hard time by the RAS in getting these two papers published. In the 1970s, when I was an elected member of the RAS council, scarcely a week went by without Lyttleton dropping by my office in the Institute of Astronomy to drone on about perceived injustices at the hands of the RAS in the 1940s. His gloomy manner must surely have adversely affected Hoyle's attitude toward the RAS.

Lyttleton was unnecessarily harsh in judging the handling of the Hoyle-Lyttleton papers by the RAS. His private letters to Hoyle are peppered with harsh words, whereas the letters from the society to Lyttleton are always diplomatic. The two papers, as published, take up over 50 pages in *Monthly Notices*, and this is after Lyttleton had done much pruning as instructed. Paper rationing meant that increasing the total number of pages in the journal to accommodate lengthy papers was not an option. The more pages Hoyle and Lyttleton filled, the fewer were the opportunities for other astronomers to get their research published. In 1984, Donald H. Sadler, who had been one of the secretaries at the time, explained at some length that "Hoyle in particular had landed the RAS Secretaries with a huge workload during the 1940s, when they found it difficult to get referees and paper."[13]

After the two papers had been accepted, Sadler contacted Hoyle: Would he like to read them at the September 1942 meeting of the society? This is hardly the gesture of a person who is seeking to squash the research. In fact, Sadler may have still been feeling embarrassed that he had been unable to squeeze Hoyle's accretion paper into the meeting of May 1941.

Hoyle accepted Sadler's invitation. When he heard of this, Lyttleton could not contain his anxiety. He prepared some detailed notes for Fred on what to say at the meeting. Lyttleton also encouraged Hoyle "to make it seem intelligible to the popular front—the back benches in this case of course." He is to take red chalk and draw a funny diagram of a red giant to amuse them. Lyttleton warns of the pitfalls of getting into any polite conversation:

> I don't have to warn you of how dangerous their seeming friendliness can be, so have as little as possible to say to them. They may plan to invite you to their after dinner gorge at the Criterion [Club]. It is this clique [i.e., the RAS Dining Club] which really controls the Society of course and at these meetings there is an opportunity for detailed "friendly" discussions with our best interests at heart.

In a mean-spirited afterthought, Lyttleton added that it would be best to mention as few names of other workers as possible. To achieve this, Hoyle should refer to "the usual formula" or "the standard theory," rather than "Bethe's energy generation." One enduring criticism of Hoyle's scientific papers has been his failure to acknowledge the work of others, and these omissions unnecessarily created enemies.

In any event, Lyttleton's exhortations would turn out to have been a waste of energy. A comparison between the published record of the meeting[14] and Lyttleton's notes[15] shows that Hoyle ignored his mentor.

On Friday, September 11, 1942, Fred and Barbara Hoyle arrived at the RAS's premises at Burlington House in good time for the afternoon tea in the library. Sadler and Thackeray cordially welcomed him, but no invitation to the Criterion was forthcoming: Mrs. Hoyle would not be allowed into a gentlemen's club. At 4:00 p.m. the company went down the wide staircase to the crowded lecture room. Hoyle spoke last, at about 5:30 p.m. The audience included the great names: Eddington, E. A. Milne from Oxford, and the astronomer royal, Harold Spencer Jones. Lyttleton did not attend. Hoyle's talk was a model of clarity, understandable by any professional astronomer of the time, and he mentioned both Eddington and Bethe by name.

Hoyle's presentation went down well. At its conclusion, Professor Thomas Cowling of Manchester leapt to his feet. He was an applied mathematician who had computed stellar models in the 1930s, including a model with convection around the core. Hoyle's rediscovery of convection excited him greatly. He openly congratulated Hoyle on the content of the paper, remarking that "Hoyle and Lyttleton have tried to build up the whole of stellar structure, setting out from the Bethe theory [of energy generation], and this method of approach is, I think, extremely valuable." Eddington behaved somewhat defensively, speaking about why he had needed to use a different approach, but he too liked the inclusion of Bethe's energy generation. However, he differed from Hoyle on a philosophical point: He wanted there to be more use of observations and fewer assumptions. Hoyle dismissed this as a "dangerous procedure" that involved arguing back from the results. With that, the meeting concluded and the Hoyles headed back to Witley.

After acceptance of the pair of stellar structure papers, Hoyle and Lyttleton switched their attention to a special category of variable stars, that is, stars whose brightness fluctuates over time—the Cepheids. These variables are especially important because they provide a means for astronomers to work out the distances of galaxies beyond our own. They are very luminous and thus can be picked out individually among the stars of nearby galaxies. The method of finding their distance makes use of a relationship between their period of variation, which is easily measured, and their intrinsic or "absolute" luminosity.

Eddington, in *The Internal Constitution of the Stars*, modeled Cepheids as single pulsating stars, puffing in and out on a regular cycle. This picture is more or less correct, but Lyttleton discovered that the underlying mechanism as proposed by Eddington would not work, and this led him and Hoyle to a different idea. They pictured a Cepheid as a binary star in which the two members orbited each other only a small distance apart inside a common extended atmosphere. In their work on the exchange of matter between the two stars, they came very close to creating the physical model that today accounts for certain x-ray stars. The final section of Hoyle and Lyttleton's Cepheid paper[16] also proposed a novel explanation of the exploding stars known as novae. They questioned how a close binary star would evolve, postulating that the two stars would get ever closer over time and would eventually merge. At coalescence a huge explosion would be certain and would explain the nova outburst. Although this model, too, is wildly wrong, it opened up new trains of thought for Fred.

Hoyle wrote: "Accompanying the fission there would almost certainly take place a streaming out between the two main components of a great deal of atmospheric gases." Three seeds were being planted in Fred's mind at that moment: an enthusiasm to work on novas and exploding stars, an idea that bodies of planetary mass might condense from the material drawn out of the colliding stars, and a speculation that such an explosive process could scatter into interstellar space heavy elements that had been made inside the stars. We noted in Chapter 4 that novas formed a topic of conversation with Walter Baade the following year. Before long, he would spend much more time investigating the heavy elements. The whole incident is a reminder of how, in science, a false trail can sometimes open up fruitful areas.

In late 1945, back at St. John's College, Hoyle returned to stellar astrophysics, particularly the chemical composition of the stars. Astronomers were slowly coming to realize that stars have substantial hydrogen content. The trail had started back in 1925 with Cecilia Payne's doctoral research at Harvard College Observatory. In her dissertation on stellar atmospheres, she correctly posited from her analysis of the solar spectrum that silicon, carbon, and other common elements are to be found in about the same relative amounts in the Sun as on Earth.[17] Helium and particularly hydrogen appeared to be vastly more abundant in the Sun than the other elements. This result disagreed with earlier theories, and Henry Norris Russell had told her that it was "clearly impossible." Deferring to Russell's stature as an astronomer, Cecilia discounted her results as "almost certainly not real," but she was soon to be vindicated. Within 3 years, Albrecht Unsöld[18] and William McCrea[19] independently established the preponderance of hydrogen in the Sun's atmosphere, a finding that Russell then accepted and further confirmed in 1929.[20] However, theorists such as Eddington stressed that these results were for stellar atmospheres, which might have compositions entirely different from the unobservable interiors of stars, and he remained wedded to iron interiors for a few more years. The next move came from a Danish expert in stellar structure, Bengt Strömgren, who computed the physical conditions in stellar interiors using the then-new atomic theory and quantum mechanics. He became "convinced that there was now no escape from the conclusion that the hydrogen content . . . was much higher than assumed in Eddington's work."[21]

Hoyle set to work assuming a completely novel composition for a normal star, quite unlike anything tried before. He noted that Russell and others agreed that stellar atmospheres were mostly hydrogen, and he reflected on the composition of the interstellar medium, the stuff from which stars form. He was impressed that Theodore Dunham, a member of the staff at Mount Wilson Observatory, had shown 6 years earlier that hydrogen is 10,000 times more abundant than "metals" in the interstellar medium, the gas and dust floating in space between the stars.[22] (Astrophysicists use the word "metal" as shorthand for all elements other than hydrogen and helium.) There is far more hydrogen around than had been thought. Hoyle felt that "we are [now] faced with a situation in which either the usual view of the internal composition of the stars must be abandoned or Dunham's results must be regarded as atypical." He plumped for the first option, to spectacular effect! He calculated the energy emitted by stars composed of 99 percent hydrogen and 1 percent metals. His curves showing that the run of a star's luminosity against its mass on this basis beautifully tallies with observations.

The conclusion of this research, that a star is almost pure hydrogen with a tiny admixture of all the rest of the elements, would soon be modified to hydrogen *and* helium as the basic composition. Nevertheless, Fred's result would have immense cosmological and astrophysical importance. Hoyle showed us that stars are made of the lightest elements, with a mere sprinkling of all the other stuff. He was by no means a pioneer in suspecting that hydrogen is the most abundant element in the universe, but he was the first to replace the iron Sun with one overwhelmingly dominated by hydrogen and helium, in which heavier elements have a minor role. Astrophysicists have followed his lead ever since.

While modeling hydrogen-rich stars, Hoyle also turned his mind to the origin of the chemical elements and, in 1946, produced a lengthy paper for *Monthly Notices* on the synthesis of heavier elements from hydrogen.[23] I shall return later to this topic in more detail, but I mention it here because it is an example of his striking capacity to work on more than one demanding topic at a time. If instead he had operated sequentially, first concentrating solely on the chemical composition research, a competitor scientist probably would have evaluated the consequences of hydrogen-rich stars more quickly than Hoyle did.

In this paper he gives a summary of stellar astrophysics, much of which has hardly changed. The content is beautifully presented. The argument is confident, well paced, and more restrained than any of the writing in the long papers he crafted jointly with Lyttleton. Here, for the first time, we read of a method for manufacturing heavy elements deep inside a star and we learn of mechanisms for returning the enriched material to the interstellar medium. This is also Hoyle's first paper in astrochemistry, a subdiscipline that he could fairly claim to have founded.

About this time, he began to supervise more graduate students. Intellectual puzzles blossomed, and Hoyle needed students to work on them. As his student, Christine Bondi tackled difficult analyses in the stellar structure models. She used the Cambridge differential analyzer, an analog computational aid developed at the Cambridge Mathematical

Laboratory in 1939. Leon Mestel joined Hoyle as a graduate student in 1948.

The start of the 1950s marks a phase change in Hoyle's career. By now he has no difficulty whatsoever in getting his papers published. The top mathematics graduates at Cambridge are clamoring to be his research students. Everywhere he goes to give talks, the lecture rooms are overflowing. This is the commencement of the middle part of his professional life. Papers and books are about to flow in torrents. And his money worries are over, thanks to abundant royalties from *The Nature of the Universe.*

In September 1952, the International Astronomical Union met in Rome, which provided the opportunity for a second meeting between Hoyle and Walter Baade. Before a formal session of presentations on galaxies, Baade asked Hoyle to be secretary of the meeting and take the minutes.[24] This gave Hoyle a ringside seat to observe a crucial turning point in observational cosmology, and he wrote:

> Baade then went on to describe several results of great cosmological significance. He pointed out that, in the course of his work on the two stellar populations in M31 [the Andromeda galaxy], it had become more and more clear that either the zero-point of the classical Cepheids or the zero-point of the cluster variables must be in error.[25]

He was saying that crucial numbers that Edwin Hubble had used to gauge the distances of galaxies were mistaken and that the age of 2 billion years at which he had arrived for the universe was too small. Baade then announced that his observations gave an age of 3.6 billion years, breaking at a stroke the logjam that had frustrated the development of explosive models for 15 years.

Hoyle and Baade got on well together. On his return to the United States, Baade invited Hoyle to spend 3 months at the California Institute of Technology ("Caltech"), to be followed by 2 months at Princeton University. Because living conditions in Europe were much inferior to those in the United States, such an invitation was a great prize. Cambridge granted Hoyle leave of absence for Lent term 1953.

At the end of Michaelmas term 1952, Hoyle flew to New York and from there took the train to Princeton Junction. At Princeton, the distinguished stellar astrophysicist Martin Schwarzschild looked after him for a few days, assisting with the purchase of an old Chevrolet car. A week after arriving, Fred set off on one of the great motoring trips of the world, driving 2,900 miles from the East Coast to Los Angeles. His route took him down to Washington, then on to what is now the Blue Ridge Parkway. I can easily imagine Fred being thrilled by the changing landscapes as he drove 300 miles through the Appalachians—the Shenandoah and the Great Smoky mountains. Then he headed west through Tennessee, crossing the Mississippi at Memphis. Soon he joined old Route 66 for the long drive across the prairies.[26] East of Flagstaff, Arizona, just past Winslow, he saw to his left the ramparts of Meteor Crater and turned onto the grassy highland to view the crater.

For anyone with a professional background in astronomy, Meteor Crater is of iconic significance, a must-see experience. The visitor today is denied the experience Fred would have had, of seeing the crater before him, in the immensity of the desert, without today's development of parking spaces, shops, restaurants, and the museum. The visitor center adds greatly to the educational experience, but detracts from the mystery and wilderness the place once held.

The first glimpse of the crater is awesome: It is 550 feet deep and 2.4 miles in circumference. Fifty thousand years ago, a huge iron-nickel meteorite, hurtling at about 40,000 miles per hour, struck the rocky plain of northern Arizona with an explosive force greater than 20 million tons of TNT. The meteorite is estimated to have been about 150 feet across and to have weighed several hundred thousand tons. In less than a few seconds, it excavated a crater over 4,000 feet across and originally 700 feet deep. Large blocks of limestone, some the size of a small house, were heaved onto the rim. Flat-lying beds of rock in the crater walls were overturned in fractions of a second and uplifted permanently as much as 150 feet. It is the nearest approximation on Earth to a lunar crater and was used by NASA for training Apollo astronauts. Fred had a deep interest in what is now called earth science and viewing the crater would have made a lasting impression.

On leaving the crater, he pottered along in his old car toward Flagstaff, passing the San Francisco Volcanic Field and the Lowell Observatory, where Pluto was discovered on February 18, 1930.[27] He swung north, for the Grand Canyon (South Rim) where he was going to celebrate Christmas Day in fine style. He took the mule ride down the North Kaibab Trail and spent the day at Phantom Ranch, on the floor of the canyon. He even managed to win a few dollars at cards with the party of mule drivers.[28] The Grand Canyon would have impressed the earth scientist and cosmologist in Hoyle very greatly indeed, perhaps even more than Meteor Crater.

Of course, like all first-time visitors, Hoyle must have marveled at the majestic grandeur of the canyon. But beneath this aesthetic appreciation lay a profound awareness of the history, and indeed the cosmology, of the place. At the base of the canyon he would be standing on Vishnu Schist, the root of an ancient mountain range and 2 billion years old. Soaring above him was layer upon layer of sedimentary rock, beginning with the Cambrian deposits of Bright Angel shale from 600 million years ago and capped off by Kaibab limestone, 200 million years old.

With Christmas Day over, Hoyle motored on, past Williams, Arizona, over the Cajon Pass to San Bernardino and then through orange groves to Pasadena.

At Caltech, Hoyle had several lectures to deliver. His speaker's notes, on unruled American quarto, have survived.[29] He covered advanced topics such as stellar hydrodynamics and convection in stars. Of course, his audiences had cosmology: Newtonian cosmology, the Einstein-de Sitter universe, and the steady-state theory. He nowhere in his notes used the expression "big bang" and had ceased to speak of his version of steady-state cosmology as the continuous creation theory. These notes show him venturing for the first time into radio astronomy. Caltech also arranged for him to give a public lecture on steady-state cosmology, not on the campus but in a huge auditorium at nearby Pasadena Junior College. Sure enough, Hoyle packed it to overflowing and the lecture was a huge success. Characteristically, he began boldly and to the point:

> Some scientists hold that the universe began. Other scientists, and I am of this school, reject that the universe began. They postulate instead what might be termed a balanced universe, one which retains its uniform nature throughout eternity. Last year the British Astronomer Royal, Sir Harold Spencer-Jones, gave an endorsement of this theory at a scientific meeting in London.

In the concluding section he spoke of his anticipation for great results from the Palomar telescopes, the 200-inch and particularly the 48-inch Schmidt. These were the best telescopes in the world, which by probing the depths of the universe could discriminate the rival theories of cosmology. He looked forward to the optical survey results from the Schmidt, expecting these to be much more convincing than the radio astronomy surveys.[30]

On the research front, his three fruitful months of being guided by Walter Baade added to his knowledge of stellar populations and of the new science of radio astronomy. He also made an astounding discovery in nuclear physics, to which we return later. He certainly developed a liking for the southern California climate and the physicists at the Kellogg Radiation Laboratory.

Hoyle struck up a remarkable friendship with Edwin and Grace Hubble. Grace, the daughter of a wealthy Los Angeles banker, first visited Mount Wilson in June 1920, staying in the Kapteyn cottage, a simple wooden lodge located some distance from the observers' bedrooms in the Monastery but close to the 100-inch dome. Her sojourn on the mountain arose through her friendship with Elna Wright, her hiking companion. Elna's husband, an astronomer at the Lick Observatory, had been granted an observing run on the 100-inch. Working at night and sleeping by day, he had to endure distancing himself from his wife and tucking into the monastic formal dining in the men's dormitory. She meanwhile was immured in the Kapteyn cottage, nestling under the fragrant pine trees. The abundance of mountain wildlife, particularly the marauding packs of coyotes, had made her nervous of staying alone, so she persuaded a reluctant Grace to be her roommate. Albert and Elsa Einstein would add their names to the list of temporary occupants in 1930. My wife Jacqueline, as a user of the 100-inch, would spend several nights there in 1976, accompanied by myself and our infant daughter Lavinia.

One afternoon, Elna and Grace wandered over to the laboratory, and as they entered the room, Grace noticed a striking man of military bearing who was engrossed in the study of a newly developed photographic plate, taken the previous night with the 100-inch. Major Edwin Hubble, following a hand-shaking introduction, strolled over to their cottage in the early evening, where he royally regaled the ladies with tales of star lore and great discoveries. When his observing run was over, he drove the pair back to Pasadena. At this time, Grace was married to Earl Lieb, a geologist from one of California's best-connected families: His father was certainly the wealthiest citizen of San Jose, and he later became president of Stanford University. Within months of meeting Edwin on the mountain, Grace's husband died in a tragic accident, overcome by carbon monoxide while exploring an idle coal mine. Grace and Edwin became romantically entwined the following year, and in 1924, after a discreet courtship, they married.

George Ellery Hale, founder of the Mount Wilson Observatory and driving force behind the 200-inch telescope at Palomar, had died on February 21, 1938. One month later, Edwin Hubble had received a telegram announcing that he had been chosen to fill the vacancy created by Hale's death on the board of trustees of the Huntington Library and Art Gallery, a legacy of Henry Huntington who made his fortune from electric railways and trolley cars serving Los Angeles. Fred Hoyle profited from Grace and Edwin's established routine of spending Sunday mornings at the Huntington Library, strolling the perfectly manicured lawns, graveled walks, and botanical gardens, spread over 250 acres. Fred admired the Huntington's fabulous art collection, with its eight Gainsboroughs in the main gallery. Grace and Edwin were fascinated by Fred's Englishness and unassuming approach to conversation. In early 1953, he spent almost every Sunday with Grace and Edwin Hubble, who also invited him to dinners at their lavish home in San Marino.[31] Edwin died of a heart attack on September 28 that year. For many years afterward, Barbara and Fred Hoyle kept in touch with Grace. A letter from her, written in 1958, to congratulate Fred on promotion to a professorship, demonstrates that the Hoyles continued a warm social relationship with widowed Grace Hubble.[32] When he was in Pasadena, Fred sometimes invited her to join him for dinner.[33]

His 1953 California adventures ended with a seminar at the astronomy department of the University of California, Berkeley. The department head, Otto Struve, impishly introduced him by saying that he was the author of a dubious cosmological theory, and then with a flourish, he presented the honorarium of $50. Hoyle did not dally in Berkeley: Directly after the talk he set off to drive over the Sierra Nevada to Reno. East of Sacramento, he ran into a blizzard but pressed on nevertheless, with drifts at the roadside the height of his car. Suddenly, he skidded badly, turned through 180 degrees, and was decisively off the road. He started to walk, an act of utter foolishness, but miraculously he came across a fellow with a tractor after only 10 minutes. He parted with some of Struve's $50 and was soon on his way down a drive he recalled as "the second hairiest descent of my experience."

From Nevada, Hoyle headed once again for Flagstaff, this time with an important mission, to go to the Lowell Observatory on Mars Hill. He started with a guided tour of the beautiful historic buildings, where Percival Lowell, the founder, thought he had spied canals on Mars. Hoyle admired the 24-inch Lowell refractor, used by the eponymous planetary astronomer to inspect the surface of Mars. In 1901, Lowell had attracted Vesto M. Slipher, a self-taught astronomer from rural Indiana, to the observatory and set him to work with a new spectrograph that Lowell had commissioned from the brilliant instrument designer John Brashear. Three years before Hoyle's birth, Slipher had found that most "nebulae"—what were subsequently realized to be galaxies—were hurtling away from the Sun at speeds as high as 1,100 kilometers per second (700 miles per second). At the Lowell Observatory, Hoyle was in the very place where the idea of the expanding universe began.

Baade had sent Hoyle to meet Harold Johnson, one of the most influential observational astrophysicists of the twentieth century, whom Baade frequently advised.[34] Johnson was an expert at measuring the magnitudes and colors of stars, color being a proxy for surface temperature, and he had measured the colors and magnitudes of stars belonging to several clusters. This was important because members of a star cluster are all of the same age and chemical composition, having been borne in the same cloud of condensing interstellar gas. Therefore, the run of color against magnitude among the stars in a cluster can be caused *only* by the differing masses of the stars; any differences in this relationship between one cluster and another stem from the clusters not having the same age. Here was an amazing astrophysical tool: By comparing several clusters of different ages, it was possible to deduce how one star of a particular mass changed as it grew older, which was one of the main goals of theoretical astrophysics at the time.

The visit hugely stimulated Hoyle, and he wanted to work without delay on Baade's two stellar populations. Johnson persuaded him to stay overnight at Flagstaff, giving him an opportunity to dine with Vesto Slipher, who had worked at Lowell for half a century. Hoyle could not resist the invitation. In 1912, Slipher had measured for the first time the velocity of an object beyond the Milky Way when he found that the great spiral galaxy in Andromeda is traveling *toward* the Milky Way at 300 kilometers per second (180 miles per second). It was the highest velocity for any celestial body known at that time.[35] Slipher's observing program soon revealed that Andromeda was an exception: By 1917, he had reported 21 galaxies receding from the Milky Way. Later, in 1929, Hubble reinterpreted and extended Slipher's results. For cosmologist Hoyle, the walks with Hubble and then the dinner with Slipher would have been like spiritual encounters. Hoyle wrote, in 1955, of "the staggering discovery made by V. M. Slipher and by E. P. Hubble and M. Humason . . . [that the] universe is expanding. . . . This is the purport of their discoveries" he was speaking from the heart about American scientists he had met.[36] And he mixed with them at a time when visitors from British universities were rare birds of passage on the American academic landscape.

Hoyle's own account of Slipher is touching and sensitive: Why, in the autobiography, written nearly half a century later, would he have so much to say?[37] Fred, always generous to others on first encounter, must have found the dinner a moment to be savored, an episode in history. He speaks up strongly for Slipher, asserting that he should be given far more credit: "The remarkable thing about Slipher's career is that he started as a young assistant . . . with instructions to find canals on Mars, and he ended up discovering the expansion of the universe. I remember him as a slender man . . . a little embittered perhaps." I hope Hoyle might also have met another charming astronomer at Lowell Observatory, Clyde Tombaugh, who had discovered Pluto on February 18, 1930.[38]

Dinner with Slipher delayed Hoyle by 1 day. He needed to drive to Yerkes Observatory, at Williams Bay, Wisconsin. Today, this is a drive of 1,700 miles, taking about 26 hours. Fred nosed the Chevy down Mars Hill. At an intersection in the center of Flagstaff, two lads in U.S.

naval uniforms asked for a ride. Fred, who had hitchhiked a lot in the 1930s, told them to hop in. They were just back from the Korean War. Having started in San Diego, they needed to get to Oklahoma City, 880 miles away. This was on Hoyle's itinerary, and by sharing the driving they covered that distance in a single day!

Three days after the University of Chicago opened, the street car magnate Charles Yerkes offered to purchase what was then the world's largest telescope—a 40-inch refractor—and donate it to the university. His largesse extended to a giant observatory to house the telescope. Offers of land poured into the university, which chose a site 75 miles north of Chicago, on Wisconsin's Lake Geneva. The nearest electric lights were 7 miles away. The university's architect Henry Ives Cobb was fond of ornamentation rooted in classic mythology. He chose a Romanesque style so frequently used for churches and monasteries in Europe: a Latin cross with three domed towers. The tan-colored brick interior is decorated with intricate terra-cotta carvings. Cobb let his imagination run riot: Everywhere in the structure, both inside and out, the viewer finds hundreds of ornate, often playful representations of animals real and fanciful, signs of the Zodiac, phases of the Moon, and many other embellishments. The building and its contents constitute a fascinating example of the architecture and technological accomplish-ments of the late nineteenth century.

A great tower at the western end of the building houses the 40-inch telescope, its steel tube 6 feet long and weighing 6 tons, pointing skyward like a giant cannon. Overhead stretches the 140-ton dome, 90 feet in diameter, an inverted bowl running on 26 wheels, with its observing slit 13 feet wide. In this dome, Edwin Hubble began his astronomical training in October 1914, taking his first photographic plates of "nebulae." (To Hubble they were always nebulae: He never used the term "galaxies.")

The Yerkes staff demonstrated to Hoyle the telescope's operation. To make the telescope accessible at all angles of elevation, the entire hardwood observing floor, 75 feet in diameter and over 35 tons, "rises like an opera set at the tough of a button," creaking and shuddering like the deck of a ship. For Fred it was all far more striking than the dome of the 26-inch reflector at the Greenwich Observatory, which has a similar moving floor. I regard it as the most attractive historic observatory in the United States.[39]

Hoyle would, I think, have found the architecture striking—very different from any nineteenth-century buildings in Cambridge. He was certainly taken by the array of annual staff photographs inside the massive entrance doors.[40] (Many years later he would introduce this tradition into his department at Cambridge.) His mission had nothing to do with sightseeing, however. He was to meet and stay with Subrahmanyan Chandrasekhar. In 1930, "Chandra" (as he was universally known) had studied the behavior of the dying stars called white dwarfs, which led him to discover the maximum mass that such stars can have (now known as "the Chandrasekhar limit"). Chandra, who had studied under Milne at Cambridge, left for a position in the United States after Eddington had publicly crushed him in a discussion at the Royal Astronomical Society. We can safely assume that in 1953 Chandra would have been regarded as one of America's leading experts on the structure and evolution of stars, and he won the Nobel Prize in physics in 1983 for his work. He would have plenty to discuss with Hoyle, who gave a talk to notch up another $50 honorarium.

On the third day of Hoyle's stay, Chandra and his wife left the house very early for Chicago, leaving Hoyle to sleep in and follow later in his own car. Surely, we have all been in the situation of locking ourselves out of an unfamiliar room or building. Hoyle, wearing only pajama bottoms and unlaced shoes, nipped out to his car to retrieve a razor he had purchased the previous evening. The front door of the house clicked to ominously, and the lock engaged. The weather being bitterly cold, the double windows with their large outer shutters were all closed, and all doors stoutly locked. For the second time on the trip, Hoyle was stranded in freezing weather and the wrong clothes. Should he drive straight to Chicago to get Chandra's key? He quickly realized he would have no time for that because he had a lunch appointment with Enrico Fermi and a seminar to give. No, he had to think of a local solution.

Teeth chattering and shivering uncontrollably, he defrosted his car, set the heater running full blast, and drove to the observatory. Fortunately, a staff member was on hand. And even more fortunately,

Chandra had deposited a spare key, which was quickly found. Hoyle made his lunch date but was disappointed in the event because Fermi would not talk any physics; perhaps the Italian felt it improper to engage in shoptalk over lunch. In the afternoon, Fred pocketed another $50 for a talk on "plasma oscillations," then a hot topic but not something he knew a great deal about. From Chicago, Hoyle set out for Princeton, 800 miles due east, where he would spend 2 months.

At Princeton, Hoyle presented his California lectures for the second time. His research there was in collaboration with Martin Schwarzschild, a very good partnership from Hoyle's point of view: They were birds of a feather. Martin had a broad and imaginative creativity, based on a thorough understanding of physical principles, and a passion for detailed and precise numerical work to determine the consequences of physical models. He also had much more respect for observations than Hoyle ever had. Throughout his career he remained focused on stellar evolution, a field in which he surely equaled Chandrasekhar, but he, like Hoyle, would never win the Nobel Prize.

Hoyle and Schwarzschild set out to model the evolution of giant stars. They wanted to account for the observed differences between Baade's two stellar populations: type I with compositions like the Sun, and type II with much lower concentrations of heavy elements. Hoyle had never forgotten the moment he learned of these populations during his 1944 visit to Baade. With Schwarzschild he would now discover why there was such a difference between the brightest giant stars in the two populations.

Hoyle's former student Leon Mestel has given a detailed technical appraisal of the Hoyle-Schwarzschild results, which appeared in a joint paper published in 1955,[41] and the impact they had. The paper "is without doubt outstanding among Fred's contributions to stellar evolution," he says.[42] The pair looked at the evolution of type II stars having a mass 10–20 percent greater than the Sun's. They consider the unfolding situation once all the hydrogen in the central core has been turned to helium. The methodology of this investigation is interesting because they emphasize that their paper is only "a preliminary reconnaissance of the problem." The aim is to understand the physics without getting bogged down in precise mathematical calculations. (There were still no digital computers.) The paper has a clear statement that they have not hesitated to make simplifications if they felt the approximation would not affect the outcome too much. Hoyle's English critics, of course, had been loudly complaining that all his papers seemed to cut corners.

Astrophysicists were thrilled when they read the paper because it tuned in so precisely with the observed color–magnitude relationships. According to the paper, the life of a star that has used up its hydrogen fuel proceeds as follows: First the helium core shrinks, which raises its temperature, heating the hydrogen immediately outside the core. This hydrogen burns in a thin shell, dumping more helium onto the core as it does so and being replenished by infalling hydrogen from above. The addition of helium to the core increases its mass and its temperature. Eventually, the core gets so hot that its helium can fuse into carbon. According to the models, the star now has two energy sources of more or less equal importance: a helium-burning core surrounded by a hydrogen-burning shell. Meanwhile, convection propels the outer layers of the star outward, greatly increasing its size. The star is now a (red) giant. The observational distinctions between type I and type II stars are attributed surface conditions due to the presence or absence of metal atoms in the two types of star.

The paper is striking for the clarity of its arguments and the superb step-by-step unfolding of the mathematics to reveal the physics. It is an impressively honest account in which every shortcoming or fudge is transparent. In their conclusion they look forward to something novel—computers! To improve on their approximations, an entirely new line of attack is desirable, and "we feel that the new line may turn out to be a fully automatic representation, using large electrical machines."

One indication of the positive impact that Baade, Hoyle, and Schwarzschild were having on advancing stellar evolution was a spectacular international conference convened at the Vatican. The Pontifical Academy of Sciences organized a study week between May 20 and 28, 1957, with the title, "The Problem of Stellar Population." Participation was strictly by invitation only: Just 17 astrophysicists took part,[43] and the proceedings filled 600 pages of the Academy's *Scripta*

Varia.[44] Hoyle was the only representative from the UK and it fell to him to produce both a summary of the proceedings and a schedule of future research. Fred was again in lovely surroundings. The Academy is housed in a beautiful sixteenth-century villa. (An immaculate restoration was completed in 2004.) However, the Hoyles did not linger in Rome's heat while he drafted his report. With three other delegates he spent a week in a small hotel in Amalfi to complete it.

In the late 1950s and early 1960s, a succession of brilliant research students joined Hoyle to work on problems in stellar evolution. His position in British theoretical astrophysics was now supreme and his output absolutely staggering. At first, his students explored the details and ramifications of the Hoyle-Schwarzschild model, which the Vatican conference found entirely satisfactory. From 1959, Hoyle teamed up with computer programmer Brian Haselgrove; together they started to use digital computers to model the structure and evolution of stars.

Computing at Cambridge started with the foundation of a mathematical laboratory in 1937. The first devices were analog; then, after the war, Maurice Wilkes headed a project to build a stored program digital computer, named EDSAC (Electronic Delay Storage Automatic Calculator). The service operated from early 1950. Researchers ran their own programs. EDSAC-1, which filled a room, was built with surplus thermionic valves left over at the end of the war.

Fred was trying to use the computer to follow the evolution of a star and his first ambition was to reproduce the results that had been obtained already by manual calculation. Research students who tried their hand at building a star with this computer included Roger Tayler and Joyce Wheeler. I shall turn to Fred himself for a description of the EDSAC computer from the user's point of view. He has left us descriptions in his science fiction novels: the Cambridge computer features in *The Black Cloud,* while *A for Andromeda* includes a specification for the really powerful computer he hoped the future would hold.[45] The computer was in the old Anatomy School, reputed to be haunted. Input and output to Hoyle's fictional computers is through five-track punched paper tape and teletype machines—this continued in Cambridge until about 1970. Typically, the tape had thousands of holes, and if a single one were out of place or blocked by chad, the program would not run. The high-speed input reader and the output punch were on opposite sides of the room, which appears to have been a conscious decision to avoid confusing input tapes and output tapes. Three thousand valves churned out 12 kilowatts of heat, fans roared away, and the teletype machines clattered.

Haselgrove was set the task of solving equations that could not be dealt with by algebra, using computer programs written in machine language. This is a difficult task, as the narrative in *The Black Cloud* makes clear. The storage capacity of the machine was so small that intermediate results were rattled out on 30 feet or so of punched tape, which was then taken across the room and fed back in through the input reader, for the next step in the calculation. To turn the ultimate result into something a human could read, the final output tape was fed through a teletype printer. According to Joyce Wheeler,[46] Hoyle mainly worked on his research in the late evenings. We know that he was crushed by his teaching workload and with drafting papers during the day.[47] In 1956, Haselgrove and Hoyle demonstrated[48] the principle of making stellar evolution calculations on EDSAC,[49] but they were frustrated by the slowness of the machine. Each "run" of their model would require many hours. The computer could not be spared just for him—James Watson and John Kendrew needed it for modeling DNA's crystal structure. But the programming technique did improve. By 1958, Haselgrove and Hoyle had comprehensive sets of calculations for ordinary hydrogen-burning stars,[50] and Fred could see that the next generation of computers would transform astrophysics.

Hoyle's interest in stellar evolution remained a favorite theme throughout his long career. Together with Schwarzschild, he was the first to consider how a star evolves once the main supply of hydrogen is exhausted. In 1994, Hoyle and Schwarschild won the Balzan Prize for this work, on the nomination of Roger Tayler and Leon Mestel. Hoyle's share of the award was 175,000 Swiss francs (or about $100,000 in purchasing power today). At Cambridge, interest in the lives of stars had started under Eddington. Hoyle was instrumental in rekindling the field, at first through his own research, and then later through a string of outstanding graduate students who modeled stellar evolution on powerful computers.

CHAPTER 7

Clash of Titans

Fellows of the Royal Astronomical Society (RAS) of a certain age will agree that the most abrasive relationship in British astronomy in the second half of the twentieth century was that between Fred Hoyle and Martin Ryle. Their academic arguments, conducted in the most public manner imaginable, lasted nearly three decades. Until both became older and wiser in the mid-1970s, astronomy at the University of Cambridge went through a period of strong polarization as a result of these two prima donnas failing to coexist more harmoniously. The destructive and competitive forces that they unleashed harmed the standing of astronomy and cosmology at Cambridge, particularly when their public disagreements became widely reported. To put this dramatic and highly significant interlude in context, it is essential to understand the background and personality of Martin Ryle.

Ryle engaged fully with all his students.[1] Always available and seldom traveling, he admired the laboratory, his staff, and his students. His door was always open; invariably, he took his morning coffee and afternoon tea with the rest of his group. Unlike Hoyle, Ryle regularly added his name to his student's research papers, not to grab personal credit but to give weight to their future curricula vitae. Hoyle's immense output includes only a handful of papers authored jointly with students, and he seldom read or corrected their drafts. Ryle was the complete opposite in this respect.

A yawning gulf separated the social and family backgrounds of Hoyle and Ryle. We have seen that Fred was the son of a cloth trader crushed by the economic slump after the First World War. By contrast, Ryle's father was royal physician to King George VI, Regius Professor of Physics at Cambridge, and later professor of medicine at Oxford. Gilbert Ryle, the philosopher, was an uncle. Martin had a privileged education, first from a private governess, and then at a private college, followed by 3 years in the grandeur of Christ Church, Oxford, where he graduated with first-class honors in physics in 1939. He joined Jack Ratcliffe at the Cavendish Laboratory to begin work on Earth's ionosphere, but within weeks he and Ratcliffe moved to the Air Ministry Research Establishment.[2]

Ryle, already a radio "ham," thus found himself in the midst of a brilliant team of electronic and radio engineers, from whom he learned tricks of the trade that he would brilliantly exploit back in the Cavendish Lab at the end of the war. Initially, he worked on antenna systems, then moved on to design prototype test equipment: a signal generator, wave meter, power meter, pulse monitor, and so on. The tragic crash of a Halifax bomber on a test flight killed key members of the scientific team, and in the resulting staff shuffle, Ryle, age 23, became group leader, taking charge of electronic countermeasures and deception. Members of Ryle's group analyzed parts of captured radar equipment, including the Würzburg equipment from the Bruneval raid.[3]

On July 13, 1944, by an almost incredible stroke of luck—good for the British but very bad for the Nazis, that is—the pilot of a Junkers 88 bomber mistook England for Germany and landed in England with his aircraft undamaged. It carried two wholly new radar instruments for detecting aircraft at night. Ryle examined the entire system and worked out the precise specification for a warning receiver that could be installed in RAF fighters. Ryle's instrument became remarkably successful at alerting pilots to enemy fighters.

As head of the group, Ryle's extraordinary inventiveness, his scientific understanding, his management skills, and his capacity for sheer hard work came to the fore. But the atmosphere was full of desperate anxiety. He was a tall individual who could be dour and aloof. Under the stress of urgent operational requirements, he was intolerant of those who did not share his immediate insight. According to a senior colleague, Ryle could be "highly temperamental and not an easy man to work with." There is no doubt he would speak loudly, clearly, and passionately when provoked. This tendency to flare up when frustrated or thwarted, lashing out at colleagues and hurling objects at people, remained with him all through his professional career. (More than once, I had to duck to avoid a blackboard eraser hurled in my direction!) This character trait gave an angry sharpness to his disagreements with Hoyle.

After the war, Ryle returned to the Cavendish, deeply affected by the tragedies of war, notably the terrible effects of British air raids on German civilians. Having worked on countermeasures to protect British bombers, he felt deep personal responsibility and this affected his outlook on his future career. He wrote:

> By the end of the war we were all very tired. Few of us knew precisely what we wanted to do. I was very tense and . . . I certainly knew what I didn't want to do. I wanted nothing more to do with military equipment. I was not one of those who would be content designing bigger and better radars in preparation for the next war.[4]

At the Cavendish, Ratcliffe gave him an entirely novel scientific puzzle, which had arisen from an incident in the war. Ratcliffe told him that on February 12, 1942, two German battleships and a light cruiser had made a daring run through the English Channel, scurrying from out of Brest, apparently hoping to make it to a safe haven in Norway.[5] Astonishingly, these vessels went unharmed. The British believed that they had acquired the means of jamming antiaircraft radar. This led to an assignment for James Hey, who had learned radar science from Ratcliffe, to investigate the jamming as a matter of urgency. Within 2 weeks, Hey had produced a secret report for Ratcliffe, explaining that the jamming was not artificial but the result of amazingly strong radio emissions from the Sun, associated with a large sunspot group.[6] This marked the first detection of an outburst of radio energy from an astronomical object.[7] At that date, astronomers presumed that the Sun would not be a significant source of radio emissions. However, further research would have to wait until peacetime, when Ratcliffe suggested to Ryle that he should investigate the Sun's radio emissions, a task Ryle performed with enthusiasm.

Ryle decided to measure the intensity, angular size, and position on the Sun of the source of radio emissions, for which he would need to build the first radio telescope at Cambridge. The Cavendish Laboratory had no money and only antiquated equipment, quite unsuitable for such research, but Ryle and his colleagues had no scruples about using their wartime contacts for scrounging and looting. In this they had official support, provided they were not stealing government property. Among other treasures, their foraging expeditions produced oscilloscopes, thermionic valves, and measuring equipment. From Germany came excellent equipment, including two large and two small Würzburg antennas captured in the field and drums of priceless low-loss coaxial cable, unobtainable in Britain.

By December, Ryle, together with a young electronics expert, Derek Vonberg, had a lash-up capable of "looking" at the radio emissions from the Sun at wavelengths measured in meters. They noted that they only detected these long-wave radio emissions when there were sunspots. However, the first observations did no more than demonstrate the principle. Better equipment would be needed for a proper investigation.

Ryle needed to improve the resolution of his antenna system. The radar antennas used for war work were low resolution for the purposes of astronomy, with beams 10 degrees wide—20 times the apparent diameter of the Sun. Ryle needed a receiving beam much smaller than the Sun's diameter. His solution to the problem—to work with antennas in pairs rather than as single dishes—would launch him on a trajectory to the Nobel Prize in physics in 1974.

In 1946, Ryle set two antennas on an east-west line, separated by lengths varying from 25 to 140 wavelengths of the radio signal they were detecting. Both dishes were connected to one radio receiver, designed by Ryle and Vonberg, to make a setup physicists would recognize as an *interferometer*.[8] In essence, the two had made an instrument operating at radio wavelengths analogous to a classic experiment involving light, known as the Michelson interferometer. The two radio engineers, who knew next to no traditional astronomy, now profited from a natural accident of timing. Solar activity—sunspots, flares, and outbursts—was soaring to the peak of the 11-year cycle, with many large sunspot groups. July 1946 saw the Sun affected by massive sunspot activity, leading to brilliant auroral displays, disturbances to Earth's magnetic field, and high levels of cosmic rays. Their paper in *Nature* suggested that the enhanced radiation came from an area considerably smaller than the solar disk, presumably from a sunspot group. By 1948, they had improved their technique significantly, using the 27-foot Würzburg antennas to observe the emission from sunspots.

Working conditions at the fledgling radio astronomy observatory could be harsh. In January–March 1947, Britain suffered the coldest winter in almost two centuries, with shortages and rationing worse than in wartime. The observatory had a wooden shed to shelter equipment, but the scientists and their technicians worked in the open, digging holes, pouring concrete, erecting steel, surveying, stringing up coaxial cables in hedgerows (so that horses would not trample them), hacking down prickly hawthorn bushes, and clearing land—all back-breaking work that reduced them to jacks-of-all-trades, while trying at the same time to make scientific discoveries. Fred Hoyle, meanwhile, could cosy up to his glowing fireplace in college, use his ration book to secure a decent meal at high table, and settle in an armchair to do his research.

I do not think Ryle and Hoyle ever "got on." Throughout his entire career, Ryle regarded all theorists, apart from the men (and in those days they were all men) in his own group, with deep suspicion, bordering on loathing. He felt that theorists were parasites, ever ready to pounce on the new results from experimenters. Ryle's observations of radio emissions from sunspots provide us with an early example of a testy Ryle–Hoyle disagreement.

The situation by 1948 was that Ryle had shown that at all times the Sun gives out radio waves characteristic of an emitter at a temperature of a million degrees. He had, in fact, discovered radio emissions from the solar corona. Furthermore, his experiments showed that enhanced emissions from the active sunspots had equivalent temperatures of up to a billion degrees. These gigantic values do not literally refer to the temperature of bulk matter, but rather to the speeds of the electrons responsible for the emissions. Ryle had models for these emission mechanisms, which he probably spoke about at Kapitza Club meetings, as well as publishing them through the Royal Society.[9] Hoyle also explored emission mechanisms.

Not surprisingly, Ryle's theory papers lack the depth, mathematical sophistication, and elegance of Hoyle's presentations. Ryle's paper on solar radio emissions is long on words and speculation and short on applied mathematics. By contrast, Hoyle was a dextrous manipulator of more complex mathematics, as his 1949 monograph on solar physics demonstrates.

Late in 1949, Hoyle and Ryle disagreed openly at the Kapitza Club. At the root of the clash was a difference of opinion about how the highly disturbed region above a sunspot could generate radio waves. Hoyle made a normal colloquium presentation of his theory, covering the blackboard with neatly written equations. In the discussion that followed, Ryle felt that Hoyle's solution would generate radiation all right, but that the radio energy would not be able to escape through the outer layers of the Sun's atmosphere. Then the debate suddenly turned sour. On impulse, Ryle sprang forward and rubbed out all of Hoyle's equations, angrily shouting that he was dead wrong![10] Hoyle was a university officer, whereas Ryle was still a fellow on outside funds. Ryle did not sense any danger from his rude intervention.

A few days later, still disgruntled, Ryle ticked off Hoyle in a letter.[11] He expresses himself royally: "We were rather disturbed by the difference of opinion that was apparent . . . concerning the possibility of the [radiation] escaping." It was important to him that "we should set this point straight." Ryle reminded Hoyle of an earlier argument, in 1946, when "we agreed that the [absorption] in the overlying layers was likely to be prohibitive." Ryle allowed himself another royal "we," pointing out that the Royal Society (no less) had accepted his arguments. "We are most anxious to reach agreement on whether or not your mechanism is possible." Ryle concluded by demanding access to Hoyle's calculations "so that we may resolve the controversy one way or the other, once and for all." Hoyle's hands were too full to respond to Ryle: The public storm had just burst over his 1949 radio broadcasts.

With improvements in instrumental technique, Ryle and his group grew bolder. The radio physicists studying the Sun began to learn some astrophysics, they joined the RAS, and they started to use the expression "radio astronomy" openly for their new discipline.[12] Ryle and a group based in Sydney, Australia, turned their attention to radio sources outside the solar system. The first Cambridge survey located about 50 cosmic radio sources but the positional data were too vague to enable more than a handful of the radio sources to be matched with optical counterparts. The Australians had identified just four of their sources. One coincided with the Crab Nebula, the fossil of a massive star seen to explode as a supernova in 1054, and another with a giant peculiar galaxy, NGC 5128. These weird objects scarcely helped the understanding of the radio sources being catalogued at Cambridge.

Ryle invented a marvellous radio receiver that subtracted out the background coming from the Milky Way, leaving just starlike radio sources.[13] The first Cambridge survey, made using this receiver, showed a scattering of 50 radio sources over the whole sky distributed at random. Five could be identified with large external galaxies, but this gave no clue about the generality of the "radio stars." What were they?

In the absence of firm identifications, a general belief grew among the British astronomers that Ryle's sources were a new class of peculiar stars located in the Milky Way, and Ryle himself repeatedly pushed this scenario in public and private meetings. At this stage, he was driven by a quest to design much better telescopes rather than a compulsion to understand the astrophysical implications. That was not good enough for Tommy Gold, by now a demonstrator (a university office at Cambridge equivalent to lecturer) at the Cavendish Laboratory. Although he was not a member of Ryle's group, he took a keen interest in their observations. As an astrophysicist, he was much more immersed in the fundamental nature of the radio stars than the radio astronomy group members.

Gold loudly and repeatedly claimed that the point sources must be remote objects far from the Milky Way. That was the way to explain the random distribution over the whole sky. Ryle found this claim astounding: The sources would need to be at least 100 million times stronger than the Sun. Thus, he preferred to think in terms of a local population inside the Milky Way and close to the Sun. In retrospect, it seems odd that Ryle clung so passionately to this belief at a time when a handful of sources had been matched to nearby galaxies. Unfortunately, little written evidence has survived of Gold's clashes with Ryle at small meetings in the Cavendish.

On the second Saturday of March 1952, Ryle received a letter from an unexpected source: the Royal Society. Written by the assistant secretary, the letter announced that his name would be going forward at the annual general meeting for election to Fellowship of the Royal Society (FRS). The council of the Royal Society handles the election of fellows with great secrecy, and Ryle would have had no inkling that he would be one of 30 new fellows proposed for election.

We can safely assume that he must have been absolutely stunned by this excellent news. Ryle had to keep quiet for 2 weeks. Hoyle learned of the achievement on March 20, when the society issued a press statement on the new fellows.[14] Ryle had won his FRS at the remarkably young age of 34; the citation highlighted his invention of the radio interferometer and his war work. I can only speculate what Lyttleton would have made of this news; it probably would have depressed him deeply. I can imagine him being incandescent and probably gossiping with Hoyle at St. John's College about the unfairness of it all. At any rate, Ryle was now standing head and shoulders above Hoyle in the academic pecking order. Hoyle might momentarily be the most famous scientist in Britain, but he was not yet up to the standards of the Royal Society.

The next month a select gathering of radio scientists convened at University College London. At this so-called Massey meeting, Ryle again argued forcefully that the small-diameter radio emitters, discrete sources as they now called them, must be dark stars in our galaxy. Gold countered that the sources were dense, collapsed, magnetic stars in distant galaxies. Hoyle publicly supported Gold's line. The discussion then became highly charged. George McVittie bluntly told Gold that he did not know what he was talking about.[15] Ryle caustically vented his frustration with Hoyle and Gold. "I think the theoreticians have misunderstood the experimental data," he snapped, referring to the very data that had just won him the FRS. Hoyle now played the stormy petrel, responding with irony:

Professor McVittie and Mr. Ryle have dogmatically asserted that the discrete sources cannot be of extragalactic origin, although of the half dozen or so that have indeed been identified, five have been found to correspond with nearby extragalactic nebulae. Presumably a discrete source ceases to be a discrete source as soon as it is identifiable as a galaxy.[16]

At this period, Hoyle, urged on one imagines by Lyttleton, still had much contempt for the British observational astronomers and a growing dislike of the physicists at the Cavendish. Only later, following his first visits to California, would he come to appreciate the importance of observations to support theory. Baade would teach him that.

At the end of the day, Hoyle and Gold munched down a miserable supper, of "last week's potatoes," and then set off for Cambridge in Tommy's Hillman, a grand car that was forever breaking down. Toward midnight, and still 20 miles south of Cambridge, stomach pangs directed the hungry pair to Jock's Café, a chips-with-everything place catering to long-distance truck drivers. It seemed an unlikely venue for a bad-tempered controversy to break loose: Jock's was the sort of greasy spoon where smugglers peddled tobacco, watches, and cameras.[17]

Fred and Tommy talked over the events of the afternoon. According to Hoyle:

Controversy had been far from my mind as I had spoken . . . on a technical problem that had worried me for quite a while: the problem of electrical conduction across a transverse magnetic field pervading a diffuse gas.

This was precisely the academic puzzle they had fought over in 1949. But, for some reason, Hoyle's methodology drew the ire of the Swedish physicist Hannes Alfvén, himself a highly controversial figure at this time. The argument prickled Hoyle, who preferred to grapple with natural configurations of the magnetic field rather than the contrived, idealized versions of Alfvén. So, Hoyle was already hot under the collar when Gold had come under attack from Ryle. They sat face to face, egg and chips on their plates. What was it with Ryle, they despaired. All Gold had said was that the isotropic distribution of Ryle's radio stars must mean they were either very very close (less than 100 light-years away) or very, very distant (more than 100 million light-years). Ryle's intermediate hypothesis, that the radio stars were distributed through the Milky Way, would lead to a bunching along the galactic plane. Gold had expressed himself no more strongly than simply saying "the possibility of the sources being mostly extragalactic should be kept in mind until more data are forthcoming."

Looking back, Hoyle felt shocked at Ryle's public denunciation of Gold. He was deeply wounded by Ryle's sarcastic comment that the theoreticians had not understood the nature of the evidence. In the cut and thrust of academic debate, it is quite normal to hear a critic say in front of an audience "You are wrong," or "I don't agree with you," or "My observations give a different result." What Hoyle now utterly detested was Ryle's theatrical manner of flat denunciation, accompanied by some rider implying that "your error would be obvious to anyone with an adequate knowledge of the subject." That was how Hoyle had been treated at the Kapitza Club, and it was the rebuke meted out that afternoon. Dangerously, by speaking up for Gold, Hoyle had now effectively aligned himself against Ryle on the radio star controversy.[18]

As Tommy and Fred cleared the remaining chips, they decided the argument could be resolved by measuring the positions of a couple of radio stars with sufficient accuracy for identification with visual objects. Unfortunately for them, Ryle had no intention of letting Gold use the radio astronomy group's interferometer. However, in my opinion, Hoyle was unfair when he later criticized Ryle for this: The Cavendish crew had worked the winter in the open air to construct the interferometer, and it is quite correct that they should have the first chance to use this instrument to acquire and interpret new data.

In general, radio observers remained committed to the concept that the point sources were galactic stars and they took little notice of Hoyle and Gold. For example, Hendrik van de Hulst gave a series of lectures at Harvard in the spring of 1951 in which he dismissed the extragalactic hypothesis as unlikely.[19] Bernard Lovell expressed the general verdict in late 1951: "Three radio stars coincide with unusual objects . . . but these are non-typical stars."[20] Including results from the Southern Hemisphere, the radio astronomers by now had about 100 radio sources and identifications with a handful of previously known objects.

Ryle, together with his doctoral student F. (Francis) Graham Smith, now solved the radio star puzzle brilliantly, in a dazzling experiment that stands as one of the most significant observations ever made in radio astronomy. Smith hooked up the two large Würzburg antennas to create an interferometer, specifically to make more accurate measurements of the positions of the four brightest sources. They covertly showed their positions to David Dewhirst, a young optical astronomer at the Cambridge Observatories. The director of the Observatories, Roderick Redman, also advised Smith to write to Baade, in California, enclosing the latest positions, and he did so on August 22, 1951. Ryle's instinct would have been to keep the data within his group, but he realized that their ignorance of optical astronomy would force them to cooperate with trustworthy optical astronomers.

Dewhirst publicized his findings at a packed meeting of the RAS in December 1951.[21] He confirmed the identification of the strong radio source in Taurus with the Crab Nebula and the source in Virgo with the colossal elliptical galaxy M87. This second identification encouraged Dewhirst to search photographic plates for both galactic and extragalactic objects at the positions of the radio sources in Cygnus and Cassiopeia. On September 23, he used the largest reflecting telescope at Cambridge to match the Cassiopeia radio source to an apparently circular nebula in the Milky Way (a supernova remnant). For the Cygnus source, he scrutinized a copy of a photograph of the search area that had been secured by Walter Baade at Mount Wilson. This showed a faint smudge of nebulosity within the error margins of the Cambridge position. Dewhirst cautiously announced, "It is difficult not to regard the coincidence as significant." The double negative in this phrase translates into "it is easy to regard the coincidence as significant." In his closing remarks to the RAS fellows, Dewhirst hedged his bet, saying the Cygnus and Cassiopeia objects could be galactic or extragalactic.

Professor Dingle, in the chair as president, opened the paper for discussion. Ryle went first. "I would like to mention some arguments against an extragalactic origin." A comparison of 50 radio sources with the nearest 300 galaxies yielded no correlation, the continuous radiation from the Milky Way showed it had a population of unresolved radio stars, and so on. Today, it reads like the response of a radio engineer who knew just a little astronomy. Gold hit back with some clever astrophysics that has stood the test of time: He correctly guessed that there could be a link between the unknown objects producing high-energy cosmic rays and Ryle's radio sources. Gold tartly reminded Ryle of his own prediction, made 3 years earlier, that the radio sources were extragalactic, and he interpreted Dewhirst's paper as "the first hint that this [extragalactic] suggestion is being taken seriously," concluding with "radio *stars* are entirely inexplicable."

Ryle leapt to his feet, asserting that if the sources *were* extragalactic, they would need to generate radio waves a hundred million times more efficiently than the Milky Way. The extragalactic protagonists had proposed no mechanism for energy generation on such a scale, so they must be wrong! Here Ryle spectacularly missed the point: He really should have grasped that if he could be less dogmatic, his research might then chase down the most energetic objects in the universe. By contrast, Gold had understood the awesome implications of his extragalactic hypothesis: the existence of extremely powerful energy sources in the distant universe. Back at the Cavendish, Jack Ratcliffe summoned Gold for a carpeting over his behavior at the RAS: As a mere demonstrator he had no right to give as good as he got from Ryle in public debates. Gold was forbidden to visit the part of the Cavendish given over to radio astronomy, and Ryle became even more secretive.

Meanwhile, out at the Mount Wilson and Palomar observatories, Baade was on the case. Smith's letter of August 22 had excited him greatly and he replied 2 weeks later. He would use the 200-inch telescope at Palomar to obtain the colors of all the candidate stars closely surrounding the positions of the radio sources. His colleague Rudolph Minkowski would take spectra of all the candidate stars to see if any showed spectral peculiarities, such as emission lines (which would indicate a high degree of disturbance). This was an excellent campaign, particularly as Baade was allowing for the possibility that Smith's positions could still be seriously in error.[22]

By late October, Baade and Minkowski agreed on the identification of three optical objects associated with radio sources: the Crab Nebula, the galaxy M87, and an abnormal nebula in Cassiopeia. The Cygnus source would have to wait a little longer.

On December 14, 1951, Smith communicated the gist of Dewhirst's

RAS talk to Baade, who had already secured a photographic plate of the Cygnus search area. This photograph puzzled him greatly. It revealed a rich cluster of galaxies, which agreed precisely with the radio position: Baade had spied "a queer object" that appeared to be two galaxies in collision, something never sighted before. An intergalactic collision would result in very high excitation in the interstellar gases. Baade dashed off a letter to Smith on April 29, 1952, adding a note that Minkowski would "shoot the spectrum" as soon as the Cygnus constellation was back in the night sky. Minkowski doubted the colliding-galaxy interpretation: He bet Baade a bottle of whiskey that his spectrograms would yield a different interpretation.[23]

Minkowski photographed the spectrum in late May. After developing the small glass slide, he looked at it in the dim red light of the dark room. He had never seen a spectrum quite like this from a galaxy. The most remarkable feature was a group of strong emission lines caused by highly excited neon gas, which would indeed represent two galaxies in collision. Furthermore, the object had a large recession velocity, 15,000 kilometers (9,500 miles) per second; this implied a distance of hundreds of millions of light-years. (Two decades later, I would have the privilege of reanalyzing this spectrum with more advanced techniques.[24]) Baade sent his amazing news to Smith on May 26 but added a caution, advising that, before they rushed into print with a public announcement, he should obtain more information on the angular size of the radio source: The diameter of the radio source must not contradict the small angular size of the colliding galaxies.

In September 1952, the International Astronomical Union's General Assembly in Rome provided the opportunity for all to lay their cards on the table. Reassuringly, the radio astronomers now had better positions, so Baade dropped his earlier caution. His principal task at the IAU was announcing the revision to the age of the universe. The day before this talk Baade grabbed Jan Oort, the father of radio astronomy in the Netherlands, and ushered him into a side room, where he fished out Minkowski's spectra. Oort stared at these with fascination, as Baade, still gloating from winning Minkowski's bottle, excitedly proclaimed that the radio stars must be remote extragalactic objects. Their boisterous backslapping caught the attention of Ryle, and Hoyle wandered over to see what the fuss was about. Gold joined in as well. Confronted with the spectra, all immediately recognized that the debate on radio stars was over. Grinning Fred was pleased as punch, sour-faced Martin disconsolate and emotionally disturbed.[25] Gold became extremely angry because, a few weeks earlier, Ratcliffe had, in effect, dismissed him by declining to renew his position as demonstrator at the conclusion of a 3-year appointment. Ratcliffe had cited "your failure to get on with Martin Ryle" as the reason for nonrenewal.

Ryle soon recovered from his humiliating defeat at the hands of Gold and Hoyle. Before the optical identification of the Cygnus source, his team had started construction of a major new interferometer, which operated from 1953 to 1960. According to Ryle, "The main purpose of the new instrument is to allow a greater number of radio stars to be located."[26] For him, the Cygnus identification "was the moment when we realised we were in the cosmology game, using a crummy instrument costing 200 pounds. Suddenly it was all much more interesting for us than galactic stars." The new interferometer would conduct a survey of the radio sources, going to a much fainter level than the first Cambridge survey. The new sweep of the skies should therefore record many more extragalactic objects and probe the universe far beyond the Cygnus galaxy. What excited Ryle was the realization that, with his equipment, he could test rival theories of cosmology. The radio sources were like beacons scattered through remote extragalactic space. How densely they were distributed at different distances would reflect the history of the universe.

In Hoyle's steady-state cosmology, the distant observable universe has exactly the same appearance as the local universe. The big-bang universe, by contrast, has aged and evolved: Distant objects appear younger than their local equivalents because of the time it has taken for their radiation to reach us traveling at the speed of light—we are seeing them as they appeared long ago. In theory, these two different scenarios can be tested against one another by taking a large sample of radio sources, measuring how bright each one is, then totting up how many there are at different levels of brightness. Surprisingly, in a statistical analysis of this kind, the outcome is not affected even if the actual distances of the sources are unknown or they do not all have the same luminosity. The principles of this technique were already well known from Edwin Hubble's work on counting galaxies.

In a steady-state universe, a plot of the number of sources counted, *N*, against their brightness, *S*, should be a strikingly simple curve. What observational astronomers actually plot are the logarithms of *N* and *S* because doing this renders the "curve" as a straight line. The counts of radio sources plotted in this manner produce a straight line with a slope of 1.5 (i.e., 3/2) for a steady-state universe. [27] If the universe is evolving, the plot has a different slope or may not be a straight line at all.

The radio astronomy group was now clearly in the cosmology game: Their new survey was netting radio sources by the hundreds. Hoyle's theory would soon be tested and found wanting. As Ryle viewed the situation, cosmology had, until recently, been no more than a playground for mathematicians, protected from the real world and safe from all possible attack. But Hoyle's steady-state model made specific predictions. In this sense it was real science, not philosophy. The results coming in from the survey were showing that the distant radio universe was not the same as the nearby radio universe. Radio sources appeared to evolve, so the universe could not possibly be in a steady state.

Ryle now played his hand as if he were in a game of poker. Gold and Bondi had both left Cambridge, so he only had to contend with Hoyle. The radio astronomy group began to operate in total secrecy. When Hoyle breezed into the Cavendish tearoom after giving a lecture, they ignored him. If anyone from outside the group showed their face in the office or laboratory, Ryle would hasten to close notebooks, wipe blackboards, and tidy away work in progress. All members of the group were sworn to unswerving loyalty and absolute silence.

The faculty of mathematics continued to burden Hoyle with lecturing duties. As had been the case since 1946, he lectured almost daily in the Arts School and, as ever, to packed audiences. He felt he could do no further work in cosmology until Ryle published his survey. But while Ryle counted, Hoyle wrote. Alan Hill, the publishing director of the London trade house William Heinemann Ltd., had approached Hoyle in late 1950 to write a popular book, *The New Cosmology*, presumably a larger, illustrated version of *The Nature of the Universe*. This book was never written; instead, Hoyle offered them *A Decade of Decision*. This is a polemic in political science, wherein he sets out some forthright personal views on population control (he is for it), atomic weapons (against them), good governance (despairs of it), and economic management (finds it abysmal). Heinemann made a bold decision in accepting this political tract from a cosmologist, publishing it in 1953.

Hoyle's standing was now such that he could position himself as a pundit on almost any topic in public life. His personal experiences colored much of his judgment. On Clement Attlee's government of 1946–1951, he had this to say:

> Six years after the ending of the war, the worst government in British history since Ethelred the Unready kept the country on a wartime footing. Although the most ravaged small nations of W. Europe had by then largely recovered from the war, we in Britain were still dressed in rags and eating last week's potatoes. . . . This worst government handed over free of charge the largest empire in modern times, converting Britain at a stroke, as in a downwards quantum transition, from being a great nation into being a small and ineffectual one.[28]

He must have been smarting from the savage way income tax was gobbling up his juicy royalties.

Hoyle richly rewarded Heinemann for publishing the political rant because he delivered the typescript of a blockbuster in 1955. Instead of the planned cosmology book, he presented the educational publisher with the best popular book on astronomy to have appeared since the war. Titled *Frontiers of Astronomy*, this book covers the whole of astronomy without resorting to equations or technical language. Many professional astronomers attest that when making the transition from school to university, this book inspired them to choose astronomy.[29]

Fred's approach to book writing differed greatly from the norm in academic circles. In structuring the content, he largely ignored the accepted curriculum as well as much of the work of others. Like Eddington before him, he seldom explained the intellectual history and achievements of his predecessors. Instead, he plunged straight in, writing almost exclusively on the topics that interested him.

It is hard to imagine any other author daring to open a book on astronomy with a discussion of ice ages. On pages 7 and 8 of *Frontiers*, he sets out a superb description of the greenhouse effect as applied to the Earth's atmosphere and its consequences for climate change. By Hoyle's third chapter, we are into the physics of isotopes and beta decay. He presents his own work on stellar evolution in the eighth chapter, which is fair enough because, by 1955, astrophysicists regarded him as a master practitioner. In a chapter on the expanding universe, he presses the case for the continuous creation of matter and acknowledges that Bondi and Gold arrived at similar conclusions from a different outlook. The Cambridge radio source counts are not mentioned, although Martin Ryle contributed a photograph of his radio interferometer, which earned him a pat on the back in the acknowledgments.

Reviewers heaped praise on *Frontiers of Astronomy*. According to the *New York Times*, it was "a fascinating book, full of the wild thinking which proves that science is alive. Outraged conservative astronomers will try to spank Fred Hoyle, but the punishment will not bother him. He has the defense prepared."[30] William McCrea, still a fan of steady-state cosmology, described the book in *The Spectator* as a remarkable story in modern science that would be ranked with Darwin's *Origin of Species*. Continuing with the evolutionary thread, McCrea demonstrated prescience:

> As is the case with the different kinds of evolution dealt with in the *Origin of Species*, his account is bound to suffer drastic modification in the course of which we may get many surprises. Mr. Hoyle is not prepared for this.

McCrea allowed himself a quotation from the text: "But in science," says Fred Hoyle, "the excitement lies in the chase not in the kill."

The radio astronomers at the University of Manchester were not, however, impressed. The *Manchester Guardian's* reviewer was critical of the book:

> Mr. Hoyle has done it again. Here is the universe from A to Z, all clearly parcelled up. It is indeed clever, too clever. Perhaps the most serious deficiency of this book is that it conveys no impression of the great many uncertainties that exercise astronomers just now and which give their subject such high interest. And because the book might have been a valuable one, it is a pity that Mr. Hoyle has been so dogmatic. For, in spite of his lucid prose and his fine illustrations, much of the dogma is unacceptable.[31]

Hoyle had finished the final draft of his typescript at about the same time as a confident Martin Ryle prepared to break cover on the new counts of radio sources. By the late summer of 1954, his catalogue of radio sources comprised about 1,700 radio sources. On August 27, 1954, Ryle presented an outline of the results at a radio science meeting in Canada, but he did not include the cosmological implications, which he would release 8 months later in a blaze of publicity.

In 1955, Oxford University appointed Ryle as its 45th Halley Lecturer. The Halley Lecture was instituted at Oxford—Ryle's alma mater and home to the eponymous seventeenth-century astronomer— in 1910, one of the years in which the comet bearing his name traveled close to Earth. Being asked to give this lecture, in the presence of the vice chancellor, was a major honor. Ryle spoke on May 6, in the splendid Pitt Rivers Museum, constructed in 1884, in an architectural style reminiscent of a Flemish town hall. In his lecture, Ryle described the Second Cambridge survey.[32] He explained the concept of counting radio sources and displayed the crucial diagram. The audience saw a steeply sloping line. The counts of radio sources increased dramatically for fainter, and therefore more distant, objects. The survey had yielded a slope of nearly 3, twice the value expected in a steady-state universe. Ryle reached for the dagger:

> Now if we consider radio sources according to the steady-state theories, the average luminosity and spatial density [of radio sources] must be constant. . . . It seems quite impossible to account for a slope of 3. This is a most remarkable and important result, but if we accept the conclusion that most of the radio stars are external to the Galaxy, and this conclusion seems hard to avoid, then there seems to be no way in which the observations can be explained in terms of a steady state theory.

He never mentioned Hoyle, bundling his opponents together as "Bondi and others." He closed his lecture stylishly, reminding the audience, "Already it seems possible to make a distinction between the two main groups of cosmological theory." The audience broke into thunderous applause. Naturally, the Cambridge theorists were not at all happy with this conclusion from radio astronomy, which came as a great shock. The charismatic and patrician Ryle had become the first person to state openly that the steady-state theory had been tested and found wanting: It was wrong.

A further sharp confrontation took place one week later at the RAS.[33] Ryle introduced his paper on the Cambridge radio telescope, mainly a technical presentation on the nuts and bolts of the instrument. Ryle's research student, John Shakeshaft, read the next paper, on the scientific results. They had detected an astounding number of radio sources, 1,936 in all. He acknowledged heroic observations made by David Dewhirst of the Cambridge Observatories. He had made some interesting identifications with the Cambridge 17-inch Schmidt telescope, but could not remotely match the power of the 200-inch in the hands of Baade and Minkowski. Shakeshaft kept the best bit until the end. The majority of the extragalactic sources "are extremely rare objects similar to the source in Cygnus." This provided a natural explanation for the difficulty of making identifications, because the majority of the radio sources would be too remote for the detection of their optical counterparts, even with the 200-inch telescope. Finally, the survey "also suggests that the more distant regions of the universe are different from our own, such a result being incompatible with the steady-state theory of cosmology."

The president invited discussion on the two papers from Cambridge. Still resentful at being kicked out of the Cavendish, Gold was determined not to let Ryle bask in glory. He was "greatly impressed by this magnificent survey" and comforted by the agreement of all parties that the sources are extragalactic, "as I suggested here and elsewhere, with much opposition, four years ago." He reminded the gathering that "Mr. Ryle then suggested that such an interpretation must be based on a misunderstanding of the evidence." Next he deftly turned the tables: "On the basis of Ryle's observations it would be very rash to regard the great majority of sources as extremely distant." Defiantly, he proclaimed that errors in the observations or interpretation could be the source of discord.

Bondi made a very witty response. A quarter of a century earlier, optical astronomers had tried to determine the geometry of the universe from counts of galaxies. By turns the conclusion was a highly elliptical universe, then a hyperbolic universe, and finally a flat universe. Small changes in the assumptions produced extreme changes in the interpretation! Could that happen to Ryle's counts?

Unusually, the president allowed Ryle a riposte to the negative comments of the steady staters. He repeated discussion from his Halley Lecture. His result was "not compatible with the steady state cosmological theories, but may well be accounted for in terms of evolutionary theories."

Hoyle did not accept Ryle's findings at all. By this stage of his career, he had seen plenty of experiments and observations that turned out to be wrong. What greatly irritated him was an asymmetrical aspect of the conduct of scientific enquiry: Someone could flatly denounce a theory as wrong, but criticism of experiment had to be supported by repeating the experiment or observation for yourself. The wealthy amateur scientists of earlier centuries were casting a long shadow: They could afford expensive equipment but Hoyle could not. All he could do now was sit on his hands and wait. Others would have to corroborate Ryle's result or point to the mistake.

In August 1955, support for Hoyle came from the radio astronomy group in Sydney, Australia. An international radio astronomy symposium convened at Jodrell Bank, Manchester.[34] The principal practitioners of the emerging discipline attended from the UK, the United States, Europe, and Australia, with 80 plenary papers read and discussed. Joseph Pawsey, founder of Australian radio astronomy, had source counts of his own, which would provide an independent check of Ryle's result. His graphs, based on 1,030 objects, completely stunned the Cambridge delegates: He found no significant deviation from the 1.5 slope predicted by the steady-state theory and no support for evolutionary models of the universe.[35] Hoyle was not present for the discussion, but his chum Gold was distinctly critical of Ryle, stating that instrumental effects and a failure to interpret the data carefully enough were responsible for the high slope in the Cambridge graphs. Bondi, too, suspected the Cambridge data were confused. Ryle, however, refused to give way and left the meeting dangerously exposed.

In May 1957, the Australians published a damning critique of the second Cambridge survey.[36] When the paper reached Cambridge, the abstract made grim reading indeed. Statistical analysis of the Australian catalogue revealed no cosmological effects. A comparison of the two catalogues was so discordant that one catalogue had to be completely wrong. In their conclusion, they spared Ryle no blushes:

We have shown that in the sample area, which is included in the recent Cambridge catalogue of radio sources, there is a striking disagreement between the two. . . . The Cambridge survey is very seriously affected by instrumental effects which have a trivial influence on the Sydney results. We therefore conclude that discrepancies, in the main, reflect errors in the Cambridge catalogue, and accordingly deductions of cosmological interest derived from this analysis are without foundation.

The Cambridge survey was corrupted because the resolution of its interferometer was too low. This had the effect that many of their radio sources were, in fact, blends of two or more weaker sources. Although Cambridge tried to squeeze a result of cosmological significance out of the flawed data, the attempt was little more than sleight of hand, giving a result that Sydney dismissed as impossible.[37] The immediate result of the Sydney–Cambridge disagreement was confusion. In the fog of war, Hoyle grabbed all the ammunition he needed to make a spirited defense of steady-state cosmology, which Ryle had targeted for disproof. Hoyle, unlike Ryle, was not at this stage spending much time on cosmology (he was immersed in nuclear astrophysics), and the debate drifted along for a couple of years. The Cambridge radio astronomers demonstrated staunch solidarity for Ryle's opinions. They modified the interferometer to work at a higher frequency, which would give greater resolving power, and quietly commenced a third radio survey. International criticism continued, with the consensus betting on Sydney. A further two decades would elapse before Ryle felt comfortable with the admission that "2C was notorious; what we did was wrong—we tried to analyse too deep."[38]

The period 1956–1958 was a time of achievement and recognition for Hoyle, as a writer and as a scientist. In 1956, he commenced a new part-time career as a science fiction writer. During the war years he had become an avid reader of science fiction, mainly the "star wars" genre. Sometimes he told Tommy and Hermann that he could have done a much better job: The books were very weak on science and poor at developing the plot and the characters. He had to wait more than a decade for an opportunity to write his first novel, *The Black Cloud*, which was published by Heinemann in May 1957.

The Black Cloud tells the story of a cloud of interstellar gas that approaches the solar system on a course that is predicted to bring it between Earth and the Sun, shutting off sunlight and causing incalculable damage to our planet. The effects of the impending catastrophe, together with the reactions of politicians and scientists, are brilliantly described, so convincingly in fact that the reader feels the events could plausibly happen. The book's preface has a disclaimer to the effect that none of the characters should be identified with the author. Be that as it may, the central character, Chris Kingsley, is a professor of astronomy at Cambridge, and Hoyle's handling is distinctly autobiographical. *The Black Cloud* reads authentically because Hoyle sets the action in places he knew well: Caltech, Palomar Observatory, the RAS, and Cambridge. His scientists are doing real science, not fantasy science: We see them using calculus, scientific diagrams, and the EDSAC-1 computer—all graphically described. The book became a sensational best-seller for Heinemann (and for Harper in New York) and then for Penguin, who promoted it as a mass-market paperback in 1960.

In March 1957, Hoyle won the coveted admission to the Royal Society as a fellow, so he was now on the same level with Ryle within the academy. He spent 2 months of the winter of 1957–1958 at Caltech, returning in January to resume his teaching. When he arrived home at Clarkson Close, Cambridge, Barbara had welcome news. In his now-frequent absences, she handled all correspondence with his publishers, the BBC, and the university. Excitedly, she waved a small envelope from the vice chancellor, Lord Adrian. The enclosed letter noted that Sir Harold Jeffreys would soon be retiring from the Plumian Professorship: Would Mr. Hoyle care to let his name go forward to the electors? Barbara sent a holding letter: "My husband is in California, and therefore unavailable at present, but I am sure he would be interested." One Saturday in the Lent term 1958, on a bright spring day, Barbara and Fred made sandwiches for lunch and went out to the country for a picnic. They drove back home for tea and soon after they returned the door bell rang. A university messenger had arrived on an ancient bicycle, holding another small envelope from the vice chancellor. The enclosed letter stated that Fred had been elected to the Plumian Chair: Would he accept? For him this was a defining moment. Henceforth his career would "now flow to an unfamiliar ocean," where he would be expected to conduct himself as a senior academic, perhaps even an establishment figure.[39] The hothead chapters of his life were now supposed to be a memory. Momentarily, he overtook his rival in the academic pecking order, but that would not last long: Mr. Ryle won a personal chair, as professor of radio astronomy, the following year.

Ryle's group pushed ahead with instrumental improvements, undeterred by Hoyle or the Australian radio astronomers. They believed their approach had an underlying validity. Ryle had an excellent opportunity to draw attention to the many achievements of his group in 1958, when the Royal Society awarded him the Bakerian Prize lectureship. These lectures originated in 1775 through a bequest by Henry Baker, himself an FRS, for an oration or discourse that was to be spoken or read yearly by one of the fellows "on such part of natural history or experimental philosophy, at such time and in such manner as the President and the Council of the Society for the time being shall be pleased to order and appoint." The 1958 lecturer could proudly announce on June 12 that

> a new radio telescope having a greater resolution and sensitivity than any previous instrument [worldwide] will soon be in use in Cambridge, and should provide reliable observation of weaker sources. The [number counts] may then be tested over a greater range.[40]

This third Cambridge survey, the 3C catalogue, published in 1958, had a huge impact on the future direction of radio astronomy. To this day, several hundred strong radio sources are known most commonly by their numbers in this catalogue: The Cygnus source is 3C 405 and the nearest quasar is 3C 273, for example. Ryle had done for radio sources what Charles Messier achieved for visible nebulas and star clusters: He had created a basic list of the brightest that would be of enduring value.

On August 30, 1957, Ryle's close colleague Antony Hewish presented the preliminary 3C results to an international conference in Brussels. He admitted that source surveys needed to be made more exact.[41] Unfortunately for observational cosmology, the 3C survey did little to clarify matters: Its slope of between 2.2 and 2.7 was far above the steady state's 1.5 and strongly at odds with the Sydney group's 1.65. Six months later the combatants again squared off at the RAS. Another member of Ryle's group, Peter Scheuer, presented a defense of the Cambridge surveys, in which he explained away the discrepancies with the Sydney results. He conceded that the cosmological conclusions of the 2C survey were fatally flawed because of the inclusion of "many sources which are now given to be below the level of reliability."[42] He then used a statistical argument that would have been lost on most of the audience to claim that 3C was revealing a source distribution wholly inconsistent with steady-state cosmology. Bondi, the brilliant mathematician, rejected the conjuring with statistics as a black art. The final word went to Ryle, who contended that the statistical argument was reliable. But he can hardly have expected Bondi to agree since, with Gold and Hoyle, he was now vindicated in the stance he had taken when he condemned the 2C survey as rubbish.

An important international radio astronomy meeting took place in Paris in August 1958, where the Cambridge radio astronomers gave the first detailed presentation of their recent 3C survey. They announced a reduction in the slope from 3 to 2.2, which did not impress the steady-state supporters. A radio astronomer from Manchester strongly criticized the Cambridge results, and the Australians continued to press the validity of their much lower figure. In discussions, the Cambridge side nevertheless displayed the same messianic certainty they had shown all along, saying they had ruled out the steady-state theory. Hoyle's supporters noted that, if the Cambridge and Sydney surveys had missed a small number of bright sources (perhaps through instrumental flaws), then the slope would easily fall to the steady-state value. A very large cosmological claim hinged on there being no error in the counts of brighter sources. Ryle was brusquely told at a later international meeting (one he did attend) that he should stop making profound cosmological pronouncements when a deficit of a mere 10 sources would swing the result to the steady-state value.

Optical astronomers soon provided support for 3C. The observers at Palomar found the improved radio positions enabled optical identifications that cleared up many puzzles; this marked the acceptance of 3C as a definitive catalogue of radio sources. In 1957, the Mullard Company, a leading manufacturer of electronic components, generously funded a new radio observatory at Cambridge. It was named the Mullard Radio Astronomy Observatory. With new facilities, Ryle pressed ahead with a fourth Cambridge survey, 4C, and its early runs showed that 3C was substantially correct. Then welcome news came from Australia: The latest Sydney results were converging with those from Cambridge.[43] The fourth survey, completed in 1965, eventually covered the whole of the Northern Hemisphere and logged 5,000 sources. But Ryle had as much cosmological data as he needed by 1961, and he released the result at an RAS meeting held on February 10 of that year.[44]

Early in the Lent term, Fred received a phone call from Mullard's headquarters in London. He recalled immediately that the company had made substantial donations to support Ryle's work. A male voice informed him that, at the RAS meeting on Friday of the following week, Professor Ryle would be announcing new results that Hoyle would find of interest. The company would be hosting a press conference for media correspondents on February 3 and "we would be delighted if you and Mrs. Hoyle would like to attend." In the same circumstances Tommy Gold would no doubt have told the informant to take a running jump, but Fred mustered what polite behavior he could and said they would both be present.[45]

At the Mullard headquarters, a black-suited executive led them into a modest-sized hall where a few journalists and BBC reporters were busily preparing for Ryle's presentation. The Mullard man bowed Barbara into a seat on the front row and then obsequiously ushered Hoyle to an uncomfortable seat on a raised dais, where he sat in isolation twiddling his thumbs, staring at the media mob, with a hint of stubborn determination in his face and general demeanor. The stage was set with an old-fashioned blackboard and easel, a lectern, and a screen for displaying slides.

What did Hoyle think as he sat under the bright lights, displayed like a scorpion in a bottle? Did he wonder if he should have worn smarter shoes? Or a more fashionable tie? Or whether there would be a drinks reception afterward? Did he puzzle over what Ryle was about to reveal? Surely if Ryle had results adverse to Hoyle's position, he would hardly set such a blatant trap. Maybe the results were consonant with the steady-state theory, and he was going to deliver a handsome apology?

With a swish of the curtain, a tall and confident Ryle strode onto the stage. The Mullard man announced that the material presented at the press conference would be under an embargo to prevent the story appearing before the RAS meeting. Then he introduced Ryle, who soon launched into a major technical lecture, which is not at all what the press was expecting. It was all familiar ground for Hoyle: the oft-repeated rhetoric detailing the inadequate 1C, the hopeless 2C, and the ambiguous 3C, superseded by a perfect 4C. He scarcely listened, absolutely devastated by the thought that he really had been set up.

Ryle announced that the 4C survey included far more sources than its predecessor and so the statistical errors that had plagued the previous surveys had vanished. The slope stood at 1.8, meaning that the steady-state theory was wrong. With a flourish, he turned to his morose victim, "Would Professor Hoyle care to comment?" Ryle's question had the hacks leaning forward in anticipation, sensing blood. With conflict and controversy, they could sell this obscure science story to their editors.

Hoyle did have an answer, but he could not use it. He knew that the best radio equipment, with low noise, had to be imported from the United States. Instead of doing that, Ryle's group had made its own preamplifiers. Hoyle's wartime experience had taught him that everyone who bread-boarded their own circuitry always claimed it was the best in the world. But he could not use this argument inside the Mullard headquarters of all places! Ryle would flatly assert the truism that he knew far more about his own equipment than Hoyle did. Hoyle gazed at the media faces, said as little as possible, and swept his humiliated wife from the room. Hoyle had just experienced the bitter fate that Copernicus with his controversial heliocentric theory had studiously avoided: being hissed off the stage.

In London, the *Evening Standard* infringed the press embargo by a few hours, running the story across the whole of the front page in its lunchtime edition. Fellows arriving at Burlington House for the 4:00 p.m. meeting were greeted along Piccadilly by news vendors' placards emblazoned with "Universe—Bible Is Correct." Hoyle avoided the potential for more humiliation by not attending the RAS meeting, where Ryle's presentation concluded:

No attempt has been made to explain the observations in terms of an evolutionary model, but it is now difficult to accept a steady-state model as representing the actual Universe.

This admission, that the results had not been benchmarked against the big bang predictions, tends to confirm Hoyle's impression that the science driver behind the radio astronomy group was a search for proof that their misguided theorist colleague was wrong, rather than a whole-hearted attempt to unlock the physics of the mysterious radio galaxies.

One of Hoyle's young disciples, Jayant Narlikar, read a prepared response after Ryle had finished. He gently pointed out that radio galaxies are rare (one in a million) and that it was therefore dangerous to make cosmological conclusions when so little was known about the physical nature of these radio sources. Narlikar proposed a slight modification to the steady-state theory, which brought closer agreement with Ryle's data. Ryle would have none of this. To him the subtle changes to the steady-state theory had the cumulative effect of making the theory more and more ridiculous. Bondi joined the clash, continuing to doubt Ryle's observations and supporting Hoyle. Ryle finished the discussion with a spectacular prediction: that the most distant sources in the survey would have recession speeds nine-tenths the velocity of light and redshifts of about 5. Confirmation of Ryle's speculation lay more than three decades in the future.

Not only could Hoyle not stomach the RAS meeting on February 10, he skipped the RAS Dining Club that evening, too. Since he dined at three of the next five meetings of the club in 1961, it is fair to conclude that his absence in February was a pointed one, intended to deny Ryle the chance of another venomous bite.

Back in Cambridge, media firestorms hit the Hoyles at Clarkson Close and the Ryles in Herschel Road. Martin Ryle took refuge in the home of his brother-in-law, Graham Smith, who fended off the reporters with untruths about Ryle's whereabouts.[46] Elizabeth and Geoffrey Hoyle were ragged at their schools for weeks. Fred's phone rang incessantly, Barbara fielding most of the calls. One voice she immediately recognized was Ryle's, and she passed the receiver to Fred, who was sitting in his favorite armchair, at work with a pad of paper. He put aside his pen and listened. Ryle gave a from-the-heart apology: "When I had agreed to the Mullard press conference, I had had no idea how bad it would be." As a result of the apology, the two were able to rub along together for another decade before the next big clash, on university politics.[47]

In the late 1950s, the radio astronomer Robert Wilson attended Fred Hoyle's postgraduate lectures in cosmology, given at Caltech; he found the steady-state theory most attractive.[48] In those days, hardly any talented physicist would choose a career in cosmology, so Wilson entered the field of radio astronomy. In 1962, he joined Arno Penzias, at the Bell Telephone Laboratories, New Jersey. The following year the two physicists used a vast microwave horn antenna, with maser amplifiers, as a telescope to study radio waves from the Milky Way. Their radio ear seemed to be plagued by a mysterious radio "noise" that was always present, was always the same intensity, and showed no variation with the seasons. They made a systematic investigation of the annoying microwave signal. Little by little, they determined that it was not coming from the telescope itself (the most obvious source), nor from nearby New York City. The radiation was not coming from the galaxy or an extraterrestrial source. They ruled out Cold War explanations such as a mysterious new weapon or fallout from the 1962 bomb tests. It was not even the pigeons that used the giant structure as a roost: Penzias and Wilson kicked them out and swept up all the droppings. They concluded that neither the telescope nor random noise was the source.

They contacted the astrophysicists down the road at Princeton University, where Robert Dicke's group was attempting to detect the greatly cooled remnant of the radiation released at the big bang. When Dicke heard of the observations made at Bell Labs, he resignedly said to his colleagues, "We've been scooped."[49] Within a few months, Dicke's group had measured the temperature of this background radiation as 3 degrees above absolute zero.[50] Penzias and Wilson had accidentally discovered the most significant evidence to date in support of Gamow's explosive universe.[51] For changing the course of the study of cosmology, they netted the 1978 Nobel Prize in physics. Wilson initially felt a pang of disappointment at finding dramatic evidence in favor of the big bang. Meanwhile, Hoyle stood his ground bravely, maintaining that the techniques used to measure the radiation's temperature were not good enough to conclude with certainty that it came from the big bang.

At the Cavendish, the radio astronomy group crowed over Hoyle's predicament. However, the more sensible physicists among them realized that the corrosive relationship between Hoyle and Ryle was damaging the standing of astronomy at Cambridge by providing newspaper hacks with highly negative copy about the university. Two of the younger people in the radio astronomy group, Malcolm Longair and Peter Scheuer, were particularly concerned at the deteriorating relationship. They staged peace talks in the Cavendish, with their boss Ryle and Fred Hoyle. The aim was to find some common ground. The meeting lasted 45 minutes and achieved nothing. Longair regards this signal failure as a tragedy for Cambridge cosmology.[52]

In the mid-1960s, Cambridge finally learned how to count cosmic radio sources correctly, winnowing the radio galaxies from a mixed bag of sources. They had listened to Hoyle's critiques and had learned that convincing data could only come from a systematic study of a uniform sample of objects, the radio galaxies. They would release their news at a conference on the evolution of galaxies, convened for April 18–19, 1968, at Herstmonceux Castle in Sussex, the home of the Royal Greenwich Observatory.[53]

Built originally as a country home in the mid-fifteenth century, Herstmonceux Castle embodies the history of medieval England. The setting is dramatic: the fairy-tale castle nestles in a hollow surrounded by carefully maintained Elizabethan gardens and parkland. Like many visitors, I have always found the first glimpse of the castle enchanting. In 1956, the Royal Observatory started an annual series of conferences based at its headquarters. The dreamy location was wonderfully relaxing. The conference photograph, taken at lunchtime on the first day, shows a happy Fred Hoyle and an even happier Martin Ryle, both grinning broadly and standing side by side.

Delegates stayed in small bedrooms tucked under the mansard roof, behind the battlements. Sumptuous food came from the castle kitchen. In the evening, after a half-hour stroll, you arrived at the nearest pub, which in those days served a local brew, Merrydown Vintage Cider, now an international brand. The observatory seminar room was on the ground floor, in what had been the chapel. However, this was too small for international conferences, which took place in the grand ballroom. The astronomer royal, Sir Richard Woolley, had two grand pianos in the ballroom: He loved to play duets, and he organized English country dancing on a weekly basis. The conference delegates could admire the high gloss on the oak floor, the paneling and fireplaces plundered from other historic country houses during restoration of the castle in the 1920s, and the elegant croquet lawn beyond, where Sir Richard played a wicked game, generally defeating his opponents.

On the afternoon of the first day, Ryle took the floor first. He gave a lengthy paper with strong astrophysical content, in which he argued that the appearance of a radio galaxy depends on its distance. Local radio galaxies were markedly different in their properties compared with remote radio galaxies. This implied that they evolved with age. Because of the longer travel time required for radiation from distant galaxies to reach us, they are observed at a much earlier stage of their life than the local ones. In a universe that has continuous creation, galaxies of all ages should be found at any distance. That was not at all what the Cambridge radio telescopes were seeing.

After tea and buns in the dining room, the delegates returned to the ballroom. Ryle took the chair for the final sessions that day, introducing his young research student Guy Pooley, a modest and quiet speaker, who addressed the throng on counts of radio sources. He displayed a graph that included results from the new fifth survey, 5C, as well as data from the reliable parts of the earlier surveys. Like David confronting Goliath, Pooley took on Hoyle, seated in the front row:

> We notice two things. First the well-known steep slope, 1.85, corresponding to an excess of weak sources. This excess is large, 6–8 times as many weak sources than in any steady-state model. . . . On this evidence we certainly cannot resurrect the steady-state model. Apart from squashing other people [he meant Hoyle], there are some positive conclusions. . . .

Pooley rounded off confidently. For him, Hoyle's position was untenable, and "an evolving cosmology, with quite severe constraints, seems the only answer."

Goliath struck back immediately, borrowing a rhetorical device of Lyttleton and Gold. "May I congratulate you, Mr. Chairman and Mr. Pooley on having provided extremely strong evidence in favor of the steady-state theory!" Hoyle argued that a significant part of the curve that had been displayed had a slope very close to that required for the steady-state theory.

Keith Tritton, then a lowly research student, had the unenviable task of transcribing the tape-recorded discussion for the meeting report to be printed in the *The Observatory*. He recalls:

> The clash took place in front of a large and distinguished audience. The row broke out after Guy Pooley's presentation, in which he made it clear that the radio source counts indisputably ruled out the steady-state theory. Both he and Ryle gave the impression they were quite pleased with themselves. Fred's reaction, bouncing back and provocatively congratulating them for supposedly proving the very reverse of what they claimed succeeded in putting both of them on the defensive. The mood changed dramatically: Ryle was very angry.

Tritton told me that his published report is toned down to convey the impression that all was gentlemanly in the exchanges, but in reality they were shrill. The audience could not mistake the dislike Ryle had for Hoyle, and there was electric tension in the room. Ryle behaved as an introvert who did not like the cut and thrust of a parliamentary debate, whereas extrovert Hoyle loved to answer back with off-the-cuff remarks. At the end of the session the delegates crept out silently, knowing they had been present at a key moment in the history of cosmology. They sensed that Ryle was correct and Hoyle had made a fool of himself.[54]

In the following years, astronomers made many measurements of the microwave background over a range of wavelengths in order to build up a spectrum that would be accurate enough to discriminate between a thermal origin consistent with a big bang and any non-thermal characteristics. The data were broadly consistent with thermal radiation at a temperature of 2.7 K (2.7 degrees above absolute zero), but the range of wavelengths sampled was too small to be certain. This all changed with the launch into Earth orbit of the Cosmic Background Explorer (COBE) in December 1990. The following month the first results from its instruments in January 1991 were presented at the Washington meeting of the American Astronomical Society. The spectrum had the perfect form of radiation from a thermal source, corresponding to a current temperature of 2.725 K. The 1,000-strong audience rose to their feet and gave a standing ovation to the COBE team. A cosmological parameter had never before been measured to four significant figures! This was now the most powerful piece of evidence supporting the big bang.

Most neutral observers will agree that, post-1965, the efforts by Hoyle and his collaborators to shore up the steady-state model became increasingly contrived. Nobody ever took the theory seriously, not even in 2001, when Cambridge University Press published a modified steady-state theory that was elaborated in great detail by Fred Hoyle, Geoffrey Burbidge, and Jayant Narlikar.[55]

In 1980, Alan Guth at Stanford University proposed a dramatic modification to the picture of the big bang, at that time taken as "standard" by most theorists.[56] Two aspects of the universe had been greatly troubling cosmologists: its flatness and its homogeneity. We do not need to launch into the technicalities here: It is sufficient to note that a violently explosive big bang has to be fine-tuned to an absurd degree in order to secure this outcome. Guth, and other theorists, solved these two puzzles by proposing that almost immediately after the onset of the big bang, the observable universe we see today expanded from an object smaller than a proton into something the size of a grapefruit. This inflationary scenario filled the tiny nascent universe with energy and propelled its expansion. This feature is built into the standard model of the universe adopted today.

Throughout the 1970s and 1980s, a strong argument against the steady-state theory was its prediction that the most distant galaxies should be accelerating; this appeared counterintuitive to most astronomers, who felt that gravity would slow the expansion. By the end of the twentieth century, however, Saul Perlmutter and others demonstrated that the expansion is indeed accelerating, thus confirming a key prediction of steady-state cosmology.[57] The current interpretation of this finding is that the expansion is being driven by mysterious dark energy.

The steady-state theory today is considered to be of interest only for the history and philosophy of science, but in the annals of twentieth-century cosmology, the steady-state theory demands more than a passing reference. As Malcolm Longair has pointed out, three features of the steady-state universe resonate with the contemporary standard cosmological model[58]: the constant density of the universe, its flat geometry, and its accelerating expansion, which it says stems from the creation of new matter. In Hoyle's version, the expansion is driven by what he called the *C*-field or creation field. In 1995, at a lecture in the Cavendish Laboratory, Fred confessed that he rued the day when he dreamt up the expression *C*-field. "If only I had chosen a more user-friendly or memorable term I might yet have been credited as the originator of the inflationary universe," he mused.

CHAPTER 8

Origin of the
Chemical Elements

C osmology kept Hoyle in the public spotlight for decades. In professional circles, however, he became more respected for his work on nucleosynthesis—that is, the origin of the chemical elements—an interest that preceded his cosmology. To appreciate his Olympian contributions to our present knowledge of the history of matter in the universe, we need to revisit Cambridge in the 1920s and 1930s.

Physicists at the Cavendish Laboratory had made stunning progress in probing the atomic nucleus, for which they won a string of Nobel Prizes. By the time Fred commenced research, all the components for major advances in understanding the relation of nuclear processes to the structure of stars were present in Cambridge, but no person or group seemed quite able to put all the pieces together. This was a consequence of the organization of the university in those days, which led to the dichotomy between astronomy and physics. Astronomy, a subject in the curriculum for centuries, had a higher prestige than physics, then an upstart less than 50 years old as a separate subject in the university.

In the eighteenth and nineteenth centuries, chemists isolated the elements to the point where John Dalton (1766–1844) could put together a plausible atomic theory.[1] In St. Petersburg, Dmitri Mendeleyev (1834–1907) noted similarities and patterns among the 63 elements then known, charting his findings in 1869 as the Periodic Table of the Elements.[2] At the Cavendish, J. J. Thomson discovered the electron in 1897 and Rutherford the nucleus in 1911, confirming the ideas of atomic structure. In 1932, James Chadwick identified the neutron. The contributions of Francis Aston (1877–1945) captured him the Nobel Prize in chemistry in 1922. Young Hoyle sometimes encountered the laureate, who was a fellow of Trinity, the college next door to St. John's.

When we consider the elements as arranged in rows in the periodic table, the step from one element to the next involves an increment of one in the proton count of the nucleus: Carbon has six protons while the next element, nitrogen, has seven, for example. The number of protons in a nucleus is known as the atomic number. Every nucleus apart from hydrogen also has neutrons. The total number of protons and neutrons is the atomic mass. As an example, take carbon, which can have six, seven, or eight neutrons. These variants are known as isotopes: All have atomic number 6, and the atomic masses are 12, 13, and 14, respectively. A convenient shorthand is to refer to these isotopes as carbon-12, carbon-13, and carbon-14 (the nuclear physics notation is ^{12}C, ^{13}C, and ^{14}C). I will use this convention for the isotopes of the other elements, such as beryllium-8, oxygen-16, etc.

An English chemist, William Prout (1785–1850), was the first modern scientist to ask the question, "What is the origin of the chemical elements?" In 1815, he published a hypothesis that the relative atomic mass of every atom is an exact multiple of the mass of the hydrogen atom. The discovery of isotopes in the twentieth century bore out his idea. The following year he suggested that all the elements were formed from hydrogen by some process of condensation.[3] Prout's hypothesis intrigued Eddington, who came to believe that all elements were made from hydrogen by an unknown process. In his popular account *Stars and Atoms*,[4] which Hoyle had read in 1927, Eddington suggested that four hydrogen atoms could combine to form helium, with a release of energy through the mass loss. Later, in his 1935 publication *New Pathways in Science*, Eddington went much further by speculating that the artificial transmutation of the elements, which Cavendish Laboratory physicists achieved by bombarding nuclei with protons, could also occur in stars.[5] Hoyle was familiar with Eddington's suggestions.[6]

By late 1945, astrophysicists understood that the bulk of the matter in normal stars is hydrogen and helium. In the case of the Sun, elements heavier than these account for just 2 percent by mass. Nuclear physics had made giant strides during the war. There were far more data on the properties of individual nuclei and many hundreds of professional nuclear physicists. In the period 1945–1950, astrophysicists began to investigate the origin of the chemical elements heavier than helium.[7] Fermi, Edward Teller (the so-called father of the hydrogen bomb), Gamow, Alpher, and Peierls all tackled the problem of cooking the elements from basic ingredients: protons, neutrons, and electrons. Their slightly different approaches all required a vast and intense source of neutrons, which they posited as existing at the beginning of the universe in Lemaître's primeval atom. They supposed that, in the earliest moments of the universe, a vast flux of neutrons would rapidly build heavier and heavier elements by the repeated addition of neutrons to individual nuclei.

Earth scientists and stellar astrophysicists had meanwhile contributed to the production of a table giving the relative abundances of the chemical elements in the Earth's crust and the atmospheres of stars. This table of chemical abundances can be considered as a Rosetta stone for nuclear astrophysics. It showed that roughly the same proportions of the commoner elements, such as carbon, oxygen, nitrogen, neon, iron, and so on, prevail throughout the galaxy. Clearly, the chemical elements had had a common genesis. The burning question was where and how nature had created the elements and their isotopes in the proportions we observe all around us. Why is iron common? Why is gold, element 79, so scarce, whereas three doors up the road, at number 82, we find lead, the commonest base metal? Why are lithium and beryllium extremely rare? Why is there a steady decrease in abundances with atomic number from carbon (atomic number 6) to calcium (20). Why is there a plentiful supply of the elements with atomic numbers in the vicinity of iron (26) and nickel (28)?

Those researchers who attempted all the cookery inside a hot big bang quickly ran into an insurmountable problem. In nature, there are no stable isotopes with atomic mass 5 or 8. This is a crucial point: A collision of helium with a neutron or with another helium does not produce a nucleus that lives long enough to participate in a further nuclear reaction. Therefore, the pioneers mostly gave up with this approach, although Gamow and Alpher, intrigued by the physics of the big bang, did not give up lightly and had eventually contrived a means of bridging the gap by 1950.[8] However, Fred, who could not accept the big bang ideas, had by then long since trumped them.

Hoyle assumed, in 1945, that neutron capture in the first minutes of a big bang universe would not work because of the gaps at atomic masses 5 and 8. He also took notice of the fact that, although the distribution of the chemical elements is roughly uniform throughout the galaxy, there are variations from star to star. If some of the elements were present in different amounts in different stars, surely that meant they could not have a primordial origin. Since it was accepted by this stage that stars transmute hydrogen nuclei into helium, could they not also process the helium into heavier elements by reactions with neutrons and protons?

As he studied the abundance tables, he picked up an important clue from the marked increase around iron, known to physicists as the "iron peak." From his solid grasp of nuclear physics and statistical physics, he reasoned that the iron nuclei must have been produced at very high temperatures, around 3 billion degrees, in the interiors of massive stars. He thus inferred that practically all of the chemical elements must have been assembled deep inside stars but under a variety of physical conditions.[9] In reaching this profound conclusion, he had been greatly assisted by the weekend he had recently spent in 1944 in the company of Walter Baade, whose gift of a selection of papers on supernovas enabled him to gain "for the first time an idea of just how exceedingly high stellar temperatures and densities might be during the late stages of stellar evolution."[10]

By March 1945, following the clues left by Eddington, Hoyle had commenced work in earnest on the synthesis of elements in stars. For the historical record, it is essential to note that Gamow's hot big bang universe and the steady-state theory both lay in the future. Hoyle did *not* come to stellar nucleosynthesis via the steady-state theory, which would require an alternative to primordial synthesis, but rather through his deep understanding of nuclear physics, which Baade had suggested he apply to the interiors of massive stars.

An outcome of the discussion was that Hoyle now began to think seriously about the late stages of stellar evolution. As he later put it:

> It was then a natural step to wonder what would be the very last of the nuclear reactions that take place within stars, instead of the first reactions which so far had occupied the attention of astronomers, following the publication in 1938 and 1939 of the proton-proton chain and the carbon-nitrogen cycle.[11]

In February 1945, Hoyle had been frustrated by a lack of data on the masses of nuclear isotopes. The following month he had a chance encounter with the nuclear physicist Otto Frisch while browsing in the library of the Cavendish. Frisch lent him a declassified compilation of nuclear masses just received from Josef Mattauch, who had been in charge of the Nazi atomic weapons program. Armed with these invaluable data from British-occupied Berlin, Hoyle set to work on the cooking of elements up to iron in stellar interiors.

The notebook in which Hoyle penned the first version of his ideas on element synthesis is preserved in St. John's College.[12] He writes boldly with a fountain pen in black ink. There are few corrections, which suggests he copied from loose sheet drafts that have been lost. His opening is dramatic: a description of the Crab Nebula, the supernova explosion observed by Chinese astronomers in 1054.[13] He explains that this as an excellent example of the breakup of an unstable star. In the manuscript, he took 30 solar masses as the initial mass, but later reduced it to 16 solar masses in the version published in the *Monthly Notices*.[14] What follows immediately is a calculation of the behavior of a rotating massive star that has just exhausted its supply of hydrogen. Shorn of the hydrogen fuel, the star must contract under its own weight, a process that increases the temperature at the center. With increasing temperature, the nuclear particles will fly around at ever higher speeds, greatly increasing the scope for nuclear reactions. Hoyle derived a central temperature of 4 billion degrees at the onset of collapse.

Already in 1939, Hans Bethe had suggested that at high temperatures there might be a nuclear reaction in which three helium nuclei fuse simultaneously to make carbon-12. Hoyle assumed this to be the case, so that he could then use carbon-12 as the source material for the synthesis of elements in a series of chain reactions. The precise mechanism for making carbon-12 in the first place would be found some years later. He used a statistical approach to model the equilibrium of nuclei, protons, and radiation in the seething inferno.

The 1946 manuscript has a staggering chain reaction that fills an entire page. The sequence of reactions starts with the synthesis of oxygen-16, when carbon-12 captures a helium nucleus. On the next line, oxygen-16 repeats the capture, forming neon-20. The chain continues through to the fusion of argon-40 to form calcium-43. In the published version, Hoyle appears to show an uncharacteristic lack of nerve, because he omits most of the steps he wrote in the notebook. He comments: "There is not sufficient laboratory evidence available to give a known chain reaction for every pair of values of [atomic number and mass]."

The seminal *Monthly Notices* paper is much longer than his draft manuscript, and it continues the element-building process right up to iron. The combined mass of all of the elements with atomic numbers greater than iron's amounts to less than 1 percent of the total mass of all the elements from lithium up to iron. Hoyle's cosmic alchemy could therefore account for the existence of most of the common chemical elements.

On November 8, 1946, Hoyle read a summary of his paper to a packed audience at the Royal Astronomical Society (RAS). At that time the director of the University of London Observatory made a point of introducing his young staff members to the monthly meetings of the RAS. This is how E. Margaret Peachey, the future E. Margaret Burbidge, came to be in the audience. She recalls:

> Listening to Fred's presentation, I sat in the RAS auditorium in wonder, experiencing that marvellous feeling of the lifting of a veil of ignorance as a bright light illuminates a great discovery.[15]

In 1943, Margaret received her Ph.D., the thesis of which was on an unusual class of hot stars. She then engaged in war work (on a range finder for aircraft) until 1947. But the concept of element synthesis in the stars now captivated her. Earlier work had established first that

Baade's Population II stars have low metal content and second that the spectra of some hot stars with strong and variable magnetic fields (Ap stars) were greatly oversupplied with certain chemical elements, at least near their surfaces. She felt that these departures from the normal distribution of the chemical elements invited investigation.

In 1948, Margaret married Geoffrey Burbidge, a research student in theoretical physics 6 years younger than herself; she thereby introduced to astronomy a future enthusiast for cosmology. In the period 1951–1953, both had positions at the Yerkes Observatory, where in spring 1953 they won time on the 82-inch telescope at the McDonald Observatory in Texas to obtain high-resolution spectra of Ap stars, with a view to understanding their bizarre chemical compositions. They worked night and day on making the observations and carrying out the reduction of the data before leaving for Cambridge, England, where Geoffrey had a short-term research fellowship in Martin Ryle's radio astronomy group. During a 2-year Cambridge stay, in 1953–1955, Geoffrey encountered Ryle's paranoid secretiveness. Sometimes the group gave him the cold shoulder, excluding him from certain internal meetings. The sojourn did, however, introduce Burbidge to what is now known as high-energy astrophysics. Throughout the Cambridge period, husband and wife worked together on the abundances of elements in the atmospheres of stars, aided by the excellent spectra secured at the McDonald Observatory.

At a seminar in Gonville and Caius College for the Del-Squared V Club, Geoffrey Burbidge gave an account of their baffling results for a peculiar star that has both a variable magnetic field and a variable spectrum.[16] In their analysis, they had compared the proportions of chemical elements in the peculiar star with those in the Sun and with the star Gamma Geminorum. The resulting composition seemed to be a drastic departure from the norm. Silicon was 10 times more abundant in the peculiar star than in the Sun. By contrast, magnesium and calcium were scarce, falling short by factors of 25 and 50 compared to the solar values. Zirconium, a transition metal, scored 20 more than the solar value. Lead appeared to exceed the solar presence by a factor of 1,500! The group of 15 elements that chemists term the rare earths (the series runs from element 56, lanthanum, through to element 71, lutetium) were spectacularly overrepresented. They are described as "rare" precisely because they are very scarce on Earth and in the solar system generally. In the peculiar star, several of the rare earths had excessive scores: lanthanum registered an unbelievable 830 times that found in the Sun.[17]

The pair of young astrophysicists appeared to have discovered a star with nuclear reactions taking place at the surface! Geoffrey Burbidge guided his audience through the nuclear physics. The strongly magnetic star must have the equivalent of sunspots. But these stellar spots had a much more intense magnetic field than the Sun. Electrically charged particles—he had protons in mind—could, he suggested, be accelerated to extremely high energies and could thereby change the chemical mix through nuclear reactions. However, he and Margaret felt frustrated by the lack of laboratory data on collisions between protons and nuclei, which meant they could not estimate how likely (or unlikely) a particular reaction might be.

At the end of the seminar, a cheerful physicist bounded up to Geoffrey and introduced himself: It was Willy Fowler, an experimental nuclear physicist from the Kellogg Radiation Laboratory at Caltech. Fowler was spending a year in Cambridge as Fulbright Professor. The Burbidges' results greatly intrigued him. His particle accelerator at Caltech could fire hydrogen and helium nuclei with energies like those the particles have in solar flares and sunspots. He joked that he only worked on the light elements, but Fred Hoyle in Cambridge had worked on the synthesis of elements up to iron. Perhaps the four of them should get together and work on the reactions of stellar nuclear physics. Thus, he introduced the Burbidges to Hoyle, thereby creating the famous quartet. However, because of his teaching load, Hoyle could not participate in the research immediately.

Fred's 1944 trip to San Diego and Mount Wilson had given him a taste for California, making Caltech a natural choice for his sabbatical leave in the Lent and Easter terms of 1953. His office was in the Robinson Building, home of the Department of Astronomy. He interacted with the nuclear physicists in the Kellogg Laboratory, became acquainted with Richard Feymann, and renewed his important contacts with the astronomers at Mount Wilson. He could do business here and build a network of collaborators, away from the curmudgeons in Cambridge. His Pasadena lectures for postgraduates included the nuclear synthesis of the elements in stars, a topic he had developed considerably since the 1946 paper. The course he gave had the title Experimental Cosmology, and it was well attended by faculty members as well as Mount Wilson Observatory staff. Ward Whaling, who was present, memorably recaptured the spirit of the time almost half a century later:

> Hoyle's lectures attempted to explain how all the elements could be built up from the hydrogen produced in his continuous creation model. His model was not popular with astronomers here, and it seemed to me they were a pretty hostile audience with lots of nasty questions.

In his presentation, Hoyle made a lot of assumptions about stellar properties, such as abundances and temperatures. The astronomers present pounced on him: "We've never observed anything close to that." Hoyle would brush off the critics by saying, "We'll continue at this point next time." Then he would work at it and return with a contrived scheme to get around the objections. Whaling had the distinct impression that Hoyle was making things up as he went along.[18]

When Hoyle returned to Cambridge, he paid dearly for his absence of two terms: His college and the mathematics faculty both required him to undertake a massive teaching load. At that time he was also hammering away on the stellar evolution paper with Martin Schwarzschild. Hence he could not immediately collaborate with the Burbidges and Fowler. There were just not sufficient hours in the day to undertake anything more demanding than watch the three from the sidelines in the period 1953–1955. At the end of summer 1954, Willy Fowler's visiting professorship came to an end. He returned to Pasadena, where the Burbidges immediately joined him at Caltech. Margaret had a postdoctoral fellowship in Fowler's group at the Kellogg Radiation Laboratory, while Geoffrey had won a Carnegie Fellowship.

While at Caltech in 1953, Hoyle worked on an important piece of unfinished business that arose directly from his lecture course: the synthesis of carbon-12. Regardless of whether the light elements are made in stars or in the big bang, the origin of carbon-12 remained mysterious. There is no *stable* isotope with atomic mass 8, so carbon-12 cannot be made by colliding an eight-particle nucleus with a four-particle helium nucleus. Hans Bethe's classic 1939 paper on energy production in the carbon–nitrogen cycle had hinted that maybe three helium nuclei could collide simultaneously to make carbon-12.[19] In this context, helium nuclei are known by their alternative name of alpha particles, a legacy from the discovery of radioactivity, and the merger of three is called the "triple-alpha" process.

Hoyle had first determined to investigate this problem in 1949. He felt then that the nuclear physics of energy generation in normal stars was "largely solved" and the synthesis of the elements in the late stage of stellar evolution pretty much explained. "The issue was to fill out the intervening steps."[20] He set a research student the task of calculating how much of the extremely unstable beryllium-8 would exist at any moment in a situation where pairs of alpha particles collide to make beryllium-8, which then promptly decays back to two alpha particles. Armed with this statistic, it would be possible to compute the likelihood of the beryllium-8 interacting with a third alpha particle *before* decaying. This would yield a production rate for carbon-12.

As luck would have it, the student in question was the only one ever to leave Fred halfway through the doctoral course. When they parted company, the question was, who now had ownership of the intellectual puzzle? The convention in Cambridge at the time was that the supervisor had gifted the problem to the student irrevocably. The problem remained the student's intellectual property until such time as he or she or an independent third party published. This pettifogging attitude of the "authorities" at Cambridge irritated Hoyle enormously, but there was nothing he could do apart from sitting it out.

At Caltech, Willy Fowler hired a research associate, Edwin Salpeter, in 1951, setting him to work on nuclear reactions in stars. He decided to tackle the fusion bottleneck at atomic masses 5 and 8. Fowler's laboratory was able to make beryllium-8 by colliding the stable isotope beryllium-9 and stripping out a neutron. Salpeter noticed an interesting feature of the beryllium-8 nucleus: It had an enhanced energy level above its normal state. Every nucleus has a certain number of protons and neutrons, making a composite system. Depending on circumstances, such a composite system adopts different states or configurations according to how much energy is vested in it. There are a variety of ways in which the protons and neutrons in any nucleus can be arranged, leading to a hierarchy of energy levels. Nuclear physicists term these enhanced energy states "resonances." As is the case with the electrons in an atom, the rules of quantum mechanics only allow certain discrete states, like rungs on a ladder.

Salpeter calculated reaction rates for beryllium-8 in its enhanced state. To his surprise, he found that the excited form of the isotope lingered long enough to be able to react with an alpha particle and make the prized carbon-12. In fact, in the core of a collapsing star of five solar masses, he reckoned 1 nucleus in 10 billion would be beryllium-8, which would react to carbon-12.[21] Caltech had trumped Cambridge—possibly.

After studying Salpeter's paper in the *Astrophysical Journal*, Hoyle was not prepared to let the matter rest. Salpeter's reaction rate was so slow that all the carbon-12 would be transmuted to oxygen-16 in a reaction between carbon and an alpha particle. Fred said to his associates: "Since we are surrounded by carbon in the natural world and we ourselves are carbon-based life, the stars must have discovered a highly effective way of making it, and I am going to look for it."

Salpeter's paper contained an important clue: It featured an excited state of a nucleus, in this case beryllium-8. In a flash of inspiration, Hoyle hit upon a brilliant idea. In a bid to circumvent the slowness of Salpeter's reaction, he puzzled over the energy levels in helium, beryllium, and carbon. He asked himself whether there could be a judicious combination of enhanced states that would allow the reaction to proceed at a greatly accelerated rate. Hoyle added together the masses of beryllium-8 and helium-4, then subtracted the resulting sum from the mass of carbon-12. He next converted this small mass difference into an energy gain for the carbon nucleus, using Einstein's most famous equation, $E = mc^2$. If the carbon nucleus should have an excited state close to the energy released by the fusion of beryllium-8 and helium-4, the reaction would proceed with alacrity. For the required consonance to occur, the two nuclei had to undergo only a gentle collision so that they would place the energy released into the balance sheet of the carbon nucleus.

Hoyle now profited from the modeling on red giant stars that he had carried out with Schwarzschild. He deduced that the center of a red giant star would have the desired combination of temperature and density to provide an ideal site for this carbon-creating reaction. The quest now hinged on the energy levels of the carbon nucleus because no energy level of the sort required was then known to laboratory physicists. By great good fortune he was in the right place at the right time because the Kellogg Laboratory had a new accelerator suited to probing the inner secrets of the carbon nucleus.

An opportunity to broach the subject of energy levels in the carbon-12 nucleus with Willy Fowler soon occurred. Hoyle had earlier been invited to speak at the midwinter meeting of the American Physical Society, at Caltech. As soon as he announced the title of his talk, "The Expanding Universe," Caltech moved the venue to the City College because Caltech had no auditorium large enough to accommodate the crowd Hoyle would certainly attract. Before the lecture, Fowler threw a small cocktail party in Hoyle's honor, to introduce him to the senior faculty members. Hoyle and Fowler exchanged pleasantries, and this was the first time Fowler had a proper conversation with him. Fred asked what he knew about carbon-12 but Willy declined to talk shop, introducing him instead to the father-and-son nuclear physicists, his boss Charlie Lauritsen and his son Tommy.

The following morning Hoyle could hardly contain his excitement when he dropped by the Kellogg Radiation Laboratory, barging into Fowler's office without so much as a "by your leave." At this moment Fowler had no idea of Hoyle's deep knowledge and strong track record in nuclear physics: He thought of him as a weird cosmologist with a pushy manner and a strange accent. Fowler and the Lauritsens were deep into a major systematic program of bombarding all manner of nuclei and did not wish to interrupt their schedules to accommodate a visiting Brit who for some inexplicable reason was obsessed with the energy states of carbon-12. Fowler pressed Hoyle for an explanation of why he thought carbon-12 had the excited configuration. He and the Lauritsens thought Hoyle's reasoning was laughable. Clearly, this man had no recent training in nuclear physics. For many years, Tommy Lauritsen had compiled reports on energy levels in the light nuclei. He had very early results on carbon-12, obtained at Cornell University before the war. At one time he had suspected an energy state at about the level suggested by Hoyle's reasoning. However, a group at the Massachusetts Institute of Technology had not confirmed this feature in 1951, and so he discounted it.

Confronted with this evidence Hoyle nevertheless insisted that the measurements should be tried again, as he often did when experimental data disagreed with whatever busy bee was buzzing inside his head. Fowler and colleagues dismissed the funny little man from Yorkshire. "We are not going to stop all this important work which we are doing otherwise to look for this state in carbon. Go away young fellow, you bother us too much!"[22] Fowler's important work was checking the operation of the carbon–nitrogen cycle for hydrogen fusion in massive stars, and particularly the synthesis of carbon-13. He was also burrowing deeply into the operation of the proton–proton cycle in the Sun. Hoyle's claim was outrageous. The cosmologist was claiming he could do what no nuclear physicist in the world could do: predict a precise energy state in an atomic nucleus.

The English visitor surely should have withdrawn quietly at this point, but Yorkshiremen have a reputation for not always behaving sensitively. Hoyle pestered the assistants and associates. Ward Whaling, fresh to the group from Rice University, found Hoyle convincing and wanted to give the idea a whirl. Fowler began to sense that, if Hoyle were right (highly improbable in his view), the consequences for element synthesis in stars would be immense. He rounded up his team for a council of war in his cramped office. Hoyle, surrounded by the world's brightest experimental nuclear physicists, suddenly became acutely aware that he could end up looking ridiculous. Fortunately for him, the nuclear physicists reached a consensus on how to run the experiment.

First, they had to make a major experimental modification. They decided to run with a small particle accelerator, using a spectrometer then attached to their large accelerator. The spectrometer, a device for measuring the energy of particles emitted in nuclear reactions, would have to be moved. Whaling's vivid recollection is of moving the spectrometer, which weighed many tons on account of the huge magnet it contained, down a narrow hallway 4 feet wide, and round two corners. They rested a large steel plate on several hundred tennis balls, slid the multiton instrument onto this platform, and set the whole in motion. A pack of graduate students feverishly fed the squashed tennis balls from the back to the front as the procession proceeded![23]

To make excited carbon-12, they bombarded nitrogen-14 with deuterons (hydrogen-2). This produced reactions in which alpha particles came flying out, leaving energized nuclei of carbon-12. By measuring the energy of the emitted alpha particles, a fix could be obtained on the energy levels in carbon-12. We can imagine Hoyle as an awkward bystander on the lower ground floor of the Kellogg Laboratory, where he watched Whaling hard at work in semidarkness, surrounded by power cables, transformers, whirring vacuum pumps, and chambers in which the atomic nuclei were hurled together.

A certain amount of folklore now surrounds this experiment and Hoyle's role. One florid account has the experiment lasting 10 days, with Hoyle nervously creeping down to the dim laboratory and then blinking in the bright sunlight as he returned to his own office.[24] In reality, the experiment took some 3 months.[25] In addition to measuring the enhanced state that Hoyle sought, the experimentalists had to quantify several additional nuclear parameters to be certain. Hoyle was back in Cambridge by the time this work had totally vindicated his outrageous prediction of an enhanced energy level. The Caltech physicists put Hoyle's name as first author on a contribution delivered by Whaling at the summer meeting of the American Physical Society. Whaling does not remember receiving any missive from Hoyle about the results. Willy Fowler reported much later that Hoyle had taken the good news calmly: He had had the courage of his conviction.[26] When the result was known, Fowler remembers that:

> We then took Hoyle very seriously, because of this triumph from our standpoint in predicting the existence of a nuclear state from astrophysical arguments. I at least took him very seriously, and we did a lot more work on his state in carbon-12. We had to show, for example, that the resonance not only existed, but that it could be formed from three alpha particles.[27]

Fowler quickly became so impressed with Hoyle's research in stellar evolution that he decided to take his sabbatical year in Europe.

In his Fulbright application he put the Bohr Institute in Copenhagen as his first choice and Cambridge as the second option. The Bohr Institute was so popular at that time that it could not take all applicants, but the Fulbright committee awarded Fowler a place at Cambridge. He and Hoyle subsequently became the very best of friends, spending much spare time in each other's company and taking vacations together in Scotland. In his autobiographical note for the Nobel Foundation in 1983, Fowler said:

> Fred Hoyle was the second great influence in my life. The grand concept of nucleosynthesis was first definitely established by Hoyle in 1946. After Whaling's confirmation of Hoyle's ideas I became a believer and . . . spent a sabbatical at Cambridge in order to work with Hoyle.

Fowler's period of leave in Cambridge would set the scene in 1954 for the crucial meeting of the Burbidges, Fowler, and Hoyle, after Geoffrey Burbidge's seminar on peculiar stars at the Del-Squared V Club.[28]

By the time he returned to Cambridge in May 1953, Hoyle had already worked out some consequences of his demonstration that helium could be turned into carbon inside stars. He could see that several of the common elements on Earth could be produced in violent nuclear collisions in which successive helium nuclei were added one at a time. Carbon-12 plus an alpha particle made oxygen-16. Successive additions of alpha particles produced neon-20, then magnesium-24, silicon-28, argon-36, and calcium-40. His paper in *Astrophysical Journal*, with the title "I. The Synthesis of Elements from Carbon to Nickel," is sweeping.[29] He details many element-building reactions in addition to adding alpha particles one by one. Two silicon-28 could perhaps bind to make iron-56.[30] Two carbon nuclei could collide and produce magnesium-24. He intended this paper to be the first of two. In the second, he expected to solve the synthesis of elements from calcium-40 to titanium-48, as well as the heavy elements beyond nickel. However, this ambitious program was about to be overtaken by his future collaboration with the Burbidges and Fowler.

Hoyle had identified two distinct processes that could build up elements heavier than helium inside stars. His first method, a so-called "equilibrium" process to make the elements of the iron peak, is also an explosive process: The requisite temperatures and densities are only achievable in supernova explosions. The second route, the alpha process, could take place deep in the interiors of evolving massive hot stars. As the different elements build up inside these stars, other element-building routes open up. However, the numerous permutations could not conceal the fact that it was impossible to make the heavier elements, such as gold or uranium. The nuclei of elements that cluster around iron in the periodic table are very tightly bound, meaning that they are the end point of nuclear reactions that *release* energy. To make even heavier elements from iron seeds is the equivalent of pumping water uphill: It requires an *input* of energy, an impossible condition in a stable star, which depends on nuclear reactions to *release* energy. At the heart of the difficulties lay electrical repulsion: Nuclei carry positive electrical charge, by virtue of their protons. As nuclei of increasing atomic number have progressively larger electrical charge, two such nuclei brought into very close proximity experience extremely strong repulsion. To bring them close enough for a merger, it would be necessary to slam them together at close to the velocity of light. Even in the hottest stellar interiors, such speeds are never reached by large nuclei. Nature just had to have another way.

A neutron has no charge, and this property allows it to penetrate a nucleus and be embraced by the strong nuclear force, a short-range force that glues the protons and neutrons together. Hoyle intended to work on a paper that would examine the means of building the middle-ranking and heaviest elements by successively adding neutrons one at a time to a seed nucleus. Sometimes the added neutron would immediately beta decay to a proton (plus electron), thus marching the seed nucleus one place along in the periodic table. This would have been his "Paper II" to follow the carbon-to-nickel paper. The idea harked back to the 1930s, when it was quickly dropped because astrophysicists such as Eddington did not believe stellar interiors were sufficiently hot to push a neutron into a nucleus. (Eddington was mistaken about that.)

To synthesize the heavier elements through neutron capture required a source of neutrons. But a free neutron is unstable: It decays with a half-life of 10 minutes. A normal hydrogen-burning star does not have a continuous supply of neutrons. Hoyle intended to invoke the spare neutrons created through carbon burning and oxygen burning. However, before he could work on this, he received, in the middle of 1953, a paper to review for *Astrophysical Journal*. He read it with intense interest and delight.

Hoyle had never heard of the paper's author, Alastair Cameron, a young Canadian physicist at the Chalk River Laboratories, Ontario, one of the world's foremost research centers in nuclear energy. Cameron became interested in element building as a result of reading an announcement from Paul Merrill (of Mount Wilson) that technetium had been discovered in several red giants stars with unusual spectra, called S-type stars. Technetium, atomic number 43, has 27 isotopes, but none are stable. Even the longest-lived, technetium-98, transforms to ruthenium-98 with a half-life of 4.2 million years, far less than the age of the stars. Merrill concluded that "S-type stars somehow produce technetium as they go along."[31] Merrill's red giants were synthesizing technetium on the fly, and this stimulated Cameron to find a nuclear process that could make neutrons faster than they could be gobbled up. That is what the paper in Hoyle's hands was all about.

The two referees appointed by *Astrophysical Journal* recommended that it be declined, but the editor (S. Chandrasekhar) was not so sure. He knew of Hoyle's impressive standing in the field and dispatched it to Cambridge for a third opinion. Cameron had proposed collisions between carbon-13 and helium-4 as a way of producing free neutrons, along with oxygen-16. But the critical referees said that the reaction rate in the process he described was too small because of inadequate supplies of carbon-13. The two referees wanted to reject the paper because they could not accept Cameron's hypothesis that large amounts of carbon-13 could be made in the late stages of stellar evolution. Thanks to his collaboration with Schwarzschild, Hoyle was by now the world expert on the structure of red giant stars. He already knew that carbon-13 would be abundant in their interiors because it is a by-product of the carbon–nitrogen energy cycle that powers red giants. Cameron's process also required helium, which is plentiful in the nuclear core. The source of neutrons had been found! To make heavy elements, Cameron's neutrons would need to stick, one by one, to seed nuclei of iron. Today, this carbon-13 process is widely held to be the neutron source for element building in red giants with masses less than eight times the mass of the Sun. From Cameron's paper, Hoyle now knew that neutrons would be plentiful in red giant stars.

By 1954, research on element building by the capture of neutrons, now called the s-process (s for "slow"), was underway in three locations: Fowler and the Burbidges were back at Caltech, Hoyle worked alone in Cambridge, and Cameron continued to labor at Chalk River. The three centers were not in communication with each other. Fowler, who felt much more at home with the lighter elements, worked with the Burbidges on the buildup of elements starting with neon-20 rather than iron. Hoyle could only dabble; he had many irons in the fire and too much teaching. In any case, he was dubious about the effectiveness of the process.[32] Cameron had good access to data on many of the individual isotopic nuclei. He got deeply into calculating reaction rates, to the extent of modifying electronic calculators intended for accountancy work to operate on protons and nuclei rather than dollars and cents.[33] In 1955, he published a major paper on element building by the slow capture of neutrons, a contribution that has unfortunately been overlooked by most historians of science.[34] The s-process could make a whole range of heavy elements, including the technetium, barium, and zirconium spied by Merrill in the atmospheres of S-type red giants.

Hoyle did not return to Caltech until the spring of 1956, when he had sabbatical leave for one term from Cambridge and an appointment as Addison White Greenway Professor of Astronomy at Caltech, a visiting professorship. This allowed him to spend a total of 7 months in Pasadena where he worked exclusively with the Burbidges and Fowler. Back home, his devoted wife Barbara kept up a busy correspondence on his behalf with his book publishers and the BBC and looked after their children ages 13 and 10. Fred's second 1956 trip had to wait until late September, when he returned to Pasadena accompanied by Barbara and daughter Elizabeth; meanwhile, son Geoffrey boarded at Bryanston School.[35]

Research in astrophysics holds a special place in the history of Caltech, a small independent university devoted to teaching and research in science and engineering. The original foundation dates from 1891. For its first 15 years, it served as a community and vocational college teaching a great variety of subjects. By 1907, the college needed a fresh purpose, which was provided by George Ellery Hale of the Mount Wilson Observatory. Elected that year to the school's board of trustees, he abandoned the high school model and instead developed the college along engineering lines. Hale was joined as trustee by Arthur Noyes, America's leading chemist, and by the physicist Robert Millikan. This triumvirate spent their World War I years in Washington, building a powerful network of contacts. From 1919, the school obtained handsome endowments. Millikan enticed the cornflake king to found the Kellogg Radiation Laboratory, and he relied on both the Rockefeller and Carnegie foundations. From this time, physics was to the fore. Millikan began a visiting scholars program, attracting a galaxy of talent from Europe. Einstein's visits in 1931, 1932, and 1933 put advanced physics on the map of Southern California. The physics department was essentially refounded in 1949, with heavy emphasis on the new field of particle physics and the installation of the electron accelerator. By the early 1950s, Caltech was the world's center for nuclear astrophysics.

In spring 1956, Hoyle lodged at the Athenaeum, the faculty club on Hill Avenue, where the facilities would be familiar to any Oxford or Cambridge don. Hale envisioned the Athenaeum as a London club: Mediterranean styling, gracious public rooms furnished with antiques, and elegant study bedrooms. The benefactors handed over a gift of stocks in 1929 to establish the club. These stocks were converted to half a million dollars in cash literally on the eve of the great stock market crash. In the evening, Hoyle dined informally on the patios, shaded by palm trees. If he wanted to treat a friend, he could use the paneled dining room, where Noyes, Millikan, and Hale looked down benignly from a portrait.

Everything was now in place for one of the greatest single collaborations in twentieth-century physics, the solution to the origin of the chemical elements by Burbidge, Burbidge, Fowler, and Hoyle, published in a paper invariably referred to as B^{2}FH.[36] With more than 1,300 citations in the primary literature, as measured by the Institute for Scientific Information, this is Fred Hoyle's most frequently referenced paper. The paper opens with a demonstration of a little learning—two quotations from Shakespeare:

> It is the stars,
> The stars above us, govern our conditions
>
> (Shakespeare, "King Lear," act IV, scene 3)

and

> The fault, dear Brutus, is not in our stars,
> But in ourselves
>
> (Shakespeare, "Julius Caesar," act I, scene 2)

In 2002, Margaret Burbidge commented in a lecture at Cambridge that these astrological fragments, which Willy Fowler had inserted, are still apposite. New observations with telescopes and equipment undreamed of in 1957 are continually building on the foundation in B^{2}FH, and the story is still unfolding.

The grand synthesis constellated a critical mass of disparate discoveries and data. The Cold War was more or less thawing: There had been a meeting of the nuclear powers in Geneva, aimed at pooling knowledge that could be useful for the peaceful development of nuclear energy. Part of the contribution from the United States to this atoms-for-peace initiative was the public release of classified data on reactions between low-energy neutrons and a wide range of elements. In the spring of 1956, the American authorities released nuclear data from their second hydrogen bomb explosion, which had vaporized Eniwetok, an island in the Bikini archipelago on March 1, 1954.

The results from the fusion bomb test were dramatic: The massive neutron flux engendered had synthesized elements beyond uranium. In 1949, Maria Goeppert-Mayer, working at Argonne Laboratory, had published the shell model of the nucleus, for which she shared the 1963 Nobel Prize. This theory enabled nuclear astrophysicists to understand how particles and forces interacted inside the atomic nucleus. Paul Merrill had discovered the unstable element technetium in S-type stars, thereby proving that stars made elements. Indeed, spectroscopists were routinely finding marked variations from star to star in the abundances of heavier elements, which surely proved that their production arose from a variety of local events rather than a single universal process. Finally, in 1956, Hans Suess and Harold Urey jointly published a seminal paper on the abundances of the elements based on meteorite data, combining Urey's knowledge of meteorite chemistry and Suess's ideas on how abundances should be interpreted.[37] Although great liberties were taken with available data, the peaks in the resulting abundance curves were the inspiration for the Caltech quartet.

At the Kellogg Radiation Laboratory, Fred, Margaret, and Geoff worked in a windowless visitors office next to the target room of an old electron accelerator. Fowler's cramped office was in the same building, round a corner. He had windows! As a step toward sketching out a theory on the blackboard, the three astrophysicists rehearsed the state of their knowledge of stellar evolution. For massive stars, it was obvious that once the hydrogen had gone, a helium core would build up. The existence of carbon-rich stars demonstrated that the triple-alpha process occurred. Some stars showed carbon-13, indicating that oxygen-16 could be made in red giants, and neutrons along with it. Animated conversation led to the conclusion that spurts of mixing in a giant star would mingle a carbon-rich core with a helium-rich shell, which would produce neutrons. These could be added, one by one, to an iron seed nucleus.

Fred and Geoff ran calculations on hand-cranked calculators. They found that building a large atomic nucleus was an extremely slow process. After being struck by one neutron, the target nucleus typically would have to wait hundreds of thousands of years for the next addition. That is why the colleagues named it the s-process, "s" indicating slow. Given a sufficient amount of time, all the nuclei in a star would transmute into the heaviest nucleus this process can make, element 83, bismuth. However, the evidence from the real world is that there is a whole range of nuclei in existence that can be made by the s-process. Evidently, the process froze out or choked off long before it had run its course. They assumed that when a star became a red giant, it became unstable and puffed off its outer layers, enriched by s-process nuclei.

To strengthen their theory, they needed more data on the actual abundances of heavier elements in stars. Geoffrey and Margaret identified the bright star HD46407, showing unusually high surface abundances of s-process elements, as a possible candidate for study. This is a barium star, so named because its spectrum reveals the presence of excess barium, element 56. In this partnership, Margaret was the observer. Geoffrey, as a staff member, could apply for observing time on the Mount Wilson 100-inch and 60-inch telescopes. However, the director of Mount Wilson, Ira Bowen, rigidly enforced an ancient policy that banned women from observing. Caltech and Mount Wilson were at this time having discussions on closer links. Fowler therefore had a chance to intervene. He pointed out that such discrimination was intolerable, at which point Director Bowen reluctantly withdrew his opposition. The Burbidges would be permitted to observe as a team as long as Margaret kept a low profile. They were to travel in their own car, take their own food, and live in the Kapteyn Cottage. By this tactic, they obtained spectra of HD46407. Margaret ploughed through the data, producing a table of overabundances for 12 s-process elements. They ranged from strontium to tungsten and included exotic rare earths.

One day, Hoyle stumbled on impressive evidence for the s-process. Each isotope has a certain likelihood of capturing neutrons, encapsulated in a number called the capture "cross section." The cross sections are measured by engineering collisions in particle accelerators. The larger the cross section, the more likely the nucleus is to capture a neutron. The group had made a table of the cross sections and observed abundances for 200 isotopes they believed were generated by the s-process. Line by line the table showed huge variations from one isotope to another, but something systematic seemed to be there. Hoyle noted that heavy elements with small cross sections are common in nature, whereas those with large cross sections are rare. This is exactly what the s-process predicts: If a nucleus has a small cross section, it will dwell for a very long time before catching a neutron and transmuting to the next element. Hoyle started to multiply, by hand, the cross section and the abundance for all 200 entries, a laborious task. With only a few exceptions, he found the product to be amazingly constant. A few were seriously discrepant, but when their paper was published, nuclear physicists remeasured the misfit cross sections and found they needed correction: The s-process theory had made successful predictions.

Despite huge successes, the s-process could not make everything. At the root of the problem lay its excessive slowness. The time interval between one neutron capture and the next allows beta decay, in which one neutron decays to a proton. Neutron-rich nuclei such as thorium and uranium cannot be made by the s-process. The only way round the problem is to bombard a seed nucleus with neutrons so fast that any unstable nucleus cannot decay before it is forced to absorb another neutron. In practice, this means absorbing neutrons at the rate of one a second or faster, rather than one every 100,000 years. Hoyle and his colleagues by now realized that some stars must have two completely different types of neutron-capture process. An extremely slow route produced neutron-poor elements, such as barium and zirconium, whereas a very rapid route was needed for the neutron-rich elements such as uranium.

Fowler and Hoyle, the two team members with strong backgrounds in nuclear physics, agreed upon a set of procedures for calculating the reaction rates for the rapid process. Hoyle started by adding neutrons to iron in quick succession and obtained encouraging results. By comparing reaction rates with the abundances, he found that most of the elements heavier than iron could be made in the r-process. The big question now facing the four friends was this: Where in the stellar universe could they find a source of neutrons so vast that one could hit every nucleus once a second?

Data from the hydrogen bomb explosion on November 1, 1952, provided a vital clue. The mushroom cloud contained superheavy nuclei with atomic numbers greater than uranium's. Only an intense flood of neutrons, the r-process, could have created them. Hoyle had already made the connection between the explosion of a nuclear bomb and the death of a massive star in a supernova explosion, so it was natural for him to speculate that such explosions are the source of plentiful neutrons and the elements heavier than iron.

Geoffrey Burbidge appeared to find dramatic evidence favoring Hoyle's supernova ideas. In 1937, Walter Baade had observed a supernova explosion in a nearby galaxy for a period of 2 years.[38] Burbidge found Baade's old records at Mount Wilson, examining the light curve. The brightness had faded in an intriguing manner. After 55 days, it had halved in brightness, then halved again in a further 55 days. This seemed to be like radioactive decay, with a half-life of 55 days. The Eniwetok explosion had synthesized the artificial isotope californium-254, with a half-life of about 60 days. Burbidge stuck his neck out: He suggested that the decay of californium-254 was the cause of the fading light in Baade's supernova. It is now known that this is not correct. The dwindling supernova is powered by the decay of nickel-56 and cobalt-56. However, in 1957, this imaginative notion of Geoff's was important for getting the theory of the r-process taken seriously.

The four decided to publish their work as a single encyclopedic paper, 108 pages in extent, the B²FH *magnum opus*. They could have produced a series of shorter papers instead, one on the s-process, another on the r-process, and so on. Had they done so, their work would have lost much of its magisterial quality and would have had less impact. B²FH remains a key paper. It defined the landscape for nuclear astrophysics, establishing a grammar and a lexicon, and providing an arithmetic and an algebra. For sure, the details would be superseded, errors would be discovered, and the data improved. Nevertheless, I would argue that theirs is the single most important paper in astrophysics published between the end of the Second World War and the widespread introduction of fast digital computers in the 1960s. Some historians have overlooked Al Cameron's contribution to nucleosynthesis. Working alone up at Chalk River, he quite independently covered the synthesis of the elements in much the same way. He produced an internal report, but the work was never published in an astronomical journal.

Of Fred's contribution to the published version, I can find but one fragment in the archives. This is an early draft of the general dynamics of the r-process. This manuscript is headed section 16, and it corresponds to section VII of the printed version, beginning at page 587 of the journal. Hoyle chose to write in blue ink on strange American paper: ruled with green horizontal lines and red columns! In places he has glued on typewritten sections cut from an earlier draft.[39] I suspect he wrote this in the summer of 1956, when he was back in Cambridge, but without access to a departmental typist because he was supposed to be on sabbatical leave.

Willy Fowler took responsibility for the final draft of the paper and for correction of the proofs. Hoyle was happy about that: Once he had done a fair draft of a paper, the publisher's editorial and production processes held no interest. The four decided that the order of the authors should alphabetical. A popular summary, published in *Science* in 1956, has Hoyle as the first named author.[40]

The paper starts with a stylish introduction by Hoyle, followed by a summary of the nuclear physics from Fowler. In all, eight different processes of energy release and nucleosynthesis are described. These are hydrogen burning, helium burning, the triple alpha process, Hoyle's equilibrium process for making the elements around iron, the s-process, the r-process, the p-process, and the x-process.

Although the s- and r-processes can make most heavy nuclei, they cannot make them all. Some heavy elements have rare isotopes that are neutron poor, for example, tin-112 and mercury-196. Their seed nuclei must have gorged on protons at some stage. Hoyle and his colleagues invoked a proton-capture scenario, or p-process. It operates in a hydrogen-rich environment and requires a temperature of a billion degrees, conditions probably found in supernova explosions.

B^2FH has no solution for the creation of the lightest elements— lithium, beryllium, and boron. All are extremely heat sensitive. Stars destroy them. The colleagues proposed a mysterious x-process, without explanation. Helium, too, presented a formidable challenge. The hydrogen-burning stars, such as the Sun, produce immense quantities of the second lightest element. Hoyle, however, realized as early as 1955 that there is far more helium in the universe than stars can make! The modern explanation of the origin of the light elements is that they were made in the first few minutes following the big bang. The expansion of the universe proceeded so fast that they "froze out" before heat could disintegrate them.

The work in 1956 on B^2FH led to a lifelong friendship between Fred Hoyle and Willy Fowler, and in May 1957 the Vatican conference on stellar evolution brought Fowler and Hoyle together once more. By now, Hoyle was testing the patience of the University of Cambridge to destruction. He had been forced to take leave without pay to complete B^2FH at Caltech. The Pope's jamboree fell in the middle of the exami-

nation period, when Hoyle would need to mark a mountain of candidates' papers. The director of the Vatican Observatory unwittingly
placed Hoyle under further pressure by asking him to write a summary
of the astrophysical conclusions of the meeting. To do this, Hoyle
decided to join Fowler at a hotel in Sorrento. Barbara and Fred drove
south in their Humber Hawk, while Willy and his wife Adrianne took
the *rapido*. Fowler, a huge enthusiast for steam trains, improbably
claimed to have sweet-talked his way into the driver's cab and taken
over the controls for a short distance![41] From Sorrento, they went south
to the stylish resort Amalfi, where Fred finished his report. Partly to
ease the tedium of the long drive back north, they gave a lift to Georges
Lemaître. For his part, he had a long list of places to visit, which distracted Hoyle even more from the pressing business of getting back to
Cambridge. Barbara Hoyle vividly recalls this car journey and the side
trips:

> It was strange to have two men in the car with such differing views on
> cosmology. It was very friendly, and while Lemaître amply consumed pasta
> and wine for lunch, Fred had bread and cheese. In the afternoons Fred
> drove while Lemaître slept off lunch in the back of the car. It was the first
> time I had encountered a priest who did not want fish on Fridays and who
> was not asked to pay for his meals in Italy.[42]

The solution to Fred's conflict with the university came the following year with his elevation to the Plumian Professorship. Like all
holders of chairs at Cambridge, he had to cease teaching undergraduates on behalf of his college, and his lecture load lightened
considerably. On the other hand, large managerial demands would
soon be placed on his shoulders.

For the 1958 symposium of the International Astronomical Union
in Moscow, Fowler organized a session on nuclear astrophysics. He
invited Al Cameron to talk on the s-process, and thereby set up the first
encounter between Hoyle and Cameron. When they arrived in
Moscow, Soviet nuclear physicists invited Fowler, Salpeter, Cameron,
and Hoyle to visit Dubna, the secret nuclear weapons establishment
north of Moscow. Cameron treated the Soviets to a talk on the yields of
fission fragments from uranium-235. In Dubna, the four visitors were
lunchtime guests of Director Igor Kurchatov. He was the key individual

who had received secret intelligence reports from spies the Soviets had planted in the Manhattan Project, although the visitors did not know that. They found Kurchatov to be a cultured gentlemen of the old school, speaking impeccable English. By then, he had turned against nuclear weapons and, like Hoyle, was arguing for an end to atomic bomb tests.[43]

Hoyle and Fowler founded a new subdiscipline in 1960, nuclear cosmochemistry, although it would be Fowler rather than Hoyle who would develop the subject. The r-process creates heavy isotopes that are unstable to radioactive decay, and study of these can be used to time cosmic events from the past. Geologists have long used the decay of radioactive potassium-40 to date rocks. At formation, the rock holds no argon-40. However, one decay path for potassium-40 produces argon-40. By measuring the proportions of argon-40 and potassium-40 in a rock, a geologist can deduce when the rock solidified. Potassium-40 has a half-life of 1.3 billion years and can be used reliably to date rocks in an age range of 100,000 to 5 billion years. Hoyle and Fowler extended the principle from Earth to the stars. In their study, Fowler and Hoyle looked at the ratios of uranium-238 decaying to thorium-232 and uranium-238 decaying to uranium-235. They used ancient terrestrial rocks and were able to demonstrate from the ages that the uranium had been created in supernova explosions long before the formation of the Earth.[44]

During 1958–1859, Hoyle's models of stellar evolution pointed to an age of 12 billion years for the oldest stars in the galaxy, a result he had presented at the IAU symposium in Moscow.[45] By now, he had access to an IBM 704 computer, generously provided by IBM, who wished to develop the academic market. Continuing this line of research in collaboration with Fowler, he obtained a similar result: The oldest stars showed that the age of the galaxy lay somewhere between 10 billion and 15 billion years, with 12 billion the likely value. Subsequent work by Fowler's school tended to confirm this early estimate, which has stood the test of time and remains unchallenged.

The collaboration with Fowler on nucleosynthesis even led Hoyle to dabble with the big bang universe for a while. Despite the great success of B²FH, there remained a nagging doubt about the origin of the lightest elements, particularly lithium: It cannot be made in stars, so it must come from another source. In 1964, in conjunction with Roger Tayler, Hoyle had started an attack on the whole question of primordial nucleosynthesis. Just what could the universe cook up in the big bang? Was it solely hydrogen and helium or could other elements be made? This development informs us marvelously about Hoyle's approach to research problems. He was always happy to compartmentalize his work so that, although he felt the steady-state model to be much superior to the explosive model, he nevertheless took pride in looking at the underlying physics of the rival model.

Bob Wagoner took on the lion's share of computerizing a vast network of nuclear reactions that might occur fleetingly in the primordial fireball. Willy Fowler commenced an exhaustive investigation of the kind he loved, with Hoyle in tow. The three of them pushed to the limit. They discovered ways to make deuterium, helium-3, and lithium-7 and made a huge effort, unsuccessfully as it turned out, to produce carbon and nitrogen. Beyond that barrier, Wagoner pressed on anyway, working through an immense array of cases to yield the heavier elements, from carbon to nickel. This effort fizzled out, at which point they all returned to B^2FH except for the lightest nuclei. Hoyle had found out how these would be made during an event he did not believe in.

As we approach the fiftieth anniversary of the writing of B^2FH, we can see that its completion was Fred's finest moment as a research scientist. The lasting friendship with Fowler would prove to be a significant motivation for establishing the Institute of Theoretical Astronomy at Cambridge in 1967. Geoffrey Burbidge would become a loyal champion of a modified version of the steady-state theory, which he continues to advocate in the present century. A decade after B^2FH, Margaret and Geoffrey Burbidge would make important observations of quasars, highly energetic objects on which Fred held deeply controversial opinions, as we shall see.

CHAPTER 9

Matters of Gravity

F ollowing completion in California of the 1957 nucleosynthesis
paper, Hoyle developed a strategy for spending some weeks each
year at Caltech. In general, he did not reside there during the
university summer vacations, which he could have done without
seeking permission from authorities at Cambridge. There were good
reasons for staying in the British Isles in the summer months. Barbara
had taken a dislike to Los Angeles and Fred absolutely detested its smog,
which drove him indoors when it was bad. Fred, Barbara, and the
children enjoyed old-fashioned seaside holidays in their caravan, rather
than expensive visits to California.

When Fred returned in 1957, the family had arranged to move
from their country house in Little Abington to a fine home being constructed at 1 Clarkson Close, on land belonging to St. John's College.
However, the builders were running 2 months late, which forced the
family to live in their caravan for several weeks. As soon as the school
holidays started, Fred's Humber Hawk hauled them to Scotland up the
old A1 road (which was not then a divided highway). From there, they
crossed to Northern Ireland, then traveled over the border to the
Republic of Ireland. For 6 weeks of glorious weather, they parked in a
narrow lane hemmed in by stone walls, sited 400 feet above the sea.[1]

Over the years, Fred used caravan trips of this kind to compose much of his science fiction. He could take the caravan anywhere he wanted, which was sometimes to Cornwall, sometimes north to Scotland, and occasionally to Ireland. As a result of the Irish vacations, Hoyle and his family became close friends of the statesman and writer Conor Cruise O'Brien.[2] The summer writing expeditions typically lasted about 6 weeks, during which time he was out of touch with Cambridge. He would often walk in the morning, sifting ideas for the development of the plot, then write longhand in the afternoon. Barbara subsequently typed the manuscript, handled all the correspondence with the publisher, and checked the proofs. Hoyle seldom had anything to do with typescripts or scientific papers once they had been accepted. He shocked one of his editors, who wished to make revisions, with "the creative aspects of this enterprise have now been concluded—do with it what you will." By the time a proof came, his agenda had moved on and he had no taste for detailed corrections. In addition to science fiction, his literary output included several murder and detective stories, mostly set in Cambridge, with titles such as *The Cambridge Madrigal Murders*.[3] None of these found a publisher: It is a demanding genre.

From 1958, Fred arranged his annual California trips by eating a little into the Christmas or Easter vacation and snitching 4 weeks from the university term. This was not convenient for his graduate students who had to manage for several weeks without his advice. In the summer, Fowler would visit Cambridge. These semiannual meetings became immensely important: During the 1960s, Fowler-Hoyle was the most famous astrophysical collaboration in the world.[4]

The first fruit of this pairing was the publication in 1960 of a large and influential paper on the synthesis of chemical elements in supernovas.[5] B[2]FH, blockbuster that it was, stayed mute on the details of where the various synthesis processes operated in the universe. After B[2]FH, Hoyle and Fowler quickly latched on to the idea that the sudden fusion of a nuclear fuel might be the source of the energy of a supernova explosion. In the middle of the twentieth century, astrophysicists regarded supernova explosions with awe. At the time, they were the most powerful events observed anywhere in the universe. The fact that no supernova explosion had been sighted visually in the Milky Way galaxy since the invention of the telescope only added to their mystique. Furthermore, the number of explosions that astronomers had observed in some detail amounted to only a few dozen. In his 1955 textbook *Frontiers of Astronomy*, Hoyle makes much of the Crab Nebula supernova of A.D. 1054, where the explosive ejection of matter could be seen so convincingly. In this text he notes that Minkowski had identified two kinds of supernova: type I with spectra deficient in hydrogen, and type II with hydrogen-rich spectra. He also gives a vivid impression of the phenomenal power of a supernova: "The energy released by nuclear reactions in only a second of time is as much as the nuclear reactions inside the Sun yield in 1000 million years."[6] He continues:

> The amount of energy released is sufficient to endow the exploding outer parts of the star with velocities of 2000 to 3000 kilometres per second, and is sufficient to enable the star to radiate at 200 million times the rate of the Sun for a period of about two weeks.

Professional astronomers who had been physics or mathematics undergraduates in the late 1950s and early 1960s claim that graphic language and the sense of immense excitement conveyed through these descriptions of supernova explosions stimulated them to become astrophysicists.[7]

Frontiers of Astronomy merely speculates on element synthesis in supernovas. In the period 1958–1960, Hoyle and Fowler created the astrophysical theory needed to understand synthesis. Their working methods can hardly have been more different. Fowler started any new problem by carefully reading all the literature, looking for clues to a solution. Hoyle made a point of reading the *current* literature but not the older papers; this sometimes led to sharp criticisms that he ignored the work of others. When the two were together, Fowler would listen to a blackboard lecture as Hoyle tried to work out a problem from first principles. Sooner or later, he would get stuck and they would terminate the conversation. Within a day or two, however, Hoyle would have found a solution, not by reading back issues of *Astrophysical Journal* in the library, but through his own pen-and-ink efforts. His fiercely independent streak would not allow him to rely on the prior results of others.[8]

Their 1960 paper is the first to note the importance of supernovas in making heavy elements. They start by asking, "When a supernova goes off, *what* explodes?" At the very end of the life of a massive star, the material in the nuclear core has reacted to make the elements of the iron peak, which are the ultimate products of energy-releasing fusion reactions. All fusion reactions of nuclei heavier than these require an *input* of energy, so the iron core can draw on no further source of energy to replace the vast amounts being emitted at the surface of the star. Inexorably, the gravitational force is going to overwhelm the dying nuclear furnace. As the core cools, its internal pressure falls, and its ability to remain stable against the inward crush of gravity diminishes. The tip moment comes very suddenly: Hoyle and Fowler showed that once the nuclear fusion peg is kicked away, the central regions of a massive dying star implode catastrophically in just one second!

To answer the question of *what* explodes, they turned to the layers of the star outside the core. These layers are rich in the light elements hydrogen and helium. With the collapse of the core, these layers have no support and hurtle inward. The energy generated by this infall is immediately converted to heat, which manifests itself by greatly increasing the speeds of the protons (hydrogen) and helium nuclei. Suddenly, there is a nuclear explosion that processes as much of these light elements in one second as the Sun consumes in a billion years. Explosion of the outer layers of the star follows the implosion of the core, which in turn releases a vast amount of gravitational energy in its collapse. This is the standard picture of supernova explosions today, first described by Fowler and Hoyle.

During pioneering studies of supernovas in 1940, Rudolph Minkowski had used the 100-inch telescope at Mount Wilson to take spectra. He could immediately see that the spectra fell into two main categories. Broadly speaking, what he called type II showed hydrogen was present, whereas evidence of hydrogen was missing from type I. Fowler and Hoyle identified low-mass stars as the progenitors of type I supernovas and high-mass stars as the source of type II events. They pointed to both Hoyle's e-process for making the iron peak elements and the r-process as being associated with supernovas, thereby linking these exploding stars to the origin of the elements. The situation today is that astrophysicists recognize three subgroups within type I—a, b, and c. Type Ia supernovas are believed to be explosions of white-dwarf stars in binary systems, whereas types Ib, Ic, and II result from the catastrophic collapse of the cores of massive stars. All types eject heavy elements into interstellar space.

As a result of the Fowler-Hoyle work, we now know that most of the stable elements heavier than iron, of which there are 66, right up to uranium, are made inside exploding supernovas. The time required for this process is seconds to minutes at the most. The explosive pulse is the source of many of the trace elements essential to our health and welfare. Four years were to pass before the duo would again publish on supernovas, following Fowler's sabbatical year in Cambridge in 1961–1962. Discussion of further research on supernovas commenced on April 17, 1962, when Hoyle and Fowler were with six dozen delegates at the Sixth Herstmonceux Conference. Hoyle's main task was to deliver a majestic review of the mechanisms of star formation. Both Hoyle and Fowler participated in debates on the best way to use a computer in astronomy: Should data be stored on punched cards (Hoyle's preference) or paper tapes (the choice of Cambridge radio astronomers)?[9]

Like theorists in the United States, Hoyle was by now using an IBM 7090 computer for theoretical work. He explained that, in his experience, skill in manipulating mathematical expressions into forms suitable for computer analysis was more important than detailed programming knowledge, which could be acquired by a technical assistant. Evidently, outside the formal sessions, Hoyle and Fowler were considering the evolution of very massive stars that Hoyle felt should be formed in collapsing interstellar clouds.

In 1963, the American Astronomical Society awarded its prestigious Henry Norris Russell Lectureship to Fowler, who presented the lecture on July 23, 1963, at its 114th meeting, held in Alaska. He blazed a new trail in astrophysics, describing the work he and Hoyle had just completed on processes involving neutrinos in very massive stars. The lecture was rich in heavy-duty nuclear and particle physics, material that would have been beyond the comprehension of many of the astrophysicists at the meeting.

The paper they produced on this work, with 10 sections and 3 appendices, extends to 119 pages in the *Astrophysical Journal Supplement Series*[10]; evidently, the editor felt it was much too long and detailed to appear in the main part of *Astrophysical Journal*. The University of Chicago Press subsequently issued this exhaustive (and exhausting) paper as a book. As with B²FH, Fowler added a literary quotation to set the scene:

> O dark, dark, dark. They all go into the dark,
> The vacant interstellar spaces, the vacant into the vacant.

> (T. S. Eliot, in *East Coker*, III, 1940)

The main problem solved in the paper is the role of neutrino processes in the evolution of stars exceeding 10 solar masses. They found that, although the emission of neutrinos sped up the evolution, the faster reaction did not trigger an implosion until most of the core had burned to iron. In writing the 1964 paper, they realized that, in equilibrium, the core burns to nickel-56, *not* iron-56, as they had previously thought. Iron in the universe outweighs all of the common metals and is the tenth most abundant nucleus.[11] It was this striking dominance that inspired Hoyle's 1946 formulation of stellar nucleosynthesis. Nickel-56 is radioactive, decaying with a half-life of 6 days to cobalt-56 and then to iron-56 with a half-life of 111 days. These decays produce gamma rays, which are detectable during supernova outbursts, thereby providing direct evidence that nickel-56 is created in the explosive envelope of a supernova.

While working on this paper with Fowler, Hoyle encountered serious difficulties with computing facilities. The national research funding agency had awarded him a grant to buy "time" on an IBM 7090 in London. This was the most powerful computer in the world and the only one situated in Europe at the time. Hoyle's research students would commute to London for the privilege of using this American computer a few hours a week. Then, out of the blue, Whitehall implemented a "Buy British" policy: Where possible, contracts with foreign suppliers were to be canceled. The country was faced with a balance-of-payments crisis, a run on the pound, and an incompetent government. The Cambridge theorists were instructed to support British enterprise in the form of the new Atlas computer at the

University of Manchester. The Ferranti company had worked with the universities of Manchester and Cambridge on computer development from the late 1940s. Their new Atlas machine opened with a great fanfare in early 1962: the world's first computer to use transistors, the fastest machine in the world—it was British technology at its best. It doubled the UK's total computer power at the flick of a switch! There was plenty of hoopla. Sir Basil de Ferranti, both chairman of the company and a government junior minister, and Lord Hailsham, the minister of education, performed the opening in front of a media frenzy. The magnetic tape decks whirred, panel lights flickered on and off, champagne corks flew, and the results of a mighty computation spewed out. Chests swelled with British pride. Newspapers reported a miracle of British technology. The Cambridge research students departed for Manchester in high spirits.

What they found in Manchester was a farce. Training consisted of pencil-and-paper exercises in a lecture room, with no trial runs on the machine. After a few days of this nonactivity, research student John Faulkner became exasperated enough to challenge one of the more knowledgeable assistants. He allowed himself to pull rank: "Excuse me, but would you mind telling me what's going on? I, and my friends here, are all graduate students of Fred Hoyle's. We've come all the way from Cambridge for this course. We haven't even seen the machine operating. What is going on?"[12]

The glum assistant took Faulkner to one side: "It's all a bit embarrassing. Atlas is not working properly and reliably yet. Look, it didn't do those supposed calculations at the official opening: We couldn't trust it to operate properly on the big day. We arranged for it to look as though it was computing, but the answers had already been stored for printing."

What a letdown. Faulkner was outraged. On hearing this sorry tale, Hoyle made an official protest to the politicians, in which he railed against the Buy British policy. His excellent students had been forced off a fine American computer to work on a wonky British machine not yet capable of producing a sustainable level of computation.

Shortly after publication of this letter, a scandal emerged at the heart of the Ferranti industrial empire. The computing side of the business was out of control and hemorrhaging cash. Ferranti was over-charging the UK government for work in progress on a surface-to-air missile, slipping the proceeds under the table to shore up the comput-ing side of the business. As a junior member of the government, Sir Basil de Ferranti had apparently encouraged the agency responsible for science funding to pour yet more money into computing research, a move that benefited his company. In the end, he resigned as minister and the Ferranti computing business collapsed, being sold for a million and a half pounds, a song. For 6 months, Fred's students had no computer. Eventually Whitehall, stung by Hoyle's protests, permitted them to go back to the IBM 7090.

The Hoyle-Fowler speculative work on supernovas had momen-tous connections because it had led the pair to work on the physics of very massive stars, up to a million solar masses. By the time of Fowler's 1961–1962 sabbatical in Cambridge, radio astronomers were well aware that certain radio sources posed an immense theoretical puzzle: How did they generate their energy? Hoyle and Fowler found a solution to this problem in August 1962, proposing the explosion of million-solar-mass stars. Ryle's radio astronomy group characteristically adopted the party line and rejected Hoyle's model as preposterous.

By the late 1950s, positional measurement in radio astronomy was sufficiently precise to allow the identification of a few strong extra-galactic radio sources with distant galaxies, the exemplar being Cygnus A (or 3C 405) which coincided with a disturbed galaxy. Optical astronomers could derive the redshift from a spectrum of a galaxy, in this case 0.057.[13] Then, by using the Hubble value for the expansion of the universe, a "cosmological" distance could be computed, which is about half a billion light-years for Cygnus A. The computed distance could then be used to calculate the intrinsic energy output (luminosity) of the radio galaxy from its observed brightness. The arithmetic is straightforward and the calculation easy. The only assumption is that the redshift is caused by the expansion of the universe, in which the radio galaxy is assumed to be a participant.

The radio power radiated by a radio galaxy is produced by inter-actions between electrons moving at almost the velocity of light and magnetic fields. The effect of the magnetic field is to force the electrons into a spiral path. This motion along a curved trajectory leads to the emission of photons by the electron, in a process known as synchrotron radiation.[14] In 1959, Geoffrey Burbidge made estimates of the total energy requirements of several radio galaxies. Lacking any information on the strengths of their magnetic field, he contented himself with a calculation of the *minimum* energy requirements to account for the luminosity.[15] Hoyle and Fowler, like the radio astronomy community in general, found these estimates absolutely staggering: Burbidge's minimum energies for radio galaxies came in the range 10^{51}–10^{54} joules, with Cygnus A at the upper end.

To put the energy requirement of Cygnus A into context, we can consider the energy content of the Sun. If you could completely annihilate all the matter in the Sun, you would get 10^{47} joules of energy. So, in a radio source with a store of energy of order 10^{54} joules, the equivalent of 10^7 Suns must be turned into pure energy. Furthermore, if the source of energy is through nuclear processes, mainly the conversion of hydrogen into helium, a further factor of 100 must be inserted because nuclear reactions are less than 1 percent efficient at turning matter into energy. The energy requirement appears to need the processing power of a billion Suns. To Fowler, "the problem was a very serious one, a very puzzling one, and a very interesting one."[16]

Fred and Willy set themselves to work on the basic source of the energy: Where did it come from, where did the electrons get their energy, and where did the energy come from that produced the magnetic field? Geoffrey Burbidge had suggested in 1961 that a million (or so) supernovas going off simultaneously in the center of a galaxy could provide the energy, a view that Martin Ryle would fruitlessly cling to for a decade.[17] Hoyle and Fowler decided that it would be implausible to organize a domino effect to trigger simultaneous explosions, so they decided to look at the properties of a single object, which they called a supermassive star, something in the range of 10^4 to 10^{10} solar masses. This was a bold step: No evidence suggested that such objects could exist. However, it was very much in the nature of their collaboration to take risks and make unsupported assumptions. Their peers expressed much skepticism because ordinary stars contain no more than about 100 solar masses.

They ignored Burbidge and other experts and plunged into the physics. To Fred's delight, he found that the structure of a supermassive star is very simple. The heat energy released by nuclear reactions at the center supports the star against gravitational collapse. The structural equations were simplicity itself. They found that the luminosity of a supermassive star is simply proportional to its mass. Curiously, they discovered that supermassive stars have more or less the same life span of about a million years, regardless of mass. This coincidence particularly excited Fred because radio astronomers had long argued from a statistical point of view that the powerful emissions of a radio galaxy lasted about a million years. The two thought this could not be a coincidence; supermassive stars must be powering radio galaxies. In August 1962, they drafted a fairly detailed paper for *Monthly Notices*.[18] This contained an important error: They assumed that the main energy source would be nuclear reactions.[19] However, by the end of 1962, they had convinced themselves that the final gravitational collapse of a supermassive star would yield much more energy than nuclear processes. This lengthened the life span to 30 million years, still an interesting result even if it did somewhat spoil the lifetime coincidence. Then Fred remembered that the Royal Astronomical Society (RAS) takes ages to publish and he may have suspected that spies were at work in the editorial office. So to secure priority for supermassive stars, they dashed off a short note to *Nature*, taking the opportunity to mention gravitational collapse as the likeliest source of energy.

Events now moved fast. *Nature* published their paper in January 1963.[20] The following month *Nature* published Maarten Schmidt's groundbreaking discovery of quasars (quasi-stellar objects), a completely new class of extragalactic objects, located at immense distances.[21] Willy was the first to hear the news, directly from Maarten Schmidt at Caltech. He immediately penned a postcard to Fred at Cambridge, with the telegraphic message "Whooppee! They have found our objects."[22]

To see how Fred's research related to the discovery of quasars, this narrative must now backtrack by a couple of years and introduce some new colleagues of Hoyle's. It was Fred's first involvement in a truly international collaboration.

In 1959, the Department of Scientific and Industrial Research (later the UK Science Research Council) appointed Professor Fred Hoyle, FRS, as the chairman of the astronomy section of its main research committee. Ostensibly, they fingered Hoyle, a theorist, because they wanted a chair who would not be trying to allocate equipment grants to himself.[23] Soon Hoyle had on his desk an application for a grant of £50,000 from the radio astronomy group at Jodrell Bank, Manchester, led by Bernard Lovell. At the time, Hoyle's astronomy cupboard was bare, all funds for radio astronomy had already been allocated to Martin Ryle and Lovell, who supported each other's applications. Hoyle liked the unfunded application, which was for an extremely long interferometer (by the standards of the day). Jodrell Bank wanted a movable telescope about 100 miles away to work in unison with Lovell's 250-foot dish for measuring the angular sizes of radio sources to much higher precision. The idea was to use the pair of telescopes to determine the sizes of the patches of radio emissions that constituted the radio sources. Hoyle, no doubt motivated by the fact that this new interferometer would be far more powerful than anything available to Martin Ryle in Cambridge, persuaded the Research Committee to make an exceptional grant of £50,000 from contingency funds. He was again mindful of Baade's advice: Good-quality observations are needed to underpin theory.

It took a couple of years to build the interferometer, and it began to bear fruit in 1961, when a grateful Robert Hanbury Brown on the staff at Jodrell Bank invited Hoyle and his young research student Jayant Narlikar to visit the radio observatory. Hanbury Brown showed them his results on the sizes of radio sources in Ryle's 3C catalogue. Mostly, they had resolved the radio sources, but a small proportion were too small to yield to the techniques then available. Hanbury Brown colorfully told Hoyle: "We have this family of chaps, less than a second of arc across. We don't know what to make of them." A worrying thought immediately crossed Hoyle's mind: Could these be Ryle's radio stars that he and Gold had rejected in 1951? Could Ryle, after all, be partly right in that old controversy? Emotionally, Hoyle had to dismiss this thought. He immediately jumped to the conclusion that the point sources must be remote objects. In 1961, this was a strange and unorthodox stance. But Hoyle could not help thinking, even then, that explosions of compact massive objects of the kind he and Fowler were toying with could be the solution to Hanbury Brown's "chaps."

Meanwhile, in California, Allan Sandage of Mount Wilson had obtained the first optical spectrum of the radio source 3C 48. It was one of the very few compact radio sources for which an optical identification had been made. But he could not make sense of the jumble of emission lines. His colleague Jesse Greenstein likewise thought that the spectrum was intriguing and, stimulated by Sandage, took further spectra. He, too, found the pattern of spectral lines baffling, and the precious glass slides moldered in his office for nearly a year. He and Sandage felt that, if only some more of the Hanbury Brown sources could be identified optically, it might be possible to gain further spectra of similar objects as a step to resolving the puzzle.

At Jodrell Bank, Cyril Hazard, in a flash of insight, speculated that a few radio positions might be improved by a clever trick of his invention. He noticed that the Moon crossed the positions of a couple of the mysterious radio sources, momentarily blotting out reception of their radio emission. Hazard realized that, by noting the precise time at which the moving Moon occulted the pointlike radio source, as well as the moment of its reappearance, the position could be obtained from the known motion of the Moon with a precision of about 1 second of arc, far exceeding the accuracy of any other method available in 1961–1962. He demonstrated this novel technique with the 250-foot telescope at Jodrell Bank by making observations of 3C 212. Soon afterward, he headed overseas to a research post at the University of Sydney. In Australia, he set his sights on 3C 273, a lunar occultation candidate. To make the observations, he would need an allocation of time on the 210-foot Parkes dish, located in rural New South Wales. This instrument was in the hands of government scientific civil servants based in Canberra who did not see eye to eye with the university physicists in Sydney. Fortunately, fate intervened on Hazard's side. The Australian authorities had recently appointed John Bolton of Caltech (and a buddy of Fred's) to be director of the radio observatory. Bolton cheerfully allocated the amiable Hazard the necessary time. Together with M. B. Mackey and A. J. Shimmins, he observed the Moon's sweeps across 3C 273 in April, August, and October 1962.

By the standards of the day, Hazard achieved an exceedingly accurate position for the radio source 3C 273. He detected a point radio source and a curious wisp of radio emission. Bolton communicated the position from the August 1962 observation by airmail letter to the head of radio astronomy at Caltech, Tom Matthews. House rules at Caltech required that the information should go to Maarten Schmidt, the accredited extragalactic observer. A photograph of the radio position showed a star of magnitude 13 together with a wisp or jet extending to 20 seconds of arc. Schmidt's initial suspicion was that the star must be within the Milky Way and that the faint wisp was a peculiar galaxy. The object was so strange that he sensed astronomical glory awaited him as the first observer to solve the puzzle.

Immediately after Christmas 1962, on a sharp frosty night at Palomar, Schmidt secured his first spectrum of 3C 273 with the 200-inch. Unaccustomed to observing such bright objects (a thirteenth-magnitude "star"), he overexposed the plate. Nevertheless, the spectrum intrigued him because it showed emission lines in unexpected places. Two nights later, on December 29, 1962, he got the exposure time correct, securing spectra with seven emission lines. But this fingerprint could not be matched to any suspect. Schmidt later recalled that the puzzle resolved on the afternoon of February 5, 1963.[24] Cyril Hazard had written to him suggesting that the optical results should be published in the same issue of *Nature* as his radio position. Schmidt thought that a good idea and, while writing the manuscript, he reached for the glass slides of his spectra. For some reason he then decided to take the ratio of the wavelengths of the emission lines against the spectrum of atomic hydrogen.

The result stunned him. It was immediately obvious that he was looking at the spectrum of hydrogen, redshifted by 16 percent. This was a crazy result. Stars of thirteenth magnitude, those common or garden denizens of the Milky Way, are not meant to show such gigantic redshifts! Why had this star such a large redshift? And what was that wisp? Was it none other than a jet, just like the one that springs from the nucleus of the giant elliptical galaxy M87? The object was strange indeed. The starlike appearance meant it had to be of small angular scale, yet it exhibited a property of an enormous galaxy—the jet—and was extremely distant by record books of the day.

Any scientist faced with this kind of discovery immediately wants to share it, so, minutes later, he told his colleague Jesse Greenstein what had happened. Greenstein pored over the manuscript of a paper he had just prepared on the wavelengths of emission lines in 3C 48. In a matter of minutes the pair now derived a redshift of 0.37, implying a huge velocity of recession, almost 40 percent the speed of light. Interpreting these large redshifts proved a great challenge. They quickly rejected an explanation in terms of a gravitational effect and instead ascribed the results to large recession velocities. This in turn implied huge distances, enormous luminosities, and tremendous energies; it was all highly speculative, but they could find no strong arguments against it.

On March 15, 1963, *Nature* published Hazard's position and the spectroscopic results for 3C 48 and 3C 273 in four consecutive papers. Greenstein's paper took all the credit for the 3C 48 results, much to the irritation of Sandage who expected his colleague to offer joint authorship. The date of publication was about 2 months after Fowler and Hoyle had predicted the existence of highly condensed supermassive objects. The new class of cosmic bodies quickly became known as quasars, the name being derived from a contraction of quasi-stellar radio source. But the trail of discovery that had commenced with the grant of £50,000 to Jodrell Bank for the interferometer would soon lead Hoyle into deep controversy concerning the nature and location of quasars.

In 1963 in the UK, the *Daily Telegraph* of March 13 and the *Sunday Telegraph* of March 17 both publicized the Hoyle-Fowler theory that supermassive objects inside galaxies might suddenly and dramatically collapse inward. Sunday's readers were told:

> huge gravitational forces, which act towards the centre, would cause the whole construction to collapse catastrophically. During the performance, powerful radio waves would emanate, and it appears possible that the Hoyle-Fowler effect is acting on a distant radio source known as 3C 273... . It is identified visually as a very distant galaxy with a flare of light coming from one side. The combination of the radio waves and the flare suggest a galaxy is in the act of implosion, with the flare representing part of the debris breaking away.

Unfortunately, Hoyle and Fowler made no attempts to develop their model further as the observations of quasars and radio galaxies improved. One could speculate that, had they stuck to their agenda, they might have produced what is now the standard model for an active galactic nucleus: a rotating supermassive black hole surrounded by an accretion disk of infalling matter, with two jets of matter streaming away along the rotation axis at the speed of light. Instead this model was to be developed at Cambridge by Donald Lynden-Bell and Martin Rees. But the pair did launch a new discipline, "relativistic astrophysics." In December 1963, an international symposium convened in Dallas, Texas, just days after John Kennedy's assassination in the same location. It brought together experts in general relativity and high-energy astrophysicists to discuss the Hoyle-Fowler effect in relation to the newly discovered quasars.

Hoyle was by this time very famous in the UK as a result of the enormous success of *The Black Cloud* and two science fiction series he had written for BBC television. Virtually anything controversial he had to say quickly became a big news story. At about the time of publication of the first Hoyle-Fowler paper, the scripts department at the BBC had become interested in making a TV series loosely based on *The Black Cloud*. However, when they contacted Fred, they learned that the film rights had already been sold to an independent producer (who never did anything with those rights). Nevertheless, Hoyle was intrigued by the notion that he might have a future as a scriptwriter, and he told the BBC that he had lots of ideas for story lines. This led to a meeting on June 8, 1960, between Hoyle, John Elliot (assistant head of the script department), a scriptwriter, and a script editor. The three BBC colleagues made a 90-minute tape recording in which Fred sketched out the plot for a seven-part series.

Three weeks later the BBC offered 250 guineas (£262.50, about $700) for seven 30-minute scripts, on which Hoyle would work with John Elliot. Hoyle would supply the scientific ideas and Elliot would script the dramatization. Hoyle laid the BBC's offer to one side for 2 weeks before responding, and when he did, he approached the offer like a scientific puzzle:

> This reply has been delayed by my wish to think over the whole ques-

tion of financial terms. It has seemed to me that the overall sum of 250 gns is unsatisfactorily low. My own computations would suggest 1000 gns. I estimate that such a sum would still be only 2–3 per cent of the receipts by the BBC from licences. Such a percentage still appears to be low—it is considerably less than rates available in the US.[25]

Hoyle's knowledge of American fees was acquired during abortive negotiations in late 1958 with a Hollywood film company that wished to make a film version of *The Nature of the Universe*. His careful reply did the trick: The BBC increased its offer to 700 guineas (about $2,000), which Hoyle accepted. The series became *A for Andromeda*. The drama is set in Fred's homeland, the Yorkshire Dales, in 1970. A group of astronomers receive radio signals from space, which they interpret as a computer program from extraterrestrials. Once decoded, the message gives them the ability to build a supercomputer, which demonstrates its power by creating an android, Andromeda. One of the scientists then decides that the computer must be destroyed, which terminates this series. *A for Andromeda* beautifully combined Hoyle's science with Elliot's writing skills to produce a character-driven science fiction story that is both believable and exciting. The production of *A for Andromeda* resulted in Fred literally discovering a previously unknown star of first magnitude. The key figure is Andromeda, the cool humanoid built according to the computer instructions from an alien civilization. With an eye on the ratings, Elliot wanted to cast a very beautiful girl in this role. Hoyle agreed, but wanted an actress who had never been seen before on screen. Where on Earth could they find such a person?

Somewhat in desperation, they went to final-year performances at the Royal Academy of Dramatic Art and London's Central School of Speech and Drama. At one of these shows, Fred instantly recognized a cool, beautiful, and very reserved young woman as the person he wanted to cast. He grabbed Elliot's arm and said: "*She's* the one. I want *her.*" This was Julie Christie: Hoyle's *A for Andromeda* made her well known and ultimately a star.[26] She was the iconic cool British blonde of the 1960s. The series aired in October and November 1961, initially to a small audience, which grew at the rate of a million people a week, so that the final episode had 80 percent of the TV-viewing audience in Britain watching. The groundbreaking series would have become an all-time classic but for an act of folly at the BBC, which threw out the films in a clear-out in the 1970s.

A for Andromeda did not use all of the story lines from the 90-minute tape recording, so the script department invited Fred to prepare outlines for another series, before *A for Andromeda* had aired. Characteristically, Fred dashed off some ideas in a hurry, recording 60-word outlines for seven programs. The script department did not like the new offering one bit. According to an internal source:

> I have read the transcript of Hoyle's proposed sequel to *A for Andromeda* and I think we must tell him it will not do. He has failed to provide us with what we expected from him, namely a new and exciting science fiction basis for a serial. What we have instead is an intellectual exercise on cops and robbers. We must tell Hoyle this will not work. We must meet him and ask him to find an entirely new idea.

John Elliot met Hoyle at his Cambridge home on the afternoon of December 19, 1961. Elliot agreed to assume a larger role in script development, to meet the criticisms being leveled at Hoyle. In the resulting sequel, *Andromeda Breakthrough*, the heroic scientists follow the trail of Andromeda to the Middle East and discover the real force and meaning of the messages from space picked up by the computer. The BBC's announcement in the *Radio Times* for the first program, transmitted on June 28, 1962, concluded with the following statement: "Film commitments make it impossible for Julie Christie to play Andromeda again, so the part is taken by Susan Hampshire." The reason for this change in casting was financial rather than creative. The BBC had refused to give Julie Christie a £300 option payment to retain her services for the sequel, so she had signed with another producer. When the BBC had made its decision on financial grounds, it could not have foreseen the exceptional ratings for the first series. Fred Hoyle thought the second series was a near disaster: He greatly disliked the political dialogue that Elliot concocted.[27] The second series failed to live up to the promise of the first in the ratings war. Fred's brief scriptwriting career was now over.

All the fame and glory now surrounding Hoyle did not play well with his academic colleagues in Cambridge. Arguably, he had attracted too many research students: In 1960–1961, he recruited five new students to add to three ongoing ones. The new students got off to a gentle start, partly because Hoyle was away for the Michaelmas term. To cope with his absences, Hoyle used grant funding to hire Roger Tayler in 1961, in part as a mentor. Tayler was a fine and capable supervisor who became a lecturer in 1964. Some other students, initially attracted by Fred Hoyle, actually worked instead with Leon Mestel or Dennis Sciama: Stephen Hawking arrived in this context in 1962.

Although Hoyle arranged cover for the teaching of graduate students, he failed to appreciate that his frequent and prolonged absences in Pasadena cut him off from the politics of university life, which in Cambridge can be devious and dirty when the going gets rough, and his travels were resented by a couple of students who felt fobbed off. Hawking is on record as saying that, on being assigned to Sciama rather than Hoyle, he felt deeply disappointed. However, when he saw how little time Hoyle spent with students, he changed his mind about that.[28] The university had opened the new Department of Applied Mathematics and Theoretical Physics (DAMTP) on October 1, 1959. Until then, the 18 academic staff, including Hoyle of course, had conducted their research in college rooms. Consequently, the staff had only limited contact with each other at a time when scientific progress was becoming increasingly reliant on networks and collaborative efforts. A novel provision in the arrangements for the new department was that its head should be appointed for 5 years at a time, chosen from among the senior staff of the department. This followed best practice in the United States and was a deliberate departure from the British habit of appointing the head with life tenure. The university regulations did not specify that the post holder *must* vacate at the end of the quinquennium, but they obviously allowed the university to renew the incumbent if the senior staff felt so minded.

The first holder of the headship was George Batchelor, then reader in fluid mechanics.[29] Hoyle was enormously put out at this choice: He and Paul Dirac were full professors but had been passed over. So badly did he view the appointment that he even went through a patch of regretting that he had accepted the appointment to the Plumian Professorship, which the university assigned to DAMTP. In Hoyle's eyes, Batchelor was a control freak when it came to deciding who could carry out what research; the department was being run as a personal fiefdom, its resources channeled away from the astronomers whenever possible and into the hands of the fluids experts. Whether these charges were valid is neither here nor there: The fact is that this is how Hoyle felt.

Nevertheless Hoyle remained loyal to Cambridge in the early 1960s when he could easily have secured a prestigious appointment at an American university. He liked the English summers and cricket too much. Besides, the clock was ticking and, in 1964, Batchelor could be rotated off the headship. Or so it seemed.

The relationship between Batchelor and Hoyle went from bad to worse. Hoyle had become very disenchanted with the department. He went in to give lectures and to see graduate students but, apart from that, settled in his armchair at home, writing on his knee. He some-times preferred to see his graduate students at home rather than in the department. They just needed to check beforehand that a home visit was not inconvenient. Barbara would not let unannounced visitors interrupt Fred's research, but she was always famously warm and hospitable to visiting students: tea and cake would appear, later to be followed by strong martinis and hearty dinners. She relished promot-ing the interactions and engagements between the young students, the brightest of their generation, and her husband. These occasions also brought out the very best in Fred, who was a man of the greatest gener-osity, charm, and humor when not under attack. The Hoyles ran an old-fashioned academic household as a place of hospitality, relaxation, and scholarship.[30]

In the department, Batchelor maneuvered for reappointment as head in 1964, arguing that the university had drafted the regulations to permit such a renewal. Hoyle would not accept this. He dug out his copy of the minutes of university meetings in 1959 and marched off to the vice chancellor to voice a grievance. Conveniently, the master of St. John's College happened to be the vice chancellor that year. Hoyle jabbed his finger at a minute recording George Batchelor's view, expressed 5 years previously, that the headship should rotate. The vice chancellor sympathized but gravely explained that a reappointment would not contradict university regulations. Hoyle "disliked the atti-tude that clear verbal agreements could be ignored." He reminded himself that, in his childhood, the cloth trade of Bradford proceeded on a handshake, without legal documents. If businessmen could be made to stand by their word, why should those running an ancient university be exempt from such courtesies?[31]

In any event, the department decided to hold an informal election. There were two candidates, Batchelor and Hoyle. Leon Mestel acted on behalf of half a dozen colleagues to get personal statements from both candidates. He wrote to them on May 6, 1964, soliciting their views on three points: the long-term development of the department, its relation to other university departments, and the role of the head of the department.[32]

Hoyle responded the next day, blue fountain pen on ruled foolscap, having first worked feverishly on a heavily corrected draft in pencil.[33] On long-term development, he proposed to reduce the workload of teaching staff and undergraduates, commenting that "the pressure on undergraduates is now absurdly great. . . . Teaching should be addressed to the average student." Hoyle felt that constraints on staff should be as small as possible. Considered as an election manifesto, the document is decidedly lightweight, lacking in punch and passion, and bereft of any exciting policy. By contrast, Batchelor's paper was like a policy document from a university officer well versed in the Cambridge committee system and expert at arguing a case on paper. On development, he looked forward to the department moving to a large building that the university press was vacating on Silver Street. With more office space, he looked forward to an increase in the number of researchers in general relativity (Stephen Hawking had just joined as a research student) and plasma physics. In an aside directed at Hoyle, he commented that the new building would enable him "to balance the centrifugal tendencies which are a natural consequence of our different scientific interests." Batchelor concluded with a provocative statement that "an enforced change of Head every five years would be wasteful of resources."[34]

In the ballot, Hoyle lost massively by 15 votes to 5. His feeble showing instantly set alarm bells ringing in the university, which feared that Hoyle would resign in a huff. The university confirmed Batchelor as head on May 25, 1964. The theoretical physicist, John Polkinghorne, wrote a note to Mestel that must have been typical of how everyone felt:

> I have been trying to get in touch with you today to find out how Fred was reacting to yesterday's decision. I should of course be very sorry if he decided to leave Cambridge. The brunt of trying to persuade him to be sensible will fall on the Astrophysics group. If there is anything I could usefully do to signify my very real esteem for his scientific talent I should be glad to do it.[35]

Hoyle was deeply annoyed with the vote, which he misinterpreted as a vote of confidence. He resolved to resign his chair at the end of September 1964.[36] Next he scribbled a furious letter to Batchelor:

> Let me say that my own opposition springs, not from any single issue of a departmental kind, but from a deep-rooted divergence as to how people should be treated. Over the past years I have watched the situation in the department with some care and have come to the conclusion that you have manoeuvred and moulded the views of the people around you to a degree that I regard as inadmissible. . . . If on Monday your party engages in a legal but distinctly unpleasant device you must not expect me to believe in protestation of innocence on Tuesday.[37]

Hoyle cleverly planted an article favorable to himself in *The Sunday Times* on July 19, 1964. Although the published story is attributed to an in-house columnist, it matches word for word a surviving draft in Hoyle's handwriting![38] He also wrote in vivid terms to the senior academic administrative officer in the university, complaining that Batchelor had strong-armed the younger members of the staff to get their votes. He concluded: "Because of my deep affection for Cambridge I find the whole affair painful and agonizing in the extreme." With that, Hoyle stormed out of the department, membership of which he had effectively resigned. Henceforth the Plumian Professor would work at home or in the library of the Observatories. In any case he had an escape route, of sorts.

When Hoyle had taken up the appointment of Plumian Professor, the prestige of this Cambridge chair ensured that he would be offered membership of influential national committees. The Department of Scientific and Industrial Research, a government agency then charged with funding scientific research, appointed him chairman of the sub-committee on astronomy. (It was in this role that he had secured the grant of £50,000 for the interferometer at Jodrell Bank.) Early in the 1960s, he conceived the idea of a national center for theoretical astronomy. His American experience had taught him that, if the UK wished to start competing with the United States, only an institute at the national level could hope to rival the astronomy schools at Harvard, MIT, Chicago, Berkeley, and Caltech.

His concept was that a group of up to 25 theoretical astronomers should be supported by government funds but be resident at a single university. When he had these thoughts, UK government support for astronomy was biased in favor of observational astronomy, partly because of its responsibility for the historic Royal Greenwich Observatory, but also because the older universities with Victorian observatories supported astronomy through the telescope more vigorously than cosmology on the blackboard. When Hoyle had embarked on theoretical astronomy, the number of professionals was a handful. In the late 1950s, it remained true that applied mathematicians held most posts in theoretical astronomy.

At Cambridge, Hoyle gathered support for an institute from an important friendship dating from the late 1940s: Alexander Todd, a distinguished chemist who single-handedly scooped the Nobel Prize in 1957 for his work on nucleotides. He was knighted in 1957 and elevated to the House Lords, as Lord Todd, in 1962. In 1959, Todd chaired an advisory committee on scientific policy for the British government. He contacted Hoyle, requesting a proposal for improving the national situation in theoretical astronomy. Hoyle characteristically made a short and clear response: Government assistance should be made available to establish such a group, freed of teaching responsibilities, at a university, perhaps Cambridge. Todd's committee immediately supported this, so Hoyle expected action.

However, the committee was merely advisory: It had no money of its own to fund Hoyle's pipedream. To get the cash, he would first need to go to the National Committee for Astronomy. This was a talk show under the aegis of the Royal Society, chaired by the astronomer royal, Sir Richard Woolley, a distinguished but old-fashioned scientist. Woolley did absolutely nothing for 6 months. Was he suspicious that a national center for theory would come to rival the Royal Greenwich

Observatory? We simply do not know, but Hoyle suspected the inaction was designed to fend him off.[39] Eventually, the national committee did meet in late 1960. When it did so, it made the classic committee decision: Let's set up a subcommittee to advise us. The subcommittee did its work very thoroughly, although Hoyle did not see it like that. They conducted interviews with theoretical astronomers all over Great Britain. The outcome agreeably surprised Hoyle: There should be a national institute with a budgetary provision of £3 million, an immense sum in 1961. The committee asked Hoyle to ascertain if the University of Cambridge would welcome the setting up of the institute there. Hoyle immediately commenced the task of convincing his friends and opponents. He fired off a memo on September 29, 1961, asking: "If funds for such an Institute were to be made available, would Cambridge University be agreeable to providing a home for the Institute?"[40] He pointed out that, although several universities would actively welcome the institute (Sussex very much wanted it), the national committee felt Cambridge to be "the most appropriate location, in view of the existing active astronomical groups."

The decision on whether to offer a permanent home for the institute in Cambridge rested with the university's General Board of the Faculties, which had to consider the long-term funding. As far as Hoyle was concerned, the board seemed to take forever to decide the matter. Hoyle's institute was not the only difficult matter before it in 1963. Professor Ray Lyttleton, together with an influential group of scientists, was lobbying to change the rule that undergraduate applicants to the university must have a qualification in either Latin or Greek, with no exceptions. This requirement could easily be satisfied by candidates from elite private schools but was proving more difficult for candidates from grammar and state schools. Lyttleton liked to say, "What use is it to a mathematics student to read a lot of stuff about golden bulls?" One of the influential scientists instructed a personal secretary to send Lyttleton the batch of papers concerned with the General Board's deliberations on Latin and Greek. By mistake, this secretary went to the wrong part of the filing system and sent Lyttleton highly sensitive and confidential papers on the General Board discussions on the proposed Institute of Theoretical Astronomy. It is possible that

Lyttleton later invented this account in order to protect a colleague who had in fact leaked the documents to him. Either way, I can well imagine the great glee on Lyttleton's face when he read the explosive contents, which he immediately disclosed to Hoyle.

Thus, when the secretary of the General Board summoned Hoyle to its administrative offices at the Old Schools in the heart of Cambridge, he already knew the bad news, which the secretary now delivered in a sepulchral fashion: "Hoyle, the university will have nothing to do with this institute, no way, full stop, end of story."[41] The batch of papers that improperly fell into Lyttleton's hands had informed Fred that the distinguished mathematician Sir William Hodge had spoken at length. As physical secretary of the Royal Society, a member of Todd's national committee, and the holder of a chair in astronomy and geometry, he spoke with authority. His colleagues from the arts and humanities listened intently as he argued that such an institute would surely become a massive financial drain on the university despite the generous external funding in the early years.

The minutes of the meeting held on February 10, 1962, give a summary of the 90-minute presentation Sir William made to the board. He argued persuasively for giving greater support to observational astronomy, and he spoke against any special favors for theory. He had chaired the Cambridge Observatories Syndicate[42] for 30 years, so he knew what he was talking about when it came to the need to support observational astronomy. He had a high reputation for making difficult decisions objectively and fairly. As the college chapel clocks struck the eleventh hour, the board agreed on two points: that the remaining domestic apartments in the Observatories should be converted for research use and that they would "recommend that the university should not encourage the establishment of an Institute of Theoretical Astronomy in or near Cambridge," because of fear that the costs would eventually fall on the university.[43] In an act of self-deception, an angry Hoyle wrongly attributed to Hodge an agenda of dirty tricks, composed out of sheer spite.[44] He could tolerate his situation no longer.

At the end of September 1964, Hoyle submitted his resignation to the vice chancellor. Barbara and he then set off to the Lake District for a week's holiday. When they returned home, the vice chancellor's response was on the doormat and the phone was ringing. The vice chancellor could not report the matter to the General Board until the beginning of the Michaelmas Term. Would Hoyle regard the letter as being in abeyance until then? Pretty soon, Hoyle had representations from respected older friends: Dr. (later Dame) Mary Cartwright (Mistress of Girton) and Sir Alan Cottrell.[45] Would he reconsider if the obstacles to bringing the institute to Cambridge were removed?

Lord Todd trekked over to Clarkson Close with Sir John Cockroft. Todd, recall, never wavered in his support for the institute in the period 1960–1963. Both men had talked to the vice chancellor: The university would welcome the institute provided it was scaled down. This suited Hoyle fine: He always believed that the Royal Society had elevated the scheme to a grandiose level. The amount of money needed was about £750,000, which Todd and Cockroft would secure. The Nuffield Foundation chaired by Lord Todd, the Wolfson Foundation chaired by Sir John, and the new UK Science Research Council would provide the funding in approximately equal shares. The Wolfson Foundation would pay for the new building, Nuffield would pay the salaries, and SRC would buy the computer. On October 16, 1964, the Labour Party under the leadership of Harold Wilson, a West Yorkshire man a year younger than Hoyle, won the British general election in a campaign that promised funding for a "white-hot scientific revolution." With this pledge of new funding, Hoyle could not now go against the advice of respected colleagues. He withdrew the resignation and flew to Los Angeles for a 2-month stint at Caltech.

By January 1965, he was fretting again. Nothing had happened while he was in California and there was a simple reason for that: Hoyle needed to make official applications, an exercise he seems to have found uninspiring. Actually, the application to the Nuffield Foundation was easy: He dusted off a submission he had prepared in 1960, updated the salaries, and popped it in the post. The SRC paperwork was trivial also, amounting to no more than a letter specifying what computer would be needed.

Hoyle took the design of the proposed building remarkably seriously. The building he desired would replicate the Institute of Geophysics and Planetary Sciences at the La Jolla campus of the University of California. Hoyle's earliest sketches have survived. He planned a building 360 feet long and 60 feet wide, with a 3-yard central corridor. He specified double glazing (then a great rarity in UK construction), huge windows facing east and west, large blackboards, excellent sound-proofing and silent floors, a library to store books but with no facilities for studying, and a decent seminar room. On one side were 21 identical offices, each designed for one staff member, but each with the capacity to double up with a visitor in the summer. And the assistant staff would have IBM golfball typewriters, capable of handling complex math; these had much impressed him at Caltech.[46]

The matter of the building's location needed careful thought. The university suggested the sheep-grazing field in front of the Observatories as a construction site. Fred surveyed this plot on a cold Sunday morning in February 1965 but found that the building he had designed was too large for the site. His son Geoffrey pointed out a nearby strip of spindly copse running parallel to the Observatories' boundary. They trudged through the neglected woodland with its slender larches. Most of the trees would have to go, but the building would fit nicely.

The university had bad news when Fred suggested this location. The real estate was encumbered with covenants held by Trinity College, from whom the freehold would need to be purchased. The next morning Fred had an meeting with the college bursar, who displayed a wicked sense of humor. "Yes you can have the site but the college Fellows are attached to that woodland. It will take a lot of extra fine port to keep them happy. The exercise is going to cost you £10,000." With this information, Hoyle set off to the university finance office, whose normal approach to balancing the budget was just to say "no" to such requests. In any event the assistant treasurer listened to Hoyle's plea, swallowed hard, stared out of his office at King's College Chapel, and said that since the university had already agreed to provide the land for the institute, he would find the £10,000.[47]

At the Department of Applied Mathematics and Theoretical Physics, meanwhile, Batchelor ordered a circling of the wagons. Seemingly endless discussions went on in university committees throughout 1965. Working groups packed with knights and lords considered the constitutional and management issues. Batchelor's concerns were understandable: The department would lose staffing positions through transfers to the new Institute of Theoretical Astronomy, and it would be impossible for the remaining staff to shoulder the teaching burden. Batchelor spearheaded a noisy campaign to ensure that the constitutional arrangements for the institute would not undermine the teaching of mathematics.

Through 1965, Hoyle fought a losing battle over the constitutional issues. He wanted the institute to be a university department but, at the same time, to stand outside the university's statutes and ordinances. He wanted a small board of six trustees to hold the director accountable, not an interfaculty committee of 15 representatives. There was never any chance that the university would set up an institution that it could not control through the statutes and a management committee. Hoyle greatly antagonized his enemies and exasperated his friends by holding out for a unique constitution that would enable him to direct the institute by post from Pasadena for several weeks a year and from his home the rest of the time.

However, he had no doubt that he would win the day. In April 1965, the university conceded that it would cover one-fifth of the salaries, and by May 1965 he had offers of £800,000 from the three donors. With the money on the table, he flew to the United States in June on a recruitment campaign, with the full knowledge of the vice chancellor. His method of hiring was to go to people he respected and offer a job on the spot.

He could not tempt Fowler or the Burbidges to take permanent positions, but they agreed to lend support through regular summer visits. John Faulkner and Peter Strittmatter accepted, according to Hoyle, although neither now recalls such an event happening as early as 1965.

Hoyle came back from Pasadena for the Lent Term of 1966 in a feisty mood, buoyed up no doubt by the strong enthusiasm Willy Fowler had for the institute. On January 20, he complained to the vice chancellor about the management arrangements for the institute. He complained bitterly that draft regulations for the institute that had been agreed with him were changed in committee while he was in Pasadena. As he put it, "The revised draft regulations, instead of being approved by the General Board, have been mauled over by various other bodies with the result that we are back now with a worse state of affairs." His bluntness (which he attributed to "my Yorkshire background") was apparent in a set of four demands, for which he said the only response acceptable to him would be "a plain yes or no." His conditions were that the ponderous discussions of committees must cease; the institute should be governed by six trustees, not an interfaculty committee; the university must ignore the fuss being whipped up by the mathematicians; and the clear yes or no must come by the end of the term.

We do not know how the vice chancellor reacted to this tactless letter because, 4 days later, Hoyle raised the stakes to a stratospheric level with an even more inflammatory missive. After he had sent his letter of January 20, someone tipped him off that the General Board had *already* accepted the hated revised draft. The individuals responsible for the acceptance of the revisions wrongly assumed that if the haughty Hoyle were presented with a *fait accompli,* he would cave in. They misjudged his doughty spirit, which now led him to make a shatteringly disloyal move.

Hoyle set to with pencil and paper, producing the most heavily corrected draft of any of his letters. The crossing out is so heavy that the indentations run through several pages, and every page is littered with balloons and arrows as he moved the text around. My guess is that it took him most of a morning to write. Barbara typed the letter and at least five carbon copies.

What message did the letter convey? He took it very seriously indeed that the university had failed to involve him in further amendments to the regulations, pointing out that the Science Research Council had explicitly made its grant "specific to Professor Hoyle and Cambridge University." He chose to interpret this as meaning that the conditions under which the institute would operate must be acceptable to him as well as to the university. He let the firecracker loose toward the end of the letter with: "This manoeuvre has now convinced me that Cambridge is not the right place in which to set up this Institute. The aim of the Institute was to deal with problems of some magnitude and grandeur. This cannot be done in an atmosphere of petty intrigue."

He enclosed copies of a three-page letter that he had already sent to the donors, letters that began, "You will be sorry to learn that negotiations with Cambridge . . . have reached the stage where a halt should now be called." Clearly, he had lost his temper and I suspect that is how his letters were interpreted. The vice chancellor and his officials would have been greatly dismayed because research councils and foundations were heaping grants on the university and colleges. There was potential here for a loss of face and the possibility that donors might question the trustworthiness of the university itself. The matter seems to have been resolved through phone calls and hastily convened meetings. The conclusion, reached in the summer of 1966, was to give Hoyle the management structure and constitution he wanted. He had finally defeated the opposition in Cambridge, but had created many new enemies in the process.

Construction started on August 1, 1966, with the building opening exactly 1 year later. Hoyle took an intense interest in the details. He toured brick makers all over eastern England to select the bricks; his choice led to problems of matching when the building was extended. His wife arranged the planting of huge drifts of daffodils, a tradition that has continued with dazzling displays 40 years later. Every office had an armchair just like the one he used at home. The delay caused by interminable discussions had a beneficial effect on the costs of the computer (an IBM 360/44) which decreased by £100,000. A few months after the opening, he recruited his student friend and wartime colleague Captain Frank Westwater, then a director of a major manufacturer of mainframe computers,[48] to manage the institute on a day-to-day basis.

He had to recruit the research staff. The experience of John Faulkner can serve as an example of how Hoyle populated his new institute.[49] Faulkner is one of many of his generation who, at age 12, listened to the BBC's *The Nature of the Universe*. His response to these broadcasts was a confident announcement to his bemused working-class parents that he would go to Cambridge to do astronomy with Fred Hoyle. They took it no more seriously than the commonly stated ambition for boys of his age and social standing to become the engine driver of a steam train, the pursuit train-mad Willy Fowler loved as a leisure activity. Faulkner won admission to Cambridge and took

Hoyle's electromagnetism classes in 1958. In Part III of the Mathematical Tripos, Faulkner scored the top grade, Pass with Distinction. When the results were published, Batchelor summoned the applied mathematicians to his office where he made this announcement: "Professor Hoyle would like to see the following students at his home tomorrow morning [a Saturday]: 9:00 a.m. Narlikar, 10:00 a.m. Faulkner, and 11:00 a.m. Chitre." Each of them would be awarded a full grant as departmental research students. Hoyle had "simply cherry-picked the top three finishers with professed astrophysical interests" according to Faulkner.[50]

At the end of his first year as Hoyle's graduate student, Faulkner attended a summer school at Lake Como, Italy, where Fred (by now the most famous astronomer in Europe) was one of the lecturers. It was the type of European academic meeting that allowed plenty of time to appreciate food, wine, and the local attractions, one of which is swimming in the lake, fed by melt water from the surrounding snowy mountains. Late one muggy afternoon, Barbara Hoyle returned from a shopping trip to Milan, somewhat wilted by the heat, and looking forward to a swim. Although the rest of the squad had already taken their daily dip in the lake, she plunged in on her own, swimming out 50 yards. A strengthening light breeze began to make the lake a little choppy. By now everyone had retreated indoors, apart from Faulkner, motivated by a mysterious foreboding. Suddenly a cramp gripped Barbara and she panicked. Faulkner leapt in and swam out to Barbara. First he calmed her down, and then brought her back to shore, swimming backstroke with one arm around her. Most of the onlookers thought they were engaged in a harmless sexy frolic, but one realized it was serious when Faulkner started to flag while still 10 yards from the shore; he dived in and relieved Faulkner. Saving your supervisor's wife from death by drowning must count as a unique introduction to postgraduate life. His act of bravery gained Faulkner admission to Hoyle's inner circle, with invitations to family holidays.

As a student, Faulkner found Hoyle immensely inspiring. This motivation came from Hoyle's extensive knowledge of cosmology and astrophysics, his identification of the significant problems in so many areas, the fecundity of his imagination, and his suggestions for tackling research problems. To this list Faulkner adds that the students got inspiration from "the wide and deep international network of academic colleagues who clearly held him in such esteem."[51]

How did Fred Hoyle interact with his many students? Did he take the "sink or swim" approach so prevalent in Cambridge at that time? Well, yes and no. From the start, he treated his students as equals, then with an amazingly progressive attitude. Whenever he gave a lecture in the department, he popped in to see his graduate students, but this was short of Rutherford's policy of seeing students every day, particularly in the terms when he gave no lectures. As we have seen, they were always welcome at his home, by prior arrangement. But during the Pasadena absences, they had to manage without his advice.

When he did work directly with students, they noticed his mind never stopped thinking about problems, snags, revisions, work arounds, solutions, . . . all in a restless stream. Perhaps he felt like Newton, who said, "I just hold these matters continually before me, until gradually they emerge into the light." His method was that of a parallel processor. Hoyle would be thinking of several problem areas for his several students, an amazing feat. He would work overnight, trying perhaps three or four approaches. When he and the student met the following day, Hoyle would present the problem from a new angle, sometimes assuming that the student would also have arrived at the same logical conclusion overnight. Of course, that hardly ever happened, with the result that the student would battle to keep up with Hoyle's evolving thoughts. But that is why working with Fred was such an amazing experience.[52]

Fred had a number of mannerisms over and above his Yorkshire accent that his students humorously adopted. For example, Fred had a particular way of starting a conversation or answering the phone by saying "Oohh, Helloow," while at the same time rubbing his nose and then gesturing with his forefinger. Faulkner and the rest got the imitation of this down to a fine dramatic art: They would greet each other in the street with a "Oohh, Helloow" and the accompanying body language. Faulkner and Narlikar continued this little antic into the later years of their careers, claiming that you could always tell if a Fred student was down the hallway by listening for "Oohh, Helloow." One day, though, Faulkner overstepped the mark, when a phone rang and he imitated Fred. On the other end of the line a surprised Barbara Hoyle said, "Fred, what are you doing there? You said you would be out of town!" She heard the sound of silence as Faulkner froze in embarrassment.[53]

While visiting the Burbidges at La Jolla in Easter 1966, Hoyle recruited John Faulkner and Peter Strittmatter (a former student of Leon Mestel) to join the foundation staff of the institute. Taking them for a long stroll along the beach, he outlined his plans, accompanied by the roar of crashing surf, and invited them to return to Cambridge with jobs at the new institute, commencing in September 1966. To John and Peter, it seemed a wonderful opportunity. Back in Cambridge, Hoyle confirmed his offer in two handwritten letters on blue aerogram forms, and the two young men accepted. Willy Fowler strongly urged caution, suggesting that they defer the offer for a year, until they could be sure the institute was fully functioning.

They both paid the price of ignoring Fowler's counsel. On their return to Cambridge in September, the institute's walls stood at a height of 1 foot. The university still had not managed to get its own legislation through its own committees in time. Construction only commenced when Hoyle went straight to the builders Rattee and Kett and told them to lay the foundations at his risk. Professor Redman, in the Observatories, usefully came to the rescue by allocating to the embryonic staff space in a wooden hut surrounded by grazing sheep. At least they could indulge in atrociously played cricket matches on the sheepfield.

Fred produced no contracts of employment for the fledgling staff. Instead, bodies and souls were kept together through the generosity of Peterhouse, the oldest college, which had already awarded short-term fellowships to Faulkner and Strittmatter. The first wave of summer visitors, the Burbidges and Fowler, benefited from largesse and accommodations arranged by Sir John Cockroft at Churchill College. The university finally managed to issue employment contracts by January 1967, thereby setting the tone for another argument. In his airmail letters, Hoyle had offered stipends of £1,830, the starting point for a lecturer; however, Faulkner and Strittmatter were not yet age 30 (the minimum age for a lectureship), which meant they should start on £1,400, equivalent to assistant lecturers. Straightening this out would have taken several rounds, but Hoyle landed a winning blow in January 1967 to pay £1,830. One chilly morning the pair set off to the financial offices to claim their costs in relocating from the U.S. West Coast to Cambridge. Unlike some universities, Cambridge had a generous policy on relocation costs, paying 100 percent of the expenses. The official who met them must have been in a state of despair: Would Hoyle's demands for extra money never cease? Their reasonable oral request was met with an artificially shocked, high-pitched, typically insincere and supercilious tone: "I'm sorry, but I'm afraid I simply don't understand. You were already present here in Cambridge [as fellows at Peterhouse] when your appointments began. Therefore, the question of travel costs to take up your employment simply doesn't arise." At which they really did depart empty-handed. They generously decided not to burden Fred with any further complaints about finance.

When the completed institute finally threw open its doors in August 1967, its sweeping program was finally "full steam ahead," as Fowler liked to say. By the summer of 1968, staff numbers had ramped up to 35, including assistants. The institute put itself on the international map with its first summer conference on cosmic rays, organized by Faulkner. In the official photograph he sits cross-legged, center stage; Hoyle had no problem with young Faulkner asserting the *droit d'organiseur*. This started a tradition that has continued ever since: running a world-class conference with good funding on a cutting-edge research topic. The other summer event was the visitor program. Hoyle (and his management committee) arranged for up to 30 summer positions a year, all well paid, and for up to 4 months at a stretch. This brought some of the best theorists in the world to Cambridge on a regular basis. I would argue that the visitor program was important in giving British theorists the confidence to engage in international collaborations, something uncommon then, 15 years before the Internet. As director, Hoyle left the scientific staff completely alone to establish their own programs, recruit their students, and build their networks. He was frequently in the institute, and curious about everyone's research, but he did not interfere. This hands-off approach, so different from George Batchelor in the center of town, appealed to Hoyle and his talented research staff who could do what they liked. By now, Lyttleton did absolutely nothing of significance and lectured to a wholly empty classroom. Even Fred left him alone.

The bait to attract the finest minds to staff and visitor positions was time on an IBM 360/44 computer, which in 1967 was one of the most powerful mainframes in any university anywhere. IBM's 360 series transformed commercial and academic computing: They were the first to be mass produced and came with an operating system suited to the normal user. This meant that individual university departments could consider having their own computer and operating staff, rather than relying on a central computing service. The 360/44 was in many ways the oddest in the IBM series because of modifications made to optimize it for scientific work, but the ease and convenience of an onsite computer under the institute's control proved an irresistible magnet.

Barbara Hoyle understood the importance of extending a warm welcome to the visitors, especially the ones from the United States. She personally shouldered the burden of finding furnished houses to rent for those Americans who brought over families for the summer. She trained Fred to be attentive to the practical needs of the visitors. When Don Clayton of Rice University arrived in London with his young family for the summer of 1967, Fred arranged for a VIP private limousine service to meet them at Victoria station and whisk them along the old A10 road to 15 Storey's Way, Cambridge. They reached the White Cottage at 10:30 p.m., their children exhausted. Clayton remembers:

> Even at that hour Barbara Hoyle was there to greet us and to introduce us to our housekeeper, Kathleen Bilham, who had lived in the house as housekeeper for two decades. Barbara Hoyle was bubbly with friendly support as we put the boys straight to bed. Finally, after we were totally enchanted by the cottage, Barbara Hoyle turned to us anxiously and said, "Oh, I do hope it will be all right", meaning the house and housekeeper. She had found it for us herself.[54]

The institute's instant success also derives from Fred's inspired decision to hire Frank Westwater as the manager, thus letting himself off the hook as far as personnel and financial matters were concerned.

Westwater took a considerable cut in salary to help his friend from undergraduate times. He ran all of the day-to-day affairs, much to the relief of the central offices of the university. Importantly, he maintained excellent relations with the numerous interlocking committees in the university, as well as outside organizations. He recognized right from the start that the institute's finances were shaky because the university would have to take over all of the running costs after the first 5 years. Turning on the money tap would require careful lubrication of the university machinery and management of Fred himself, to ensure that he attended the really vital university meetings.

Unfortunately, a nasty rift developed between Frank Westwater and Barbara Hoyle. Both passionately wanted the institute to succeed and both worked tirelessly to that end. Both achieved their ambitions for Fred. Westwater viewed Fred's poor relationship with the university authorities as a cloud over the future. Captain Westwater's naval instincts led him to believe that the good ship HMS *IoTA* would founder unless he issued clear instructions to Commander Hoyle, an approach that was at odds with Barbara's protective instincts. She was Fred's helpmate, interposing herself between him and the world so he could think in peace. She felt that many people were trying to undermine Fred, which was true, so she simply pulled up the drawbridge and dropped the portcullis. Inevitably this led to sharp conflicts between herself and Westwater. Barbara, for example, insisted on appointing Fred's personal assistant, who was required to make a lunchtime phone call to 1 Clarkson Close reporting on the comings, goings, and general chitchat.[55]

Whatever the office gossips had to say, Westwater always acted honorably. When Faulkner declared his intention of returning to California after 3 years back in England, Westwater asked if he would mind sharing his reasons. After a good discussion they swapped IoTA "horror stories." Always loving a good yarn, Faulkner told the sorry tale of the phantom jobs and nonappearance of salaries and expenses for himself and Strittmatter when they had first joined IoTA. An astonished Westwater was hearing this saga for the first time—the events having taken place before he joined the staff. He immediately volunteered that he could do nothing about the unpaid salaries but he could right the wrong over their relocation expenses. He suggested they both draw up a detailed account of their earlier expenses, which the university paid in full without fuss.

Within weeks, Westwater died of complications following prostate surgery, Eddington's curse. He left an institute that would blossom into one of the greatest astronomy centers in the world.

CHAPTER 10

Mountains to Climb

n the mid-1960s, relentlessly taunted by his opponents, Hoyle was a caged bird. Devious and envious colleagues, desperately searching for a final stitch-up, flapped and screeched at the gaudy parrot who had so much to say. They were blackbirds defending their territory. After Hoyle withdrew his resignation in October 1964, he still felt trapped by the pecking order in Cambridge politics. In an effort to free his mind of turmoil, he decided to graduate from hill walking in the Lake District to mountain climbing in the far-flung regions of the Scottish Highlands, 500 miles north of Cambridge. Here his mind could soar as a free spirit. The Scottish mountains are dangerous places, not to be explored alone under any circumstances, so he engaged the services of one Dick Cook, president of the Lake District Fell and Rock Climbing Club, whom he knew.

In spring 1965, they drove to Inverness, accompanied by a third climber, Norman Baggaley. A blizzard that blanketed the eastern Highlands in deep snow marked their arrival. Out west the snow was not so bad, so the following morning they headed southwest along the already ploughed road through the Great Glen. After much skidding and sliding, Fred's Humber Hawk nosed along the shores of Loch Duich, which he had visited in the summer of 1935 on his first Highland hike. His mentor Dick Cook wanted a mountain he and Norman had not tried before. Cook pored over his maps, alighting upon a mountain called Moruisg, 20 miles to the north, on the edge of a great wilderness of barren peaks. In summer, this mountain has grassy slopes, the going is easy to its peak at 3,045 feet, and there are no steep sections. Late spring 1965 was altogether different. The trio's ascent was not too challenging, apart from intense cold in the last 1,000 feet. By contrast, the descent, in the face of a furious icy wind, was awful. Hoyle's spectacles iced up (he never used contact lenses). An ice storm drove a hail of darts at his face. Grimacing in pain, Fred chewed on the cord of his anorak hood to prevent himself from biting his tongue. When he reached the road, he found he had chomped clean through the nylon cord. He had fought his way through the worst mountain wind he had ever experienced. On such an afternoon, while he battled primeval elements, the Plumian Professor could dismiss Cambridge University politics as a trivial pursuit.[1]

In the following days, they scaled several peaks, unaided by the ice axes or crampons that Hoyle carried routinely on all subsequent winter climbs. On this trip, they took advantage of the direct heat of the late spring Sun, which melted the surface ice by the early afternoon. Confidently, their boots crunched through the upper crust to the soft snow beneath. But they put themselves in considerable danger by, for example, scaling the Five Sisters of Kintail, which lie at the head of Loch Duich. To conquer these peaks, all over 3,000 feet, they had a very steep climb from the valley floor to reach a ridge. Traversing this ridge, they faced a 2,000-foot precipice on one side, with ice and projecting rocks all the way down. A weak Sun shone from a cloudless azure sky, bathing the mountains around them in a blaze of light. Fred loved days like this.[2]

One evening, after a hearty Scottish dinner, Dick Cook opened a slim book, which he leafed through carefully and silently, pausing to enter a single tick mark by pencil here and there. His curiosity aroused by this silent armchair activity, Fred asked to see the book. Dick handed over the 1953 edition of *Munro's Tables*, from an original compilation in 1891 by Sir Hugh Munro.[3] Fred learned that a "Munro" is a hill in Scotland that is over 3,000 feet (914.4 meters) high. The list, from the very beginning, has classified such points into "separate mountains" and "subsidiary tops." Where a ridge has several summits all over 3,000 feet high, it is a matter of opinion how many are mountains and how many are separate tops. That is why Sir Hugh's list is important: It tabulates the separate mountains, currently 284. I doubt that Fred Hoyle ever allowed himself to be sucked into the problems of classification as applied to Scottish mountains, but his competitive spirit was stimulated when Dick Cook explained that he was in the process of climbing every Munro. That is why he had selected Moruisg, which Hoyle noticed that he had just ticked off in his little book.

On his return to Cambridge, Hoyle bought his own copy of the book and unfolded his Ordnance survey maps. He found scores of mountains that, until then, he did not know existed. The Munros he had just scaled had provided some splendid days with magnificent views. He resolved, in 1965 at the age of 50, to climb all 284. Scaling every Munro would be his substitute for being a free bird.

Despite his doubts and misgivings about continuing his caged lifestyle in Cambridge, Fred Hoyle somehow transformed professional astronomy in the UK during 1968–1972. By example, he masterfully led the way in showing inward-looking groups and cliques how to morph into a powerful international community, well able to punch above their weight on the international stage.[4] His influence extended well beyond Cambridge, geographically as far as possible in fact, since he played a crucial role in the foundation of the Anglo-Australian Telescope, to mention just one achievement. He became, for those same 5 years, an establishment figure, with metaphorical mountains to conquer. His achievements as a leader of scientific affairs in the UK were considerable.

In the month of October, Cambridge is often blessed with cloudless, atmospheric days on which the slanting light of the Sun shows off the historic buildings and surrounding countryside to best advantage. On one such day, a few weeks after the Institute of Theoretical Astronomy (IoTA) opened, Hoyle rolled up at his office after a 10-minute walk from home, to be confronted by a face from the past— Jim Hosie. The two had last seen each other at mathematics lectures in 1935–1936. A distinguished civil service career had propelled Hosie to the directorship of the division of astronomy and space at the Science Research Council (SRC), effectively placing him at the summit of UK policy making on spending decisions in celestial realms. What business had brought him to Cambridge by steam train on a first-class ticket from London? His mission: to entice Hoyle to join the *crème de la crème* of the UK science establishment. Hosie offered Hoyle three positions: chairman of the astronomy division of SRC; membership on the astronomy, space, and radio board of SRC; and membership on the Science Research Council itself.[5] Hoyle could not resist these opportunities, which suddenly placed him at the epicenter of policy making, rather than in the penumbral location he had endured inconsequentially for the past 2 years.

Hoyle now wielded greater power for influencing the future direction of British astronomy than any professor in a university, or indeed the astronomer royal. Having created the IoTA as a national facility, he was confident about operating across the entire spectrum of astronomical activity. I feel sure he was strongly motivated, for entirely unselfish reasons, by the prospect of transforming the UK's astronomy efforts as a whole to higher levels of international distinction. In the 22 years since the end of the war, very little had been done to obtain world-class facilities for younger astronomers. Naturally, some of the brightest among them had voted with their feet in a mass exodus to American observatories. Britain lost the Burbidges, Wal Sargent, Alex Dalgarno, John Faulkner, Tommy Gold, and Peter Strittmatter, among the brightest people of their generation. These losses mattered greatly to Hoyle.

So, how did Hoyle use his chairmanship? In radio astronomy, the successes during his tenure included rebuilding the mighty 250-foot radio telescope at Jodrell Bank and providing the University of Cambridge with a large radio interferometer for Ryle. In optical astronomy, two telescopes were built in Australia for Southern Hemisphere observations: the Anglo-Australian Telescope and the 48-inch Schmidt Telescope. The UK also made a start in infrared astronomy, which Hoyle pushed strongly. The number of grants and awards available to university groups rose substantially. All this happened when the SRC had to deal with several conflicting projects, but Jim Hosie proved highly effective at channeling SRC's money into astronomy.

Her Majesty the Queen had inaugurated the Isaac Newton Telescope on December 1, 1967. The disastrous siting of this 100-inch telescope, above the marshes of Pevensey Bay, was an enormous disappointment to UK optical astronomers. Fortunately, during the years of wrangling that preceded its opening, the British made strong efforts to plan and fund a major telescope in the Southern Hemisphere. Possible locations for such a telescope were South Africa, Australia, or Chile.

Discussions held with numerous parties over many years led the Australian government to propose that Australia and the UK should join forces to build a large optical telescope at Siding Spring in rural New South Wales. The design would closely follow that of the successful 4-meter telescope at Kitt Peak, Arizona. Leading astronomers of both nations quickly united behind this suggestion and, by August 1967, a joint planning committee held its first meeting in London. The membership included Bondi and Hosie, plus the astronomer royal, Sir Richard Woolley; the concept of an Anglo-Australian Telescope (AAT) was essentially Woolley's. A man of presence and style, he deftly kept the project alive during critical periods. By the time of the second meeting, Bondi had landed the biggest space research job in Europe, the directorship of the European Space Research Organisation, the forerunner of the European Space Agency. With Hoyle as the director of the astronomy division, Hosie handed him Bondi's vacant seat, thus starting a long association between Hoyle and the AAT.[6]

Through his many visits to the United States and his extensive networks there, Hoyle was aware of design weaknesses with the Kitt Peak telescope: Its massive yoke mounting would not transfer easily to a higher latitude. According to Margaret Burbidge, who replaced Woolley on the management board in 1971, Hoyle's advice was crucial in forcing management to face up to their contractors and suppliers in order to get the design problems solved.[7] Hoyle also understood the importance of excellent instrumentation, although he disagreed with Hosie on how research grants for this purpose should be awarded to universities. Hoyle wanted people already familiar with large telescopes to design instruments (spectrometers, detectors, imaging systems, and their computers). Such people were to be found in California and

Arizona, rather than on the Sussex coast or in Edinburgh. The SRC, however, was constantly under political pressure to spend the money at home, by giving responsibility for instrument development to university staff with little hands-on experience.

The Anglo-Australian Telescope project required Hoyle to travel often and extensively almost from the moment the institute opened. An intergovernmental act of parliament, passed in 1970, established an Anglo-Australian Telescope Board with six members, effectively as the owner-managers of the telescope and its associated observatory.

The 1967 agreement on the Anglo-Australian Telescope stimulated Hosie to commission a review on how the SRC should allocate its resources for astronomy in the Southern Hemisphere. Hoyle did not participate in this review but, as chairman of the astronomy division, he eventually considered its recommendations; his most important decision would be to equip the observatory in New South Wales with a 48-inch Schmidt telescope alongside the AAT. Encouraged by this positive outcome, the SRC decided to undertake a similar policy review for the Northern Hemisphere, with Hoyle in the driver's seat. This appointment illustrates how Hoyle had by now graduated way beyond being a theoretical astrophysicist by securing a series of management positions at the pinnacles of the profession.

The chairmanship of the Northern Hemisphere Review Committee required very hard work and political skills for conflict resolution, both character strengths of Hoyle's at this stage. However, in the long run, this review would prove an unwelcome distraction from his duties at Cambridge. His committee comprised the astronomers royal (Woolley and Hermann Brück from Edinburgh, who was astronomer royal for Scotland), Bernard Lovell, J. M. Cassels (a nuclear physicist from Liverpool), the space physicist Robert Boyd, and Jim Hosie. Only Woolley and Brück could claim any professional experience in optical astronomy. The remit of the review committee included a requirement "to inform itself of the views of British astronomers, including such expatriates as the Board shall decide and the staffs of the [Royal] Observatories."[8]

Hosie's board had also sounded out Geoffrey Burbidge and the Caltech astronomer Wallace Sargent as expatriate consultants. This led to trouble straightaway. The two consultants participated fully at the second meeting on April 17, 1969, where Woolley protested strongly that they appeared to have voting rights. Hoyle neatly solved that one in a clever put-down to Woolley: Henceforth no member would have voting rights! Any action on their recommendations would lie with a higher authority. However, Woolley's shot was symptomatic of the immediate difference of opinion that opened up on many issues between the royal astronomers and the holders of university posts. And henceforth Woolley nursed a little grudge.

As secretary of the review, Hosie convened nine meetings in the space of 15 months. Burbidge and Sargent flew over for all that were relevant.[9] The fur started to fly at the fifth meeting on July 31, 1969. Hoyle had prepared a paper proposing that a national center should be created to manage any future large telescope. This should be in a university town and not attached to a royal observatory. The expatriate astronomers signed an inflammatory paper that was highly critical of the Isaac Newton Telescope: "In its present location [it] is not merely ineffective but useless." They were scathing about the incompetence of the Royal Greenwich Observatory (RGO): "It would be folly . . . to go ahead with a northern hemisphere large telescope if any part of its development or operation is to be put in the hands of the present RGO." The two astronomers royal would never agree to this snub to their institutions.

Hoyle has given us a summary of the dilemma the review committee faced.[10] The university astronomers regarded the availability of a large telescope as a great privilege and were willing to compete with each other for observing runs. The government astronomers looked upon observing as extra work for which they should be paid overtime and disturbance allowances. Hoyle wryly notes that the royal astronomers were far more concerned about their pension rights than the research possibilities of a new telescope. This makes the point that the communities based in universities had a very different set of priorities than the civil servants at the Observatories.

The report recommended the establishment of a Northern Hemisphere observatory; this should be "provided urgently on a good site in the Northern Hemisphere—which, for climatic reasons, means outside the UK." The facilities so provided should comprise "a large telescope and one of medium aperture." No one quibbled with either of these recommendations, but they came to blows on how to manage astronomy in the UK at the national level. The report proposed that new national facilities should be the responsibility of a National Centre for Astronomy, which would have a management committee drawn from the universities. The royal observatories were to play no role other than filling two seats on the management committee.[11]

Woolley reacted very angrily. He despised the continuing presence on the committee of the two expatriates. He would dissent; the resulting report could have little weight and "will not do any good to the image of astronomers." A stalemate developed: Woolley and Brück would sign the report only if it omitted the proposal for a new center and "avoids pejorative reference to the Royal Observatories." A semblance of stability was achieved by early autumn of 1970: The 40-page report said its recommendations had met with the approval of the majority of the committee, apart from the astronomers royal who submitted a minority report opposing the creation of a national center outside their Observatories. This report was never published. And Woolley's little grudge had meanwhile grown to vengeful proportions.

The situation on Northern Hemisphere astronomy remained confused throughout the 1970s. Hoyle's position as chairman of the planning committee expired in 1971, and his involvement ended. In 1975, the SRC abandoned the notion of a national center, deciding that the new observatory would be the responsibility of the Royal Greenwich Observatory. An international agreement signed on May 26, 1979, established a new observatory on the island of La Palma in the Canary Islands. This observatory opened in 1985, and its 4.2-meter William Herschel Telescope saw first light in 1987. It had taken 18 years from the start of Hoyle's chairmanship of the Northern Hemisphere Review to the realization of a telescope able to compete with Arizona, California, and Hawaii.

In the Southern Hemisphere, Hoyle saw a positive result for his efforts more quickly. By 1970, construction of the Anglo-Australian Telescope was already well advanced. Much of Hoyle's work involved interminable discussions on administrative issues: where to locate the scientific office, how to recruit staff, and what body would "own" the Anglo-Australian Observatory.

It is no exaggeration to say that telescope building distracted Hoyle far too much in the years following the opening of the IoTA. As a research establishment, IoTA was an instant success, becoming within 2 years one of the most important centers in the world for theoretical astronomers.[12] In terms of staffing, visitor programs, international conferences, and facilities, Hoyle had gotten everything just right, including surprisingly small details: the golfball typewriters, expensive carpeting to absorb noise, and provision of a Xerox copier.

From 1964, Hoyle had used the Xerox copiers at the copy center on the Caltech campus. Fowler had quickly become a big fan of the copier as an aid to academic productivity, and he arranged for assistants to carry copying jobs to and from the center. By observing Willy's team at work, Fred quickly realized how significant the Xerox would become. He decided to get one for IoTA and to grant unlimited "people's access" to it, something unheard of in universities in 1968. Summer visitors to IoTA really did plan long copying sessions carefully because they did not have to sign for or pay for any copying done in Cambridge. As a further aid to productivity, Hoyle found cash in his operating budget to pay for *Astrophysical Journal* to come by airmail.[13] This meant that the IoTA copy was always the first in Cambridge, by some weeks. Henceforth, envious colleagues in the Department of Applied Mathematics and Theoretical Physics and the Cavendish Lab had to pedal out to IoTA on the Madingley Road if they wished to keep up with the literature.

In staffing the institute, Fred had quickly put together a world-class group of people. He passionately believed in excellent creative individuals, and he paid little attention to what opinions they might hold. He did not surround himself with nodding sycophants, nor did he demand loyalty to his particular point of view. He adopted an Olympian perspective: The best could be champions. He differed markedly from other leaders in astronomy because he eschewed the concept of a large collaborative program that everyone must support. Ryle, for example, brooked no criticism from within and would not have dreamt of giving any position to someone with views contrary to his own. Hoyle treated everyone on their merits. He wanted his associates to understand the very complex universe to the best of their abilities rather than to conform to a set of ideas already accepted by the consensus.

So, Hoyle largely let the IoTA staff do as they pleased, but he did give firm leadership to one program, research on nucleosynthesis. Personally, he had little time for research in this area, but he wanted to make an important niche for nucleosynthesis in the institute. He put Fowler, as a member of the summer staff, in charge of building a research team. This was an inspired appointment. Willy threw all his energy and enthusiasm into Hoyle's dream of generating astrophysical leadership in England. Moreover, Fowler had a dream of becoming an honorary Englishman and retiring in Cambridge, so he badly wanted the academic programs at IoTA to succeed.[14]

Don Clayton was one of the first people to whom Fowler now turned. Clayton had been his research student at Caltech, where he came to the attention of Hoyle. On completion of his doctorate, he accepted a position at Rice University, Texas, where he began to write a textbook on nuclear astrophysics. On March 3, 1966, Fowler's secretary typed a remarkable letter of invitation to Clayton. Fowler wrote:

> I have heard rumors that you are writing a book on nuclear astrophysics. As you know, Fred Hoyle and I are also doing one. How about getting together? . . . It sounds like a natural hook-up to me. In regard to details—credit, financial, and otherwise—it would be alphabetical, CFH, and the sales and royalties would certainly be more than three times otherwise with the H! Is it not time for you to have a sabbatical? How about coming back [to Caltech] for a year? . . . In a year we could finish the book. Come back! Bring [the whole family]. If Rice won't pay we will.

That invitation gave Clayton much to think about. Its final sentence showed "Willy's legendary determination to make the best things happen without leaving them to chance." He and his wife felt immediately attracted by the excitement of Pasadena, which they missed in Houston. But he had not been at Rice long enough to earn sabbatical leave, and his own book was so near to completion that he could not merge it into Fowler's concept. With hesitation he wrote to decline the publishing opportunity but said he was nevertheless keen to return to Caltech "to dunk myself back in the mainstream." Fowler responded with a terse, almost telegraphic two-liner, the brevity of which disturbed Clayton: Had his judgment upset his former supervisor? On the contrary; in Fowler's eyes, Clayton had gained in stature by his professional response in declining to collaborate on the book.

Within a week, a truly excited Fowler contacted Clayton. And this time he did not write; he picked up the phone instead.

In great clarity, Fowler explained Fred's hope that nucleosynthesis should be the main thrust of IoTA's 5-year research agenda. Hoyle himself would need to scale down his research efforts. He had the institute to run. Fowler invited Clayton to spend the year at Caltech, on whatever terms he could agree with Rice University. He should give priority to finishing his own textbook, which he did, then work hard with himself and Fred on a major collaborative monograph—a professional-level handbook on nuclear astrophysics—that they could finish in Cambridge in the summer of 1967. Thrilling stuff.

Clayton booked a transatlantic call to Hoyle. The Yorkshire man, a trace of gruffness in his voice, warmly invited the young Texan to Cambridge. Over an echoey line, Hoyle asked Clayton to spearhead, with Willy, the nuclear astrophysics program as a regular summer visitor. Hoyle would be visiting Caltech in February 1967 and suggested they discuss the details there. That was all Clayton needed to know. He moved to Caltech immediately and started a 7-year summer project for IoTA. These would be golden years for nucleosynthesis.

Fowler's affection for Cambridge was so firm that he had bought and renovated a house on Oxford Road, a few minutes walk from IoTA. Clayton engineered an annual migration of a "Texas mafia": his Rice colleague Dave Arnett and a string of talented students, including Stan Woosley, Kem Hainebach, Dave Schramm, and Mike Howard. All were frequent guests to Fowler's house. They became experts on the country pubs and inns, particularly the Green Man and the Red Lion in Grantchester. They consumed cheddar cheese and pickle lunches, downed pints of warm Greene King beer and Suffolk ale, supported cricket matches (with Hoyle), and even attended sheepdog trials.[15]

Beyond Cambridge Clayton and Fowler regularly accompanied Hoyle on annual climbing expeditions, mostly to Wester-Ross and Sutherland. Typically, Fred drove Don Clayton all the way in his Lotus Ford Cortina, one of the most powerful cars on the market. He cursed the ambling caravan drivers on the twisty roads in the Lake District and the Highlands. At Inverness, Fowler joined them by train and they bought their own copies of the Munro tables. Within days they began ticking off their achievements. What did they talk about as they scaled the mountains? Science and philosophy of course. Solar neutrinos, counts of radio sources, cosmology, nuclear astrophysics, quantum mechanics, the problem of free will, the nature of time, all the topics Fred Hoyle never stopped thinking about.

Unlike most climbers, they did not rough it with tents and sleeping bags. Instead, Fred had become an aficionado of remote Highland inns, often converted hunting lodges. He knew them all. Each day the three clambered over the hills, with the aim of completing the climbs by teatime. Fred liked to finish by around 4:00 p.m. and get back to the inn for buttered scones, strawberry jam, and tea. Young Clayton would have preferred a cold beer but followed Fred's example. In the evening, Fowler always chose the wine, a right he exercised with at least as much experience as Fred used in choosing mountains. And of course the talk was of science. Never in human history was so much nuclear astrophysics discussed in such remote isolated inns.

In October 1980, Fred Hoyle bagged his last Munro, Bla Bhienn (Blaven), in the black Cuillian Hills on the misty Isle of Skye. The task had taken him 15 years, years in which he rescued British optical and theoretical astronomy from the brink of obscurity. Unfortunately, he never bothered to report his achievement to the Scottish Mountaineering Club, which is why he does not feature in their official list of Munro completers.

Several summer collaborations brought together the Texas mafia with B²FH, creating the world's most powerful concentration of nuclear astrophysicists; Willy was the godfather. The group made intense use of the free computer time on the powerful IBM 360/44. Dave Arnett found an entirely new process that had been missed in the B²FH paper: explosive nucleosynthesis. In simulations on the 360/44, he heated carbon to a temperature slightly above that at which it burns gently to oxygen and neon. The clattering of the line printer showed the detonation of a thermonuclear explosion. The debris from this explosion contained the correct concentrations of the prominent isotopes of carbon, oxygen, neon, sodium, magnesium, and aluminium to account for their cosmic abundances. What an explosive result! Another strand of the research concerned conditions in the supermassive stars that Hoyle and Fowler felt could be linked to quasars. All these faithful friends regarded the computer simulations of the nuclear physics of stars as enormous fun. Post-B²FH, they invented a completely nuclear astrophysics agenda, one that has continued to the present day under Clayton.

Everybody at the institute had to enjoy themselves. Fred was determined about that. The institute could not exactly replicate the legendary Friday night parties of the Kellogg Radiation Laboratory that followed the Caltech nuclear physics colloquia. Cambridge did not have the weather, and besides, the beer was too warm. Instead, the Cambridge astronomers treated themselves to a nice lunch on Fridays, booking the spacious private dining room at the University Centre, overlooking the River Cam with its punts, Laundress Green, and the back of Darwin College. These were truly remarkable gatherings because, for the first time ever, they brought together in an amicable forum the institute staff, the radio astronomers of the Cavendish Lab, and the astrophysicists from Applied Mathematics. John Faulkner recalls: "There would be very engaged conversations between the three groups, who were always fairly well represented."[16] The radio astronomy group would pitch in on gossip or general astronomy, but go tight-lipped if the conversation edged toward what might be going on in their group.

Long after the lunch had ended one Friday in February 1968, the radio astronomer John Shakeshaft paused at the exit door, turning to address John Faulkner and Peter Strittmatter. In his precise way, he said, "Oh, by the way, I really think you'll want to pay attention to a very important seminar that will be given next week—it will be extremely significant and you won't want to miss it!" Strittmatter asked what it was all about, but Shakeshaft, who had already said more than he should, politely replied that he was not at liberty to say more, apart from urging them to be present. Early the following week, an announcement appeared saying that Professor Antony Hewish would give a seminar on the discovery of a new type of pulsing radio source. This would be on something not yet published, a first in the Cavendish group's history. On February 20, the IoTA crowd, including Fred who was just back from Caltech, trooped down to the Maxwell Lecture

Theatre in the Cavendish Laboratory. Physicists packed the wooden benches of the gloomy hall (it had not changed since Rutherford's days). Everyone in Cambridge with the remotest interest in astronomy attended.

Jocelyn Bell, a young research student who had played the central role in the discovery, modestly assumed a backbench position. Tony Hewish, at once a charming and modest person, informed the rapt gathering of the discovery of a rapidly pulsing radio source (the term pulsar lay in the future). Jocelyn had nicknamed it LGM1, Little Green Man One, because for a few days they were not sure whether it was a natural cosmic signal or from an alien intelligence.[17] However, they soon confirmed that they had found a new kind of astrophysical object that emitted a bleeping radio signal with astonishing regularity. The audience immediately sensed that they were being treated to something momentous, a truly remarkable breakthrough, a eureka moment in the very laboratory that had already won many Nobel Prizes. But neither the audience nor the radio astronomers knew just what had been found this time. Hewish speculated that it might be the vibrations of a white dwarf star, but then teased his audience by mentioning the discovery of further objects of this kind, "all over the sky." Bell had scooped a further three, and the blips from each LGM had a different period. But Hewish did not present data about the other three, nor did he reveal that LGM4 had a period of just a quarter of a second![18]

Frenzied discussion followed. In the questions after the seminar, Fred said (imagine his Yorkshire accent), "Ah doan't think it's a whaite dwarf—Ah think it's a suupernoova remnant." Fred had preceded this gem by saying this was the first he had heard of these objects, which was true because the Cavendish radio astronomy group had withheld announcing their discovery until they could be certain they were not just dealing with locally generated interference.[19] He used the expression "supernova remnant" not in the exact sense, which is an expanding nebula, but as a general term for what gets left behind. If Hewish had revealed that LGM4 had a period as short as 0.25 seconds, my guess is that Hoyle would then and there have said that the pulsing radio sources must be *rotating* neutron stars. That would have made the seminar even more exciting: A neutron star is the burned-out nuclear reactor of a massive star, collapsed down to a spinning object a few miles in diameter.

The following weeks were thrilling indeed. Hoyle could barely contain his excitement at a meeting of the SRC the next day. As he took his seat, Sir Bernard Lovell asked him, "Have the Palomar observers any exciting news on quasars, Fred?" "Not much about quasars, but yesterday at a colloquium in Cambridge, Tony Hewish announced that he had discovered some radio sources which emitted in pulses at intervals of about a second." Lovell found this astonishing, but Hoyle assured him that Hewish had been observing the sources for months in great secrecy and had established that the signal was not caused by interference.

Nature published the discovery paper on February 28, 1968, but the Cambridge radio astronomy group continued to keep the positions secret.[20] They did release them in confidence to Bernard Lovell at Jodrell Bank and phoned them through to Allan Sandage at Palomar Observatory. Then a leak occurred: Bernard Lovell phoned the data through to Hoyle after a couple of days and, in turn, Fred called Margaret Burbidge in La Jolla with the news. Soon the theorists joined in the fun. The discovery paper had hinted that the regular pulses might be the *vibration* (they actually used the term pulsation) of a white dwarf star. Any radio observatory with a fully steerable dish was now listening to pulsars. Observers and instrumentalists rapidly cobbled together search strategies to find more of them. In July 1968, I declined the offer of a research studentship in nuclear physics at Oxford and instead accepted a place with Ryle's group. I felt the work of Tony Hewish and Martin Ryle sounded much more significant than measuring the energy levels in atomic nuclei.

Everyone by now accepted that they were dealing with stellar objects, but what kind of star? Some theorists thought the pulses were caused by the *vibrations* of a white dwarf star. Jerry Ostriker produced a model based on a rotating white dwarf.[21] Tommy Gold (at Cornell University) took this model a stage further by suggesting that they must be *rotating* neutron stars, sending out energy in a beam, just like a lighthouse. When the radio beam swept across Earth, observers registered a radio blip; a second or so later, the spinning star completed a further rotation and the next blip would be detected. This imaginative suggestion implied that neutron stars could be real-world objects rather than just theorists' toys. John Gribbin, one of Hoyle's research students at IoTA, kept the general public informed with a string of gripping news stories for *New Scientist* magazine. I would not be surprised if Hoyle had discussed his idea of a supernova connection with Gold, but the sources do not enable me to prove this. Hoyle knew, of course, that Gold directed the Arecibo Observatory, which had the largest single-dish radio telescope in the world, and it would be characteristic of Fred to want to get Tommy observing on a big dish. Hoyle could be confident that the Cornell radio astronomers would tell him of any discoveries, unlike his secretive Cambridge colleagues.

Crucially, Gold made a startling prediction: These very stable cosmic clocks would be observed to be slowing down as a result of the powerful radio emissions being generated. The energy for the radio emission had to come from somewhere, and Gold thought the conversion of rotational energy of motion would do the trick. If the clocks were running down, they had to be neutron stars. But the Cambridge LGMs had not been under scrutiny for long enough to show such an effect. Undeterred, Gold submitted the paper on his model to a conference on pulsars, held in New York City on May 20 and 21, 1969. To Gold's annoyance, Al Cameron, who was in charge of the scientific program, refused to give him 5 minutes at the podium: "Your suggestion is so outlandish that if this were admitted there would be no end to the number of suggestions that would equally have to be allowed."[22]

John Maddox, the editor of *Nature,* felt otherwise. The paper, received on May 20, was published on May 25, surely a record.[23] Maddox had a sixth sense for spotting a good story. By contrast, the New York conference on pulsars was a fiasco, dominated by a report of startling new observations that turned out to be entirely false and by theoretical models of vibrating white dwarf stars.

The next development came in October from Australia, where astronomers reported the detection of a pulsar with an 89-millisecond period, the shortest then known. Its location was a young supernova remnant, which confirmed Hoyle's initial hunch. Cornell's Arecibo Observatory, in Puerto Rico, did not have a steerable antenna but, by

chance, the Crab Nebula passed through the beam of their fixed telescope once a day. Gold, as observatory director, asked his colleagues to
investigate the Crab Nebula, site of the nearest supernova explosion in
our galaxy. Within days, they had the clinching discovery: The Crab
Nebula had the fastest pulsar then known, repeating every 33 millisec-
onds. Within a day, the expected slowdown was observed!

This discovery heralded a great new era in astronomy. A world of
very high density and very high energy concentration suddenly opened
up. One cubic inch of a neutron star has about the same mass as a
cubic mile of iron. Its surface gravitational attraction is 200 billion
times that of Earth. In this world, Einstein's theory of general relativity
reigns supreme and Isaac Newton's mechanics are irrelevant. Overnight, the theoretical astronomers realized that black holes—the last
stage of collapsed matter after a neutron star—could also be real-world
objects. For the first time, they had the confidence to start research on
black holes, which we observe today as the energy sources in x-ray
binary stars. The study of black holes would soon be a major activity
for the institute.

Hoyle stirred a hornet's nest of a different kind as the number of
known quasars increased. At first, he was especially happy to think that
the objects whose existence he and Fowler had predicted really did
exist, as million-solar-mass objects collapsed by gravity. He was in no
mood to think that the redshifts could have an origin other than in the
expansion of the universe. But that feeling did not last. As new observations poured forth from radio and optical observatories, he quickly
lost faith in the argument that large quasar redshifts are due to the
expansion of the universe.

Quite soon after the initial discoveries, quasars were found to have
properties very different from those of galaxies. There was no correlation between their redshifts and magnitudes, as happens for galaxies. A
dim quasar might have a low redshift or a high redshift. You could not
make a guess on the basis of its brightness on a photograph. Soon it
was discovered that the radio and optical energy from some quasars
fluctuated on a timescale of a few weeks or months. This is unheard of
in normal galaxies. Hoyle immediately saw that fluctuations place a
severe limit on the radiating process. The short response time meant

that the radiating region could be no larger than the solar system. In 1966, with Wallace Sargent and Geoffrey Burbidge, he claimed that the variations implied either that the radiating surface was moving at the speed of light or that the objects were much closer than the distances obtained by assuming that their redshifts are due to the expansion of the universe. According to Hoyle, the quasar redshifts could not be interpreted as simply due to the expansion of the universe.[24]

Together with Geoffrey Burbidge, Hoyle worked on a detailed paper that examined whether or not the evidence pointed to quasars being at great distances or comparatively nearby. They looked at the question from both points of view, deducing the nature of the objects for the two alternatives. Furthermore, they proposed a program of observations capable of distinguishing the two possibilities. In the case of local quasars, they proposed that these objects had been expelled at extremely high speeds from the cores of disturbed galaxies. This opened the interesting possibility that some of the ejected quasars could be directed toward the Milky Way, in which case they would show a blueshift. (Objects going away from the observer show a red-shift in their spectrum, whereas those traveling toward the observer show a blueshift.) Their paper suggested searching for blueshifted objects in the vicinity of the nearby giant radio galaxy NGC5128.[25]

Their expulsion hypothesis was so radical that they did not bother to send it to the Royal Astronomical Society (RAS) for publication, fearing out-of-hand rejection. They did get lucky though with the *Astrophysical Journal*, where the editor, Subramanayan Chandrasekhar, was an old friend of the pair. In the 1940s, Chandrasekhar had been publicly crushed by Eddington at an RAS meeting, so he may have felt motivated to help Hoyle publish independently of the RAS. After pondering the matter for a month, he gave his chums the benefit of the doubt. The University of Chicago Press published the paper, but not with alacrity. During the year it took for the paper to appear in print, Geoff and Fred were often disturbed in the dead of night by the telephone. On several occasions a mysterious voice that neither could place threatened them with the ruin of their scientific reputations.[26]

No blueshifted quasars were found, but there was a new cause for disquiet. By 1966, three quasars were known in positions on the sky close to bright galaxies. Each case could have been explained by a chance alignment of a nearby galaxy and a distant quasar. Hoyle chose to interpret this argument as very unlikely, based on the statistics. "One could easily see that the odds were against this being so," was how he put it.[27] When the positions of two extragalactic objects, such as a galaxy and a quasar, were closely matched, but their redshifts were very different, they were said to have "discordant redshifts." Was the match a freaky coincidence, or were they physically related? If a true physical connection existed, then both of the redshifts could not be due to the expansion of the universe. At least one had to have a contribution from a different cause.

Support for the existence of discordant redshifts came from Halton (Chip) Arp, an observational astronomer at the Mount Wilson and Palomar observatories, with a penchant for closely scrutinizing photographs of galaxies in a search for appendages and alignments. By about the date the institute opened, Arp had found a number of quasars that appeared to lie close to galaxies. There were symmetrical arrangements, which strongly suggested physical association of a high-redshift quasar with a low-redshift galaxy. Something other than the expansion of the universe had to be the cause of quasar redshifts.

At the institute, Fred now commenced his own search for discordant redshifts. He turned to Cyril Hazard, the only radio astronomer on the institute's staff. Hazard proposed that they should try to match the positions of point radio sources to optical objects. Some of these would be quasars, and interesting associations with galaxies could be checked. There was no point in using Martin Ryle's published catalogues of radio sources because Arp was well ahead of them in checking the 3C and 4C sources. However, Hazard, who had recently returned from Australia, could instead use the positions of radio sources recently discovered in a new survey conducted by the University of Sydney. Hazard proposed an elegant method to accomplish the identifications.

The Cambridge Observatories owned a film copy of the Palomar Observatory Sky Survey, a photographic atlas of all the sky visible from Palomar. Hazard proposed to use this to see if an optical object lay at precisely the same position as a radio source. To carry out the identification he would supply the institute's IBM 360/44 with the radio position. The computer program searched on a magnetic tape catalogue for all the stars within a specified angular distance of the radio source position. These star positions would be printed out in the form of a map centered on the radio position and at exactly the same scale as the Sky Survey film. The printer would use translucent paper. The plan involved fairly sophisticated equipment by the standards of the day.

Hoyle's computer staff soon gravely announced that the 360/44 would need a memory upgrade costing £50,000 to cope with the number crunching. The potential cost to the institute was equivalent to hiring 10 postdoctoral staff for 3 years. Hoyle knew with absolute certainty that he could not cover this from his normal operating budget. So, he contacted Jim Hosie, his boss at the SRC. Hosie gave an informal nod, but Hoyle must follow the correct procedure and process a grant application through the normal triple-layer committee structure. This would lead to a delay of 6 months. But Hoyle was in a hurry. He cut a sharp corner.

Surely this grant would be a shoo-in, he told himself. After all, he chaired one committee and had a seat on the other two. It would take IBM 6 months to prepare the upgrade. So Hoyle ordered the upgrade, confident that the research council money would come through just in time to pay for it. Overconfident, he made an enormous blunder.

In the year before Hoyle joined the research council, it had already adopted a policy that new grants for radio astronomy equipment would be confined to the Cambridge and Manchester groups. This was to avoid a clamor from other universities for millions of pounds to join the radio astronomy game. So, Hoyle's grant application for radio astronomer Hazard failed: Martin Ryle at the Cavendish Laboratory regarded it as a matter of deep principle that no other university department should be funded for radio astronomy.[28] However, Hoyle had by now taken delivery of the memory, which increased the throughput of the institute's computer by 50 percent. To pay the bill, he acted as an entrepreneur: He sold computer time to the UK Medical Research Council and the mighty IBM itself. No doubt this arrangement was advantageous to both parties. A good deal of progress was made on unraveling the structures of complex biological molecules. But his unauthorized expenditure had not created any friends for him in the central university administration, which henceforth regarded him as an unreliable manager of budgets.[29]

There came the day when a grinning Cyril Hazard stood in Fred's office proffering a stack of polaroid pictures. "Fred, look at this one," he said. To Hoyle's eye it was an uninteresting double star. "Not so," responded Hazard, "both objects are much bluer than normal stars; I think they are quasars." Through Margaret Burbidge, they contacted Joe Wampler at Lick Observatory, California. His observations showed that both objects were quasars, but their redshifts differed by a factor of 5. To Hoyle's mind, this dealt a fatal blow to the idea that quasar redshifts were simply due to the expansion of the universe, which was by now the consensus view among astronomers specializing in extragalactic research. Hoyle reckoned the chance of a fortuitous pairing at about a thousand to one.[30]

Meanwhile, Arp's results were treated with derision by everyone except Fred Hoyle, who realized that, if a proportion of radio sources appeared not to be in harmony with the Hubble law, then counts of radio sources could not be used for cosmology. One day in 1971, he phoned Arp, who was slightly surprised by the caller's unusual accent. "Hello. This is Fred Hoyle from the Caltech campus. Can I come and see some of your pictures of connected objects with different redshifts?"

Arp thrilled to the prospect that a person of such eminence wanted to see his images. Hoyle pushed his thick glasses up on his forehead and brought the glass photographic plate almost touching his face. Excitedly Arp began to explain the main features of the photograph, but Hoyle snapped at him: "No, don't tell me anything; I just want to look." He stared for a few minutes in silence, thanked his host, then departed silently.

In the following years, Arp paid a very heavy price for pursuing this line of inquiry. His colleagues in Pasadena became so disturbed and disbelieving that they recommended the director stop his program by refusing observing privileges. The director swung the axe. Arp's appeal to the trustees of the Carnegie Institution of Washington, who owned the observatories, fell on stony ground. He took early retirement and moved to Germany.

The work of Hazard and Arp slowly increased the number of strange pairings and alignments, which Hoyle felt strongly must mean that quasars are expelled from galaxies at very high velocities. But he never convinced his Cambridge colleagues or the wider astronomical community. He had plenty of opportunity to press his point of view, though. In 1971, the American Astronomical Society awarded him its prestigious Russell Prize and Lecture, for "lifetime achievement" in astronomy. Hoyle received the prize and delivered the lecture in Seattle at the society's April 1972 meeting. Arp took the opportunity to read a short observational paper arguing an extreme proposition: The excess redshifts were related to the age of the objects, and the atomic constants were changing with time! Hoyle titled his prize lecture "The Developing Crisis in Astronomy." He first showed a series of pictures of apparently close associations between quasars and galaxies. He rounded off his little slide show with the assertion that the odds they were all freak coincidences was about a million to one. Then came a mathematical argument demonstrating how atomic and nuclear masses could vary as a function of cosmic time. Triumphantly, he concluded, "This concept appears necessary if we are to understand the result reported by Arp [at this meeting] for the galaxy NGC 7603 and its appendage." So now Arp knew what Hoyle had silently concluded when he returned the photograph of this object to Arp's desk![31]

Outside the lecture hall, an aging Martin Schwarzschild scolded the pair of them: "You are both crazy!" This startled Arp, although he did like being linked to Fred. Hoyle just looked blank. *Astrophysical Journal* routinely published the Russell Prize Lectures. Shortly after the meeting, Fred sent his manuscript to the editor. To Hoyle's utter amazement, he received a referee's report demanding extreme changes to the text that was, after all, the record of what he had actually said. He was so angry at this shoddy treatment that he never responded.

Hoyle's Seattle lecture had ended with an alternative idea in cosmology: The masses of elementary particles can change with time. A different approach to cosmology characterized Hoyle's personal research effort at IoTA, although his interest had started long before then, in 1948, with continuous creation and the steady-state theory. Hoyle's goal in the 1960s and in the IoTA years went far beyond the classical gravity of general relativity. This would be achieved by involving all manner of imaginative physics, most particularly by making connections between particle physics and cosmology. Einstein's general theory is silent on particle physics, and that worried Hoyle. The situation in Cambridge today is that Stephen Hawking has long been the standard bearer for a more complete cosmology. Hoyle explored this territory hand in hand with Jayant Narlikar, in a partnership almost as remarkable as the Hoyle-Fowler collaboration.

Narlikar had been in the same cohort of mathematics undergraduates as Faulkner, whom he pipped at the post to top the Tripos list. He had attended the same summer school at Lake Como as the life saver, where Bondi's lectures on cosmology made a deep impact on him. One of his early projects had taken him to the IBM 7090 in London to model counts of radio sources, the first computer simulation in cosmology.[32] They published two papers while Narlikar was still a graduate student, which tells us something about the younger cosmologist because Fred Hoyle seldom coauthored papers with his students. During 1965–1966, by then a postdoctoral researcher, he had worked with Hoyle on the formation of a massive object at the center of a galaxy. Together they worked out the consequences for a galaxy with a supermassive black hole at its center. Those researchers interested in galaxy formation and evolution took little notice, but today, of course, there is enormous enthusiasm for the existence of colossal black holes in the nuclei of galaxies.

The main thrust of the Hoyle-Narlikar collaboration before and during the IoTA era was alternative cosmology, particularly the possibilities offered by action-at-a-distance theories, a throwback to the seventeenth century. Isaac Newton, the founder of modern theoretical physics, had shown that the motion of a planet could be understood in terms of the forces acting on it from other planets at a distance. This action propagated through space instantly. In the century after Newton, the French physicist Charles de Coulomb had discovered that the physical law describing the dynamical behavior of electrically charged matter is remarkably similar to the gravitational case, apart from the fact that electric charge is both positive and negative, which leads to both attraction and repulsion. Like Newton, Coulomb sensed that the action between electric charges propagated instantly over the distance between them. By the middle of the nineteenth century, the investigations by Carl Friedrich Gauss on electrical charges in motion led him to conclude that the action does not happen instantaneously, but moves at the speed of light. That clue enabled James Clerk Maxwell to give mathematical expression to the concept that the forces arise through the medium of a field of force rather than action at a distance. And Einstein, of course, then replaced Newtonian action at a distance with the gravitational fields of general relativity. So by the twentieth century the concept of action, as opposed to field, was a museum piece.

Or was it? Bondi's lectures at Lake Como had set Narlikar's mind racing. Suppose an action rather than a field governs the universe. What are the consequences? In action-at-a-distance physics, the interaction travels forward and backward in time symmetrically, thus apparently allowing time travel and violating the principle of causality. But in the real world, there is no time travel. So, if action at a distance is to work, the universe must be so structured as to prevent propagation into the past, in which case what are the cosmological consequences? In some sense this is a philosophical puzzle, a deeply intellectual question of the kind that appeals to academics working beyond the fringes of accepted knowledge. Fred was always happiest when beyond the curve. The pair showed that action at a distance could work in a steady-state universe but not in a big bang universe. This discovery was a powerful argument in support of Hoyle's conviction that, despite the successes of the big bang model, it must be wrong. The two would continue to work on their different approaches to cosmology for three decades, undeterred by the lack of support from consensus cosmologists.

CHAPTER 11

The Watershed

B y early 1971, the problem of continuing funding for the Institute of Theoretical Astronomy (IoTA) had become critical. As we have seen, the Nuffield Foundation and the Science Research Council (SRC) had agreed to bear the lion's share of the running costs (80 percent) until July 31, 1972, at which date it was assumed the University of Cambridge would take full responsibility for the salaries. Of course, there would have to be a limit as to how much the university could afford for the support of astronomy and astrophysics.

From the teaching point of view, the radio astronomers in the Cavendish provided physics lectures and laboratory classes for the Natural Sciences Tripos, while the teaching officers at the Department of Applied Mathematics and Theoretical Physics (DAMTP) carried a large load for the Mathematical Tripos. The permanent staff of the Observatories had graduate students but played no role in under-graduate teaching. This was not due to a lack of willingness on their part but simply reflected the fact that observational astronomers worked at night and in observatories overseas. Most of the institute staff were on short-term research contracts and therefore were unavail-able to support undergraduate teaching, although they had graduate students. In any case, the whole point of founding the institute had been to free the best minds for research, which is partly why George Batchelor had opposed its foundation in the 1960s.

At the research level, the university supported three branches of astronomy: optical astronomy at the Observatories, radio astronomy in the Cavendish, and theoretical astronomy at both DAMTP and the institute. Research in optical astronomy did not attract international attention, perhaps because the university had failed to invest in new facilities. By contrast, the radio astronomers and theorists were conducting world-class research. Any important discovery they made was likely to be reported immediately in the media, often internationally, and this brought considerable credit to Cambridge University as a whole. No other British university was making such a large and varied commitment to astronomy.

The uncertain funding outlook began to affect the staffing of the institute adversely. Although the summer visitors continued to flow from America, Hoyle could not offer anyone a permanent job. Back in 1967, Leon Mestel and Dennis Sciama in DAMTP both decided to seek opportunities that would be independent of Fred and the institute. They had their eyes on career development, which Hoyle could not offer, partly because he was not permitted to poach teaching officers from DAMTP. Sciama went to Oxford as a fellow of All Soul's College, and Mestel went first to Manchester and then Sussex. Stephen Hawking had joined the institute staff but in an unestablished position. He still lacked a permanent university office when he was elected a fellow of the Royal Society in 1974; at age 32, he was one of the youngest fellows in the long history of the society. Three years after opening the institute, Hoyle lost two of his brilliant stars, Faulkner and Strittmatter, both to California.

It is possible that, when Hoyle chose a site for the institute next to the Observatories, he had a long-term plan. In the early 1960s, the university lacked a credible policy for the future of optical astronomy, which was not the fault of the university. The nation had emerged from the chaos of the war with no strategy to compete with the large optical telescopes at the Lick Observatory, Mount Wilson, and Palomar; at home, the Royal Observatory had chosen a completely inadequate site on which to conduct optical astronomy and had then compounded that error with observing programs that lacked cutting-edge astrophysical content. No wonder the young observers with talent had moved permanently to California and South Africa. Hoyle knew that the professor of astrophysics, Roderick Redman, would retire in 1972. He may have calculated that one or two other staff members would reach their limit of tenure before that date. Furthermore, Hoyle possibly had a somewhat low opinion of the research staff in the Observatories. After all, he was ever full of praise for the Californian observers but remained silent on his Cambridge colleagues whose publications had dwindled to a trickle. But the Observatories had lacked access to a decent telescope, so how could its staff compete with the expatriates such as Wallace Sargent and Margaret Burbidge? It would be surprising if Hoyle had not been thinking, back in 1965–1966, that the university could quietly run down optical astronomy at the Observatories, thus releasing space and jobs for the future institute.

If Hoyle did have those thoughts in 1965, he could not have foreseen that he and the SRC would have transformed the prospects of UK optical astronomy by the early 1970s. As we have seen, Hoyle had played a large part in ensuring the success of the Anglo-Australian Telescope (AAT), the 48-inch Schmidt, and plans for a Northern Hemisphere observatory in the Canary Islands. By early 1970, the university realized it needed a new strategic plan for astronomy to cope with the arrival of much better national facilities for optical astronomy and the impending lack of funding for theoretical astronomy. During Lent Term 1970, two members of the university's General Board visited Hoyle on a mission: They had some news he did not want to hear.

The General Board proposed to shut down the Observatories in September 1972, on the retirement of Professor Redman, unless Hoyle would agree to an amalgamation of IoTA and the Observatories. Redman had already been under instruction, since 1967, not to fill any academic positions that fell vacant. The economic argument was that no other university was conducting astronomical research in four separate departments. Cambridge now needed to reduce the number of departments, to lower overhead, to deliver an efficiency gain, and to save money. Furthermore, the academic staff of the merged institute would have to pay greater heed to the undergraduate teaching needs of the physics and mathematics departments. But if Hoyle did not want a merger, the Observatories would close down. So, why on hearing of the possibility of shutting down optical astronomy and concentrating on theory did Hoyle not joyfully embrace the two messengers?

He had recently dropped his opposition to an amalgamation for two reasons. First, the SRC had more funding for astronomy in 1971 than in the mid-1960s, partly thanks to Hosie's skill at swelling the budget, and Hoyle hoped to get a decent grant from the SRC to secure IoTA for a further period. Second, all his work for the SRC on improving facilities for UK optical astronomy had led him to conclude, quite correctly, that the universities must start producing young professionals properly trained in conducting observations. On reflection, it seemed to Hoyle that Redman's retirement gave the university an opportunity to appoint an experienced user of large telescopes to the chair of astrophysics. Of course, he already had a candidate in mind, Wal Sargent, but he did not tell the General Board's emissaries that.[1]

The General Board responded with a classic Cambridge maneuver: They set up a small committee to advise on the merger of the two departments. This committee took the future of the Observatories as its main concern but summoned Hoyle's opinions. It also sought the opinion of Hosie at the SRC as to how the university should profit from the opportunities being offered by the new telescopes. Its findings became incorporated in a Report of the Council of the School of the Physical Sciences to the General Board, dated February 24, 1971. Hoyle, as a member of the council, signed the report without asking for any reservation or minority opinion to be added. Redman certainly worked very hard to secure a good outcome for the Observatories, without causing any harm to theoretical astronomy. Notes that he took at the meetings and wrote up later in the evenings are a testimony to his constructive efforts.[2]

The report contained one recommendation that Hoyle found deeply unattractive. At the behest of Hosie, who, remember, held the SRC purse strings, it recommended that Redman's chair "should be filled by an Astrophysicist with a strong interest in fostering instrumental development."[3] Hoyle had made strenuous efforts to keep instrument development—Hosie's big idea for the Observatories—in the background. At one stage Hoyle told the committee that the university should "stop preparing to do astronomy and really do some."[4] In other words, a university with the standing of Cambridge in astronomy should concentrate on getting first-class observers at the business end of the telescope and leave instrument development to other universities.

Was Hoyle justified in being opposed to instrument development as the new focus for optical astronomy at Cambridge? That is very hard to judge. His American experience hardly helped because the Lick Observatory, Caltech, as well as the Mount Wilson and Palomar observatories had telescopes dedicated to their host institutions that were supported by excellent workshops. The observers in those institutions had technical staff to make instruments. The UK was on the point of having access to international telescopes, but the skills of instrument development had been lost. Any large telescope needs cameras, spectrographs, and interfaces with computers. If the UK could not take some responsibility for these needs, it could hardly expect favorable allocations of observing time on international projects. Cambridge, through the Cavendish Laboratory, had an international reputation for inventing novel instruments—Ryle's interferometers, the electron microscope, the mass spectrograph, the cathode ray tube. It must therefore have appeared a reasonable bet to other members of the committee, as well as the SRC, to make instrumentation the priority at Cambridge.

The issue dragged on through the early summer of 1971. Hoyle was by now getting deeply concerned that no application had gone to the SRC to extend the institute's grant beyond July 31, 1972. His staff members became increasingly concerned about their prospects. At the institute, knots of people would sometimes assemble gloomily in the evening, chewing over the uncertain outlook.

The report of the General Board on the future of astronomy in the university saw the light of day on July 21, 1971.[5] Hoyle was not a member of the General Board, nor did he have any chums among its 12 members to speak for him. This report included a series of recommendations, the effect of which merged the institute and the Observatories into a new Institute of Astronomy, commencing on August 1, 1972. The university promised to keep its funding at the same level but not to increase it. The 17 members of the graduate staff at IoTA would need support on outside funds (i.e., the SRC) if their posts were to be renewed. There would be three professorships: the Plumian (Hoyle), theoretical astronomy (Lyttleton), and astrophysics (to be filled). Following approval of the report by the university, the grant application could be made and a successor appointed to Redman's chair of astrophysics.

If this all sounds unduly complex, it is because that is how such matters were (and still are) generally conducted in Cambridge. All changes in administration, hiring, and funding had to be routed through a Byzantine committee structure which could take months, even years, to respond. Fred Hoyle never came to terms with "the system," its attendant bureaucracy, and the need to win quiet support from colleagues.

The July 1971 report appeared to be a reasonable outcome from Hoyle's perspective. He could play along with his senior colleagues for now. Filling the chair of astrophysics assumed a higher level of urgency because the SRC could hardly be expected to approve a major grant to any institution with a vacancy in a senior post.

Boards of electors, comprising eight senior academics drawn from Cambridge and other leading institutions, fill professorships in Cambridge. In the case of the chair of astrophysics, the eight electors were appointed until September 30, 1971, after which a renewal would be more or less automatic for any elector wishing to continue with the unpaid position. This board met twice, well before September 30, to make a recommendation on Redman's successor.[6] At its first meeting, the board drew up a short list of about a dozen candidates, including Hoyle's favorite, Wal Sargent. At its second meeting, the board postponed any decision, presumably because not all of the candidates who had been sounded out had responded.

The university authorities dispatched official papers in August calling for a meeting of the electors. This led Hoyle to believe that, if the electors could meet in September, he could get his grant application to the SRC in October. But the month of September falls in the long vacation, the period when university officers are not required to be in the university. Although a September date was agreed provisionally, the earliest date on which it was possible for the electors to actually meet was in October 1971.

The Cambridge University *Reporter*, an official gazette published weekly, had carried a notice in August 1971 calling for suitable candidates for the chair of astrophysics. The announcement stipulated "an astrophysicist with an interest in fostering instrumental development." This wording was chosen after much discussion within the university and Hoyle had fought against it. In the institute, some of the staff felt that the wording had been deliberately framed to disqualify Hoyle's candidate and to exclude theoretical astronomers.

Hoyle's commitment to the Anglo-Australian Telescope took him away from Cambridge in the second half of October and the first week of November, causing him to miss the October meeting of the electors. When he returned to Cambridge, he found the minutes of that meeting, plus the papers for the next meeting of the electors, scheduled for November 26, 1971. To say he was shocked at what he read is an understatement. He was utterly devastated and momentarily at a loss.

Something irregular was going on behind the scenes, he concluded. First, the grand scheme for instrumentation was now lying at the heart of the board's strategy. They had already identified a front-runner from the list of candidates. Hoyle was clinging to the hope that he might still be able to argue for a change of plan, although he had not prepared a written case. Second, and far worse for him, the university had made two changes to the board of electors rather than renewing the pair whose terms expired on September 30. He felt that his interests had been sidelined during his absence.

Hoyle's understanding of the normal practice at Cambridge was that, if an elector's term finished after the date of the notice of a vacancy, the expiring elector would simply be renewed. Importantly, electors should not be changed once they have started on their deliberations to choose a new professor. Hoyle stared in complete disbelief at the papers now before him. A distinguished Cavendish physicist, Brian Pippard, and Francis Graham Smith (a former member of Ryle's group at the Cavendish) had replaced two of the people he thought he could count on, Sir Bernard Lovell and Professor P. Hall.

Hoyle and Lyttleton wrongly chose to interpret the changes to the electors as blatant gerrymandering, and they wrote furious letters of protest, claiming that in the entire history of the university there had never been a case where an original member of an appointments committee had not been retained until an appointment was made. They would later discover that the two electors who were rotated off received no communication from the university whatsoever, despite having attended meetings earlier in the year. Lovell's first knowledge of the appointment of Redman's successor came from reading the official announcement in *The Times*. However, the university had meticulously followed its established procedure by publishing the changes in the *Reporter* before September 30. Since neither Hoyle nor Lyttleton protested at that earlier stage, one can only conclude that they failed to read the notification.

More than two decades later, Hoyle claimed that he had expected the votes on the old board to tie at four–four, which was probably correct. He had fretted that the Cavendish men, Sir Nevill Mott and Sir Martin Ryle, would have been against him. He had also assumed that Sir Richard Woolley would have taken every opportunity to vote in a direction opposite to his own because the two had had bitter arguments over the future of the Royal Observatory. Hoyle seems to have become rather paranoid about the changes to the board of electors because, as he recounts in his autobiography, he had anticipated that the new board would block the vote against him with four members associated with the Cavendish. Adding Sir Richard to these four gave, at best, a five–three division against him. His agreement to rescue the Observatories had led to this grim situation.[7] And because his own knighthood still lay a few months in the future, he was not in an ideal position to pull rank at the forthcoming meeting.

In a highly charged atmosphere, the vice chancellor personally chaired the meeting of the eight electors on November 26. Who were the candidates before them? Because the university library has recently closed the archive on this meeting, it is not possible to give the list. However, the list as examined before closure tells us that the university had solicited applications from about one dozen distinguished candidates in the UK and the United States, drawn from both theory and observation. One British candidate had had the courtesy to ask that his name be withdrawn. Hoyle's preferred candidate, Professor Wallace Sargent of Caltech, neither applied nor did he receive a request to submit a *curriculum vitae*. About 3 weeks earlier, Sargent had told one elector, Pippard, that he was not interested in a professorship at Cambridge. Prior to the meeting, the electors would have had a duty to study the astronomical work of the candidates and consider whether that work could be expanded were they to be given the Cambridge post. Some of the electors already had deep personal knowledge of some candidates. Sargent and Donald Lynden-Bell, for example, had both worked for Woolley, who had the highest regard for their abilities.

The vice chancellor opened the meeting of November 26 with a proposal that the issue be decided there and then. He noted that a preference had emerged at the October meeting, the one Hoyle missed. Everyone present knew that Sir Richard Woolley would retire from the Royal Greenwich Observatory on December 31 and leave immediately for South Africa. This would remove one of Hoyle's opponents, so naturally Hoyle played for delay, or "time for further reflection," but he had no support for that.[8]

Hoyle had his opportunity to propose Wal Sargent, about whom he must have spoken with warmth and enthusiasm. However, there was no support for that because Sargent had already said he would decline the post because in Cambridge he would not have a big telescope at his beck and call.

Hoyle then suggested they consider the theorist Geoff Burbidge, whose wife was shortly to assume the directorship of the Royal Observatory. To Hoyle's irritation, electors quickly dismissed this distinguished candidate "as a bull in a china shop." The electors then moved to a resolution to elect Donald Lynden-Bell, a theorist who had been at DAMTP before accepting a senior position at the Royal Greenwich Observatory. They carried this motion by five votes to three. The four electors associated with the Cavendish Laboratory, plus Woolley, voted for the motion. Hoyle, Sir Harrie Massey, and William V. D. Hodge voted against. Woolley had personal knowledge of Lynden-Bell's outstanding abilities and had seen him in operation as an inspiring theorist surrounded by observers. Hodge, whom we have seen strongly supported observational astronomy, possibly felt that an applied mathematician would not have the correct mix of skills to run an observatory, so that may be why he voted against.

According to Pippard, the four electors from the Cavendish Laboratory had no discussions prior to the election itself. Hoyle became lost in a fog of self-deception over this election: There was no conspiracy to spite Hoyle. Furthermore, in considering who was the most able candidate, the observational astronomers on the board would have been aware that Hoyle by this time was on a wild goose chase, claiming that quasar redshifts were not due to the expansion of the universe. Perhaps they felt that in the new Institute of Astronomy they needed to appoint a professor who could counterbalance Hoyle's wilder ideas and take over when Hoyle would retire 3 years hence, at age 60.[9]

When the news broke, it initially caused concern among the younger and untenured members of staff in the institute. I was by now a staff member, having been appointed to a visiting fellowship. Our contracts would expire in 7 months, on July 31, 1972, and no grant application had gone to the SRC. Our immediate fear was that Fred would resign, leaving us in the lurch. He had been spending far less time in the institute, as if he sensed impending doom. However, he did not resign in November because he first wanted to ensure that the SRC would renew funding for the institute's research staff for a further year. The council could hardly be expected to support an institute that was bereft of a director.

New Year's Day 1972 brought him a wonderful surprise: a knighthood from Queen Elizabeth II for services to astronomy. That was welcome public recognition. He had to wait until February 8, 1972, for the good news from Jim Hosie that the grant application had gone through. But even then he hesitated to throw in the towel. Whatever his feelings were, he reined them in. Did he perhaps think that he should, after all, be able to work with Lynden-Bell? On a personal level, there was never any tension or animosity between Fred and Donald. Each had mutual respect for the other's achievements and a shared delight in applying mathematics to astrophysics.

February and March 1972 took Fred away to Australia on government business concerned with the future management of the Anglo-

Australian Telescope. Margaret Burbidge, who had been designated in 1971 as Woolley's successor to direct the Royal Greenwich Observatory, took his position on the AAT Board. By early 1972, the construction of the telescope and its observatory was going smoothly. But the planning for the future management of the observatory was making no progress.[10] In an effort to tackle the disagreements over management, Margaret Burbidge arranged to host a board meeting in La Jolla, California, roughly equidistant from the UK and Australian capitals.

Fred Hoyle and Jim Hosie traveled on the same flight to Los Angeles, just days after the SRC's approval of the new grant for IoTA. On the long flight to Los Angeles, they discussed many questions concerning the SRC and British astronomy. Hoyle had his important positions on many key committees and had recently commenced his term as president of the Royal Astronomical Society, so there would have been plenty of issues to talk about.

In the course of this no doubt animated conversation, Hoyle informed Hosie that he would insist on being the new institute's first director. However, Hoyle had already indicated to Lynden-Bell that he would not wish to head the Institute of Astronomy beyond his sixtieth birthday. That would have given him a 3-year stint, though university ordinances specified a 5-year appointment. As Fred later put in a personal letter to Barbara,[11] "[Hosie] responded . . . in genuine surprise, and for a few moments his remarks were entirely spontaneous." From Hosie, Hoyle learned that "Cambridge" had been in touch with the SRC about the new grant, all behind Hoyle's back. In the course of those discussions, "they" had asserted that Hoyle did not wish to be the head of the merged institute, and "they" had even asked the SRC for the go-ahead to appoint Professor Lynden-Bell as the first director of the new institute, which is what they had decided. Hoyle would later learn that the actual grant application submitted by the university had Lynden-Bell as the potential grant holder. But there is a natural explanation for this that does not involve a conspiracy: Hoyle was in the United States and then Australia and therefore not able to sign the long-overdue grant application.

While still in flight to Los Angeles, Hoyle decided that he must stage a showdown with the vice chancellor, to whom he immediately drafted a letter stating that he would not stay in Cambridge except as head of the institute. After landing in Los Angeles, he ventured south to La Jolla, where he decided to sleep on the letter for two nights, perhaps conscious that jetlag might be making him impetuous.

By the time he felt refreshed, he had changed his mind. No, he would not write with a veiled threat, fearing that if Cambridge called his bluff, his actual resignation would then lose impact in the media. It might look as if he were being petty, and he was determined to cause an explosive showdown on a national scale.

On February 14, Hoyle dispatched a three-paragraph letter of resignation to the vice chancellor. He opened by saying that he had decided to resign over the appointment of Lynden-Bell to the chair of astrophysics, but he had stayed his hand until the 1-year holding grant had been arranged. The second paragraph informed the vice chancellor of his resignation from the Plumian Chair from July 31, 1973, thereby giving 16 months' notice. The last paragraph protested that "for a substantial block of electors manifestly to have discussed a name in advance without reference to me, the present Head of IoTA, . . . was a truly monumental discourtesy."[12] The discussion to which he refers took place at the meeting of the electors he missed, where the merits of several candidates must surely have been discussed.

The same day, he sent Peter Vaugon, Westwater's successor as secretary of IoTA, an announcement for the notice board. The statement started by pointing out that when IoTA was founded, there had been an informal understanding that its future beyond 1972 would be discussed in 1970. This had not happened because "the university became concerned instead with the future of the Observatories." Hoyle's announcement continued that he had "agreed to the proposed merger . . . because I had no wish to have optical astronomy die in Cambridge." Then came the sting: He had hoped the university would appoint an astronomer with a distinguished record in optical astronomy. Finally, he gave his own interpretation of recent events: Forces within the university have seen the new arrangements as an opportunity to downgrade IoTA. The institute had become a department with an international reputation "which some people have found embarrassingly high."[13] Fred did not use such polite language in his personal letter to Barbara: He denounced them as "lying knaves" and promised to root them out.[14,15]

What really brought about his resignation and what did he mean to achieve? We have some clues in his autobiography, the letter to Barbara, and his interviews with the media. Paranoia and conspiracy are the words that spring to mind in trying to understand this puzzle. For whatever reason, at a very deep level he chose to interpret the various actions of the professors in the Cavendish Laboratory as a plot to force him out of Cambridge. He felt that "the SRC was sucked into a clever plan."[16] In his eyes, Jim Hosie and the chairman of the SRC had really let him down by not pressing the university to reach an accommodation. As he confided to Barbara, "Had the SRC told the university that only if I were given unstinting support would there be substantial backing from them I think we would have had little trouble." He promised her he would stage "the best roughhouse you have seen" and concluded with, "There is no other way to play the cards in this situation. [My secretive opponents] will not be winkled out except by a full scale explosion."

The university regarded the resignation letter and the accompanying notice as private. Nothing was released to the media, nor was anything communicated openly to the IoTA staff. The lack of action was understandable: The vice chancellor wished to have a private meeting with Hoyle to persuade him to change his mind, as he had done once before in 1964. But Hoyle would be away on his travels in California and Australia until the beginning of April, so the resignation letter was laid to one side, or so it seems today, because no reply was sent. The vice chancellor's wife was in the last stage of a terminal illness, and there may well have been confusion among his staff as to whether he or one of the officials in the General Board should respond.

The chairman of the SRC, Sir Brian Flowers, rushed to Cambridge and had hard words with the vice chancellor over the university's handling of the affair.[17] One can imagine these senior academics being concerned that fallout from the impending row would harm science funding. Donald Lynden-Bell wrote a generous letter from Sussex, expressing surprise at Fred's decision. He made the imaginative suggestion that with Margaret Burbidge (accompanied by Geoffrey) taking up the directorship of the Royal Observatory, maybe they and Fred could start a new theoretical center of excellence in Herstmonceux Castle. (But would Willy Fowler have moved from Cambridge to Sussex, I wonder?) Lynden-Bell hoped Fred would find a way to reverse his decision.[18]

At Jodrell Bank, Sir Bernard Lovell received his copy of the resignation letter together with a personal note penned by Fred. The critical point of this letter reads thusly:

> I do not see any sense in continuing to skirmish on a battlefield where I can never hope to win. The Cambridge system is effectively designed to prevent one ever establishing a directed policy—key decisions can be upset by ill-informed and politically motivated committees. To be effective in this system one must for ever be watching one's colleagues, almost like a Robespierre spy system. If one does so, then of course little time is left for any real science.

On receiving this sad letter, Lovell asked Hoyle if he would consider coming to Manchester. Hoyle grabbed this lifeline, and Lovell began to put arrangements into place. While in California and Australia, Hoyle had let slip, deliberately, that he had resigned. As he put it in the letter to his wife, "I am now much better able to make enquiries about other possibilities—I ought not to let slip the opportunity of this trip." He received verbal messages of support from his American and Australian contacts, but no firm job offer.

One afternoon in early April, Fred Hoyle left his home in Clarkson Close. He strolled across St. John's College playing fields, remembering with pleasure how Ray Lyttleton, 20 years earlier, had won a seemingly lost game of cricket by scoring a brilliant century.[19] He wandered along the Backs to Garret Hostel Lane with its humpbacked bridge. His destination was the Master's Lodge of Trinity Hall, where he had an appointment with the vice chancellor, Alec Deer. Hoyle and Deer were both elected fellows of St. John's in the same year, so they knew each other well. It was therefore no strain for Hoyle to tell his friend that the decision of February 14 remained unchanged. As they talked, they consumed half a bottle of sherry. Then Hoyle drained his glass for the last time, set it down on the table with a "thank you," and that was that. His long association with Cambridge, a 39-year stint, was over.

By late April 1972, Hoyle was back at La Jolla for another meeting of the AAT Board. On April 23, 1972, a firestorm erupted in the UK. A news item in *The Sunday Times* proclaimed: "Hoyle quits in row over bureaucracy."[20] In a transatlantic phone call, the reporter asked if he would now "join the brain-drain," which brought the response, "I am turning the options over in my mind." The following day Pearce Wright, the science correspondent of *The Times*, claimed to have contacted several prominent fellows of the RAS who were critical of Cambridge politics and the failure to keep Hoyle on board. *The Guardian*, under the headline "Hoyle and Water," pressed the view that the appointment of a theoretical astronomer to Redman's chair had been the last straw. The *Daily Telegraph* quoted him as saying, "I may be an awkward customer, but I have my own way of doing things and I can't stand bureaucracy. I have resigned without a suspicion of regret and there is no hope of a reconciliation." He made it clear that his resignation was about the system and its secrecy; he had no personal reason to oppose Lynden-Bell.

An article in the April 28 issue of *Nature* considered the impact on the scientific community. Its author pointed out that Sir Fred's resignation could leave important gaps to be filled on the SRC and the AAT Board. The final paragraph said that, with Hoyle's resignation, "British astronomy will have lost the advantage it gained last summer when Professor Margaret Burbidge was wooed back from the United States." Hoyle did not, in fact, give up any of his official positions. Margaret Thatcher, as Secretary of State for Education and Science, extended his appointment to the AAT Board so that he could see the project through to completion.[21]

By early May the institute staff fully accepted the changes and settled to supporting Lynden-Bell, who now had the funds to extend the contracts of the theorists for a further year. Some staff hoped that Professor Martin Rees, who had taken a position at Sussex, would return to Cambridge as Hoyle's successor to the Plumian chair. On May 8, *The Times* quoted Hoyle as saying that he "would not automatically take a post in the United States," to which he added that a return to Cambridge "would not be good for him or Cambridge." This article provoked the chairman of the Faculty Board to write a strong letter to

Hoyle, using a Royal Society letterhead. Concerning the directorship, he evidently decided to lance the boil:

> May I assure you that no Director has been appointed. I have no knowledge of the origin of the reports that Professor D. Lynden-Bell has been appointed [director]. The Board will not make an appointment without consultation with Redman, Lyttleton, and yourself.[22]

Hoyle responded somewhat weakly to this. The origin of the report had been Hosie's claim on the flight to Los Angeles that the university had asked if "the SRC would think it reasonable if D. Lynden-Bell were appointed Director of the Institute, since Professor Hoyle did not wish to be Director."[23] On hearing this, Hoyle had jumped to the wrong conclusion, incorrectly assuming that an appointment to the directorship had already been made. It had not. An article in *The Sunday Times* of May 14 claimed that "there is a distinct feeling at Cambridge that nothing would cause greater panic than a change of mind by Hoyle." An unnamed university spokesman volunteered that "the main worry among the majority of astronomers is that Hoyle's departure might affect the inflow of funds, for everyone concedes that British astronomy has never known such a fundraiser."

At Manchester, Sir Bernard Lovell lost no time in finding a position for Hoyle. On July 3, 1972, the senate of the university approved Lovell's recommendation that

> Professor Sir Fred Hoyle, MA (Cambridge), ScD (East Anglia), FRS, be invited to accept appointment as Honorary Research Professor of Physics and Astronomy, to hold office from a date to be arranged to September 30th following his sixty-seventh birthday.

Hoyle took up this appointment on January 1, 1973, about the time he and Barbara moved from Cambridge to Dockray in the Lake District. From that base, he frequently visited Manchester and Jodrell Bank both for discussions and lectures. He continued as a brilliant lecturer playing to packed audiences in both public lectures and scientific colloquia.

Lynden-Bell inherited a merged institute with an incredibly strong base in theory and an important history of observational astronomy. He had a little internal problem to deal with from a handful of people who did not care for the use of the expression "institute" and the absence of "Observatories" in the title "Institute of Astronomy." However, in practice, everyone worked under Lynden-Bell's direction to ensure that the new department would be successful. The university supplied a good measure of financial support, grants came from the SRC, and Lynden-Bell forged harmonious relations with DAMTP and the Cavendish. Martin Rees returned as Plumian Professor in 1973, and he would become an important magnet for research students and postdoctoral workers. The annual departmental photograph for that year (the 150th anniversary of the Observatories) is packed with bright young faces who 30 years later hold prestigious chairs and direct major departments the world over. Stephen Hawking, despite becoming increasingly frail, continued to visit the institute three or four times a week. The program of summer visitors flourished, major international conferences continued to be hosted, and there was a clamor for office space from researchers on sabbatical leave. Ray Lyttleton retreated to a large empty office that had formerly been Hoyle's. Lynden-Bell signaled his commitment to observational astronomy by basing himself in Eddington's former dining room in the historic university observatory. The resignation was a bitter blow for Willy Fowler. With no Fred in Cambridge, he had to revise his own retirement plans (he remained in Pasadena) and sell his house on Oxford Road.

Fred Hoyle gifted the university a superb institution with a world-class reputation. Everyone would agree that his successors to the directorship (which changes every 5 years) and to the Plumian chair have ensured that Cambridge has grown into one of the largest and most productive schools of astronomy in the world.

Stones, Bones, Bugs, and Accidents

An immediate consequence of his resignation was that Hoyle was no longer the holder of a mainstream university post, and therefore he would no longer have research students, post-doctoral research assistants, or research staff alongside him. Barbara Hoyle would have to function as his typist, his diary manager, and his publishing agent. The professor who believed so strongly in high-quality support for research had suddenly cut himself off: At the mundane level, he no longer had the right to use a photocopier free of charge and, more seriously, he could no longer apply to the Science Research Council (SRC) for research funding. Worse still, he had resigned 10 years before the retirement age and this would have a devastating impact on his future income.

Did he really have to do this? If instead of resigning the chair, he had simply resigned the remaining months of his directorship of the Institute of Theoretical Astronomy (IoTA), he could still have made his point. Though his Plumian chair was assigned to the new Institute of Astronomy, he could have worked from St. John's College or from home. So long as he had carried out any teaching duties required of him, the university would have left him alone. He would have been free to travel to Caltech, to write books, and to give acclaimed public lectures. Perhaps he could have developed a sideline as an anti-establishment pundit with a radical newspaper column. Instead, he opted for complete severance, which reveals how strong his feelings were. In 1933, he had come to Cambridge as an outsider, and now he would be leaving with that status.

For whatever reason, in the period from 1970, Hoyle allowed himself to be consumed by corrosive paranoia. The clashes with Ryle, Batchelor, and others should have been laid to rest, but Hoyle's action suggests he perceived that they were still plotting his downfall, and he must have been deeply unhappy (perhaps suffering from stress?) to resign a position that had been held by his hero Eddington. He probably, but wrongly, attached too much importance to threats of resignation as a weapon in Cambridge politics. Bernard Lovell recalls some advice he received from Hoyle: "In 1965 to 1970 on many occasions he urged me to resign when things were going against us, but I always persuaded him in those days to stay put and fight."[1]

In terms of official standing in the profession, he continued as a member of the Anglo-Australian Telescope (AAT) Board until May 1975, steering it through very difficult times, and he served his 2-year term as president of the Royal Astronomical Society (RAS), delivering a brilliant presidential address at the February 1973 meeting.

For that talk, he returned to a much-loved theme, the origin of the universe. His opening remarks show that he had lost none of his consummate skills as a provocative popularizer of science:

> The discovery of the expansion of the system of galaxies poses a problem concerning the origin of the Universe which after fifty years of work by cosmologists appears perhaps more mysterious today than it did at the end of the first quarter of the century. Much effort has been expended in discussing alternative cosmological models, with the hope that a particular model might be singled out through a better agreement with observation. This effort has been unproductive, however, since even five decades of accumulated data have not sufficed to distinguish with sufficient accuracy between the various possibilities.[2]

In the lecture, he explored the issue of what came before the big bang. For him the central problem was "the meaning we attach to the concept of an origin for the Universe." He unburdened himself to his audience by explaining that the conceptual difficulties of a sudden origin are "so insuperable that it seemed almost inevitable that one had to by-pass them" with a steady-state model involving the creation of matter. After 15 minutes the address became quite technical, as if he was booking his place in the history of cosmology rather than giving the fellows an entertaining talk. And yet, if the RAS had had a press officer in 1972, the address would surely have been front-page news because of his conjecture that the observable universe is but part of a much larger (and unobservable) universe. He said: "The structure we ordinarily think of as the Universe is but part of a larger whole."

The Cambridge difficulties did not deflect him from his duties with the AAT Board, where he faced a difficult political challenge. When the assembly of the telescope had started, the British and Australian members could not agree on the future management of the telescope or the location of its administrative and research offices.[3] As a board member, Hoyle threw himself into these issues, always wanting to ensure the best possible outcome for British astronomy. On management, the unresolved question was whether the board or an Australian university should run the show. The Australians favored putting the management in the hands of the Australian National University (ANU) or Mount Stromlo Observatory, both in the Australian capital Canberra. The SRC and British board members perceived a series of political maneuvers in Australia as a blatant attempt to grab control of the telescope. The SRC knew it could rely on Hoyle to resist this takeover bid, and it was for that reason that Margaret Thatcher, as the minister responsible for science, extended Hoyle's term on the board.

Hoyle, together with the other British members, Jim Hosie and Margaret Burbidge, faced a tense struggle to retain control for the board because the Australian minister for education and science, Malcolm Fraser, chose to interfere on behalf of ANU. He summoned Hoyle and other board members to his office in Parliament House, Canberra, for a severe dressing down. He shouted at the three British members: "The Australian government is already pouring cash on the support of optical astronomy at the ANU and we cannot afford to support a second, independent, institution." The ANU felt that Hoyle and Hosie were looking for an opportunity to establish "a base for a British operation of no small magnitude," a British Southern Hemisphere

Observatory complete with a millimeter-wave radio astronomy dish. Although in the agreement to cooperate signed in 1970, both governments had signed away ownership of the telescope to the board, the minister now wanted control firmly in Australian hands. Fraser gave a veiled threat that, if necessary, he would ask Margaret Thatcher to dismiss Hoyle, Hosie, and Burbidge if they would not back down.

Margaret Burbidge convened a second board meeting in La Jolla in April 1972 to resolve the issue. The Australians pressed a narrow interpretation of the 1970 agreement, saying it was an agreement about a *telescope*, not an *observatory*. Hoyle, Hosie, and Burbidge would not fall for that semantic ruse. The participants later reported that it was the most difficult, hard-fought meeting they had ever attended. Only the calm voice of Margaret Burbidge maintained a sense of decorum. In any event, the Australians lost the vote (the only one the board ever took), which meant that the board had resolved to appoint its own scientific director, together with support staff, technicians, and astronomers. Fred Hoyle had made a big personal sacrifice by attending this crucial La Jolla meeting, when he should have attended to his postresignation affairs in Cambridge. But he knew it was only Margaret's second meeting, and he wanted to be sure of being present at the vote.

The Australian astronomical community as a whole backed the solution, although the cabal trying to scoop the "scope for ANU" did not give up the fight. In August 1972, Fraser and Thatcher met in Canberra, with AAT management issues at the top of the agenda. Fraser still wanted a single integrated observatory in Canberra, firmly under Australian control. Thatcher, cool and well briefed, succinctly informed Fraser that she regarded his ministerial interference as completely unacceptable, and she certainly would not allow control of the telescope to pass to a third party. Her gritty performance meant the board could at last commence its search for a director. Hoyle and Margaret Burbidge successfully argued that the appointee must be the best available astronomer, with no restriction on nationality, and chosen as the result of an international search. Fred must still have been smarting from what he perceived as the secretive machinations at Cambridge.

A new crisis hit the board at the end of 1972. A general election in

Australia brought to power the first Labour government in decades. The day after the government assumed office, the vice chancellor of ANU wrote to the new prime minister urging that the university, not the board, should appoint the director. Shortly thereafter, the new minister for science, W. L. Morrison, summoned Taffy Bowen, then the chairman of the board, who had just been appointed to an attractive position in the Australian Embassy in Washington. Morrison demanded that ANU run the telescope, to which Bowen responded that the board was a body legally established by two governments, one of which was now on the verge of acting illegally. Morrison hit back with a threat to sack him as chairman, which produced the response that a minister had no power to dismiss a board member. At this, Morrison exploded with anger, producing his ultimate weapon: If Bowen wished to take up the diplomatic appointment in Washington, he must resign from the board because "as minister I need a Chairman close by whom I can consult on a daily basis." Bowen resigned on the spot and the board temporarily had no chairman.

The headstrong minister had failed to take account of the rules on chairmanship (his officials had not had time to brief him). With the resignation of the Australian chairman, responsibility for filling the position passed to the UK, which appointed Fred Hoyle. A promotion at the SRC next removed Hosie from the board. With both him and Bowen gone, and Margaret Burbidge's impending resignation from the Royal Greenwich Observatory, Hoyle became the most experienced board member. He still had much to achieve.

He urgently needed to appoint the first director. In his search, he sounded out his contacts in the United States, a recruitment method that had been successful for IoTA. To his dismay, he quickly found that the direct interference of Australian ministers had seriously tarnished the appeal of the AAT. In March 1973, Hoyle was back in Canberra to interview two candidates, Hanbury Brown (by then at the University of Sydney) and Bev Oke (Caltech). Both rejected the offer of the directorship because of the continuing hostility of the ANU. Fortunately, help was at hand. Morrison decided to hold a meeting of Australian astronomers to discover their views. On finding that the local community of astronomers was heavily against the position of the ANU, the government quietly forgot about the telescope. Hoyle's next stroke of luck was in persuading Joe Wampler of the Lick Observatory, California, to accept the directorship for a fixed term of 2 years. Fred had come to know Joe through his collaborations with the Burbidges and Wal Sargent. Hoyle's choice was excellent because Wampler was a leading instrumentalist. He brought to the new telescope a state-of-the-art detector, which got observational astronomy there off to a flying start.

With Wampler in place, the next task to fall to Hoyle and the board was the recruitment of scientific staff. Hoyle used his American and IoTA experience to influence the board when it determined the terms of employment. The staff astronomers would have a small share of the "time" and would be expected to support university astronomers who had won "time" on the basis of competing bids. In effect, the board's scientific staff would be concerned mainly with collaborative programs with visitors. This arrangement contrasted with the conditions at the Royal observatories, where much of the "time" went to permanent staff. Hoyle was very keen on staff–visitor collaboration, which had worked so well at IoTA.

The period 1973–1974 was one of feverish activity for Hoyle, who made several trips to Australia. The telescope saw "first light" (took its first astronomical photograph) on April 27, 1974, some months ahead of the American 150-inch telescope nearing completion in Cerro Tololo, Chile. It now fell to Hoyle as chairman to plan an official inauguration with several hundred guests. Prince Charles agreed to perform the opening ceremony on October 16, 1974. Unfortunately, the weather at the Siding Spring site in New South Wales sprang a nasty surprise: A ferocious gale lashed the mountaintop. Hoyle's duty was to welcome Prince Charles at the entrance of the telescope dome, which is more than 80 feet above the ground. Hoyle has given us a vivid account of this moment:

> The vehicles arrived at last and there came a dash from the protocol squad to open the car doors. Gough Whitlam, the Australian Prime Minister, was the first man up. He moved bravely out through the door of the building. Whitlam is a big man, and it was not hard for him to avoid being blown away in the wind, but I recall in a kind of fascinated horror as his hair instantly stood straight on end. My turn came next. Move forward, bow, shake hands with the Prince, and make a little prepared speech of

welcome, I told myself. But every syllable was blown clean away in the roar of the wind, so that I was instantly reduced to the grimaces of a primitive man.[4]

Inside the building, 500 guests were seated beneath the dome. A string quartet entertained the waiting guests with music by Purcell and Dvorak. At the start of the ceremony, Hoyle conducted some 20 distinguished guests to the platform. After a presentation to Prince Charles, Sir Fred gave the opening address, which included the memorable comment that "a telescope is a good example of the things which our civilisation does well." Gough Whitlam commended him for that. Following the ceremony, Hoyle escorted the royal visitor and other distinguished guests who wished to examine the telescope and its ancillary equipment.

The board remained incapable of agreeing on where to locate the permanent offices. There were three possibilities: near the telescope at Siding Spring, in Canberra, or in Sydney. Barbara Hoyle accompanied Fred at the inauguration, and she visited all three locations to see which she would prefer if Fred should decide to move to Australia, a possibility he now had in mind. She liked Coonabarabran, the small town 25 miles from the telescope and dismissed Canberra (boring) and Sydney (too large). The December 1974 meeting of the board, held in San Francisco with Wampler present, likewise failed to agree. The Australians wanted Canberra or Sydney, the British Coonabarabran. Wampler made it clear that he wanted no associations with any other institution, whereas the Australian members still pressed for a close relationship with a university. In any event, Hoyle did not press for a formal resolution, which was probably a wise move on his part given the deep divisions. Wampler was allowed to build up staff on a temporary basis in Epping, a suburb of Sydney, which with the passage of time became accepted as the permanent home.

Fred Hoyle chaired the board for the last time in May 1975 and stood down on July 31, 1975. Events that began for him in 1967 had led to the completion of two fine telescopes, the AAT and the United Kingdom 48-inch Schmidt. Under Wampler, the AAT got off to a cracking start, and the renaissance of British optical astronomy commenced. Hoyle could look back on his period of facilitating telescope building with very great pride. His contributions to the board meetings, as well as his blunt handling of the politics throughout the years of turmoil, really did secure the best outcome for British astronomy.

While working hard for the AAT, Fred Hoyle also had to sort out his domestic and financial affairs. Following his resignation at the age of 57, he could not collect a pension until his sixtieth birthday in 1975, when the amount he could expect would be much reduced: By leaving at 57 rather than the compulsory retirement age of 67, he had greatly diminished his future pension pot. In the short term, he set a goal of earning the equivalent of his Cambridge professorial salary: He could accept American appointments for half the year, receive handsome honoraria from public lectures, and moonlight as a book writer.

Barbara and Fred solved the question of where to live in a characteristically decisive manner. They toured the Lake District, Cornwall, northern Yorkshire, and Central Wales: Fred needed fresh air, walking country, and opportunities to appreciate natural history. It was love at first sight when they saw an isolated Lake District farmhouse at Cockley Moor, Dockray. It had a sunny outlook, remarkable views, and 3 acres of derelict paddock. Its situation is above Ullswater, at an altitude of 1,400 feet on the slopes of Helvellyn, the highest mountain in the Lake District. They snapped it up immediately and had a surplus £10,000 to bank from the sale of the Cambridge house.

Once Barbara and Fred had made the farmhouse habitable, their closest friends began to visit, staying over for a few days. Don Clayton, the Burbidges, and Willy Fowler were guests in 1974. After his visit, Fowler exclaimed, "It's heaven on Earth!"[5] Hoyle, by now approaching his sixtieth birthday, took Clayton on grueling walks and climbs, including ones up to the summit of Helvellyn. When they paused for a breather, they talked of nucleosynthesis in nova explosions.[6] On some evenings they all enjoyed quiet dinners in country inns.

In financial affairs, Hoyle's American colleagues rallied to help him. He accepted offers of employment from Caltech and Rice University for two quarters of the year, that income more or less equaling his Cambridge professorial salary. These arrangements gave him a first-class academic environment in which to do his research, free at last from the obligations to create such environments for others to work in.

Hoyle spent most of November 1973 in Texas with Don Clayton, working on papers on nucleosynthesis. This was the first of five such meetings in which the two of them investigated a paradigm for nova outbursts in which the hydrogen in the outer skin of the star explodes. Their model included a prediction that is testable by observation: The explosion should create sodium-22, which is radioactive and decays to neon-22 with a half-life of about $2^1/_2$ years. The beta decay would then cause the emission of a gamma ray with a precise amount of energy. So far, attempts to detect gamma rays of the predicted energy from nova explosions have failed, but the search continues, almost three decades after Clayton and Hoyle made their prediction.[7]

On this first visit to Texas, Hoyle stayed at the faculty club on the campus of Rice University. In the first week, he gave a public lecture on Stonehenge (a new obsession) in an auditorium filled to overflowing and, for his efforts, he received an extremely generous honorarium of $5,000. One morning in the second week, Don Clayton's phone rang at 8:30 a.m., with Fred on the line, "Don, I think I have clumsily broken my arm by falling in the shower." Clayton took him at once to his own physician who confirmed by x ray that the ulna was cracked through and displaced. The doctor explained that the best thing would be to pull it out on the spot, and he had medications to soften the pain. Fred said, "Just pull it out now. I don't need medication!" He got more than he bargained for, because the bone did not want to reset easily. Fred gritted his teeth and, after 10 minutes of torture, it set properly. The doctor placed a soft cast over it, which Fred wore on the flight back. The sight of Fred's arm on his return home was the first Barbara knew about this accident, which he had kept from her because he did not wish to alarm her. Fred was a tough guy.[8]

In 1974, the Royal Society awarded Fred Hoyle the Royal Medal in recognition of his achievements in astrophysics. This is a major award, much more prestigious than the Gold Medal of the RAS that he received in 1968 or his Bruce Medal of the Astronomy Society of the Pacific, scooped in 1970. King William IV had established the Royal Medal in 1826. The conditions of the Royal Medal are that two are awarded each year, one in the biological sciences and the other in the physical sciences, and they recognize the importance of work carried out in the previous 10 years. The award is subject to the approval of Her Majesty the Queen, thereby acquiring the characteristic of a personal gift from the sovereign. Fred Hoyle took particular pleasure in winning the Royal Medal because previous winners from Cambridge had included Jeans (1919), Eddington (1928), and Dirac (1939).

Hoyle made his next visit to Clayton in 1975. He flew first to Canada, where he pocketed useful fees for delivering three public lectures at McGill University in Montreal. Following the third lecture, he gave several media interviews at a press conference. After questioning him on the properties of pulsars, one interviewer unexpectedly raised the matter of the award of the Nobel Prize in physics in 1974. As we have seen, this had gone jointly to Martin Ryle, for the development of radio interferometers, and to Antony Hewish, "for his decisive role in the discovery of pulsars." From the moment of the official announcement, tongues wagged that the Nobel awards committee should have included "the girl," Jocelyn Bell, as a third winner, and the tongue wagging continued for months.

In Montreal, one newspaper reporter now probed Hoyle's opinion: Did he think Jocelyn should have shared the 1974 award? Hoyle had not anticipated this question, to which he responded openly and informally, in the naive expectation that the reporter would check "the story": "Yes, Jocelyn Bell was the actual discoverer, not Hewish, who was her supervisor, so she should have been included." And for good measure he added, "When Dr. Bell confirmed her initial findings, her directors kept it a secret for months—no one in the group was allowed to speak to anyone else outside." This chance remark became "the story" and ran as big news through the industry's wire services. By responding off the cuff to an unexpected question, Hoyle had unwittingly set himself up as a very public critic of the Nobel Foundation.

International coverage of his comments exploded within hours of the newspaper interview. In London, *The Times* and the *Daily Mail* gave prominence to his opinions. The *Mail* reporting on "a blazing academic row," phoned both Tommy Gold at Cornell, who backed up Hoyle, and Professor Antony Hewish. The latter reacted sharply: "What Hoyle has said is untrue, absolutely untrue. It makes me so angry. It is ridiculous to suggest that the results were stolen." Discoverer Jocelyn downplayed her role: "It was Professor Hewish who did the preparatory work; I was the individual who analysed the data—doing the spade work." Hewish went on record in *The Times* with the accusation that Hoyle had made "an astonishing fabrication for which he could see no explanation."

As the storm broke, Hoyle headed south to Texas to join Clayton, who had received a copy of the feature in *The Times* by fax. When Don picked up Fred at the airport, he immediately noticed that his visitor looked terrified. "Don, I think I've really blown it this time! The quotation in *The Times* is actionable as libel. If Ryle or Hewish launch a libel action, I could lose everything. You've got to help me—the only way out of this situation is to write a careful letter of explanation for publication."[9] *The Times* had attributed to Hoyle the following accusation: "The directors, about eight of them, were busily pinching the discovery from the girl."[10] Hoyle explained to Clayton his grounds for concern:

> Back home, defamation of character in this manner is a very serious offence in civil law. The courts take a strong line if you wrongly accuse someone of stealing, lying, or sexual misbehaviour. A plaintiff with a strong case would have no difficulty in finding a high-profile libel lawyer to take the case on a no win, no fee basis. The point is Don, I am facing personal bankruptcy because damages awards are stratospheric and include all the plaintiff's costs as well as the defence costs.[11]

Gravely concerned, Clayton drove Hoyle down to the Gulf Coast, where the two of them stayed in a rented cottage on the beach at Freeport. Over the Easter weekend, they worked together on several drafts of a careful letter of explanation for *The Times*. Hoyle dispatched the letter on April Fool's Day, but his missive was not a prank: It was deadly serious. After reviewing the circumstances of the interview, he concluded thusly:

> Miss Bell's discovery required great persistence. When she first noticed the existence of the pulsar signals they were thought, not unreasonably, to be only man-made interference. It was Miss Bell's continued investigation of the signals, and of showing their source to be an object that changed position from day to day with the stars which constituted the crucial step. Once this step had been taken, nothing that happened from there on could have made any difference to the eventual outcome.
>
> There has been a tendency to misunderstand the magnitude of Miss

Bell's achievement, because it sounds so simple—just search and search through a great mass of records. The achievement came from a willingness to contemplate as a serious possibility a phenomenon that all past experience suggested was impossible. I have to go back in my mind to the discovery of radioactivity by Henri Becquerel for a comparable example of a scientific bolt from the blue.

So what really impressed Hoyle was Bell's willingness to consider a phenomenon that everyone else said could not happen. Much of Hoyle's research post-Cambridge had that characteristic. In the final paragraph of his letter, he proffered the olive branch:

> I would add that my criticism of the Nobel award was directed against the awards committee itself, not against Professor Hewish. It seems clear that the committee did not bother itself to understand what happened in this case.

It appears that Ryle may have taken legal action, according to an interview Hoyle gave to a journalist from *Time* magazine in the 1980s. In a personal act of kindness, Ryle settled for a written undertaking from Hoyle not to criticize the radio astronomy group members ever again.[12] Ryle made it quite clear he would resort to the courts if Hoyle dared to break the gag order.

The Nobel Prize for physics in 1983 was awarded to William Fowler, for his work on the origin of the elements, and to Subramanayan Chandrasekhar for his research on stellar structure. On October 19, 1983, the phone call from Sweden completely devastated Fowler as he realized Fred had been passed over. Fowler's citation reads:

> for his theoretical and experimental studies of the nuclear reactions of importance in the formation of the chemical elements in the universe.

Chandrasekhar's is:

> for his theoretical studies of the physical processes of importance to the structure and evolution of the stars.

Stellar evolution was the common theme of the 1983 award. One member of the jury, the elderly and heretical Hannes Alfvén, had publicly argued with Hoyle in the 1950s about the origin of the solar system, so he may have felt quite strongly that Hoyle did not make the grade. Some astronomers still feel that the academy committed a gross injustice by declining to split the award three ways, thus including

Hoyle. A suspicion will always remain that the jury deemed Hoyle to have disqualified himself either because of his public disagreement with the 1974 award or because he had descended into a great deal of speculation on bizarre topics. Willy's regret at the exclusion of his friend never diminished with the passage of time, and thereby he lost some joy from his own share of the prize. However, their scientific partnership survived this test. And in the long run, Fred did share the Balzan Prize in 1994 with Martin Schwarzschild. More importantly, in 1997 he and Ed Salpeter shared the Crafoord Prize, awarded by the Royal Swedish Academy of Sciences to mark accomplishments in scientific fields not covered by the Nobel Prizes.

With the fateful letter dispatched, Clayton and Hoyle settled to their research in astrophysics by continuing to investigate novas. Hoyle had long thought about the origin of dust in the interstellar medium. All studies of our solar system's origin assumed that the newly formed Sun was surrounded by a disk of gas and dust that eventually condensed into planets. Starting with his work on accretion with Lyttleton, Hoyle realized that detailed information about interstellar dust was needed to appreciate its significance for star and planet formation.

For decades, Fred had asked himself, "Where does the dust come from?" By 1975, he and Clayton had a remarkable solution: They believed that solid particles—they called them grains—formed in the outer shells of nova explosions. In their model, they regarded the grains as carbon, specks of soot if you will.[13] A few years earlier, Hoyle, together with his research student Chandra Wickramasinghe, had argued that conditions in the atmospheres of red giant stars were favorable for the formation of grains, which they considered to be carbon.[14] It is now known that red giants make silicon carbide, a type of sand, rather than carbon, but this does not diminish Hoyle's achievement in identifying nova explosions and red giants as the source of grains in interstellar space. The great success of their 1976 paper was a prediction of the blend of isotopes of carbon to be expected in grains that were older than the Sun. Their prediction inspired searches for these ancient grains once geologists had perfected techniques for examining in minute detail the composition of meteorites. Hoyle lived barely long enough to hear that the predictions were confirmed by observations, when, in 2001, physicists announced the isolation of grains of silicon carbide in a meteorite. These grains had just the right mix of isotopes expected from formation in a nova. As Don Clayton has put it:

> The discovery of grains [formed before the Sun's birth] silenced forever the doubters that man could ever hold in his hand a solid object that existed before the earth and planetary rocks existed. Laymen and scientists alike were stunned to learn that this had happened.[15]

Each of the grains studied by Sachiko Amari and her colleagues contained more than a million million atoms.[16] This large number assured that the excess amounts of carbon-13, nitrogen-15, and radio-active aluminum-26 and silicon-30 were characteristic of the grains' environment at the time they condensed. That was the signature of the nova. Their discovery sparked intensive investigation of the stardust predicted by Clayton and Hoyle.

Fred's close friends organized a marvellous treat for his sixtieth birthday, a conference held in Venice with the title "Frontiers of Astronomy." For a venue, they selected a beautiful monastery on the Isola St. Giorgio, which is immediately in front of the Palazzo Ducale and the Piazza San Marco. The title of this conference echoed Hoyle's influential 1955 textbook. Following on the heels of a golden era for nucleosynthesis, the theme running through the program in 1975 was largely nucleosynthesis, with talks on that topic from several collaborators. Most of his chums were present, and they had a good time.[17]

At his birthday banquet, Hoyle received a folder of congratulatory telegrams from across the globe. The Prime Minister of India, Indira Ghandi, personally signed a letter, full of praise, "on behalf of the people of India." There were messages from officers of Caltech, Cornell, the Royal Society, the National Academy of Sciences, and many other institutions. Not so pleasing was the letter from Harold Wilson's private secretary at 10 Downing Street, which mistakenly referred to "Sir Frederick" and sniffily announced that the Prime Minister only sent 100th birthday greetings. The civil servant concluded with, "I shall be glad to send a message from the Prime Minister on July 14, 2015." Fred gave an after-dinner speech on the theme "He who lies among dogs must learn to pant." He first dealt with a round of thanks, starting with

Willy Fowler, "my fellow dinosaur." Next he turned to Martin Rees, to thank him for arranging that the former IoTA building should be named the Hoyle Building, adding that "my departure from Cambridge was one of the saddest moments of my life." Hoyle's banquet speech included a reflection on his personal attitude to scientific research:

> The hard thing about being a scientist is the need for new-style problems. You become excited for a few years about some form of investigation, but after a while the demand grows for some change. Writing papers in the same field becomes less interesting and a search for something different begins. In my own case I have never been able to rid myself of my early interests. Whenever I attend a discussion of nuclear astrophysics I suffer an attack of intense nostalgia.[18]

The BBC celebrated his sixtieth birthday with a half-hour radio program in which the editor of *Nature*, John Maddox, held a conversation with Fred and his friends. Maddox asked Hoyle for his latest views on the universe, which brought the response:

> Many of the past generation believed they were very close to the ultimate structure for the universe, and that it was only a question of time before extra work would fix the final details. I don't believe this at all myself. I think what we see is a tiny fragment of a much bigger structure. The big advances in astronomy come when there are big advances in physics. We shall find it difficult to arrive at a unique answer for the universe because we see only a part of it.[19]

By this time, of course, a younger generation of cosmologists—in Great Britain, people such as Stephen Hawking, Martin Rees, and Paul Davies—were speculating that the visible universe is a mere bubble inside a much larger entity. Hoyle's response to Maddox shows that he remained thoroughly modern in his outlook.

The continuation of research in nuclear astrophysics was by no means Hoyle's only collaborative effort in the 1970s and 1980s. Jayant Narlikar visited the home on Cockley Moor on eight occasions between 1973 and 1987, to continue working with Hoyle on gravitation and cosmology.[20] Together they wrote a major book, *The Physics-Astronomy Frontier*, which they felt "should be easily within the scope of anyone who has a strong high-school background in general science."[21] Perhaps Fred was trying to repeat the success of his *Frontiers of Astronomy* (1955). If so, he must have been profoundly disappointed because the new book failed to make any impact, despite its imaginative content. It is, in fact, at a much higher level of presentation than high school science. Besides, the market for popular science books had changed beyond recognition: Teenagers of the television age expected color photographs, elaborate graphics, and an unchallenging text.

The twilight years also led to a collaboration of a unique kind with Chandra Wickramasinghe. This former research student had secured an appointment as professor in the mathematics department at Cardiff University. Following Fred's departure from Cambridge, Chandra persuaded Cardiff to offer an honorary professorship to Hoyle, which Hoyle accepted of course, in addition to his honorary place at Manchester. In the quarter of a century 1977–2001, Hoyle and Wickramasinghe coauthored almost 200 research papers and reviews, a truly astonishing record on any count. Bear in mind that Hoyle had no normal university job, or proper salary for that matter, after 1973. Nevertheless, free of charge to the taxpayer, he dove into a succession of intellectual puzzles that on one measure—the papers published— account for two-fifths of his entire output, a remarkable statistic. Curiously, Fred Hoyle devoted only two paragraphs—fewer than 200 words—to this aspect of his career in *Home Is Where the Wind Blows*.[22] Furthermore, his brief comment was not set in a research context: It simply prefaced his shortest chapter, on climbing the Munros, and read as an afterthought—did he add it at proof stage?

In the decade 1975–1985, the big problem they attacked was the origin of life. The pair came to think that life is a cosmic phenomenon and not the lucky outcome of a number of highly improbable events that took place locally on Earth. They resurrected a very old idea, panspermia, the idea that life seeds are distributed widely throughout the cosmos. What were the life seeds? Hoyle and Wickramasinghe proposed that they are bacteria, an idea ridiculed at the time because bacteria are too easily destroyed by ultraviolet radiation in space. Together they found a clever way to beat this objection.

They focused their attention on dense interstellar clouds, which were already known to contain abundant supplies of simpler organic molecules, the most basic molecular building blocks of life. Large portions of such clouds are composed of dust and grains. Hoyle and

Wickramasinghe imagined that biochemistry could start among the prebiotic molecules and grains. This process eventually led to bacteria, which cannot survive for long in space. However, in a dense molecular cloud, there will be billions of comets, formed by the condensation of volatile material such as water, methane, and ammonia. The bacteria could survive inside comets. Indeed, the larger comets would have warm liquid interiors where the assembly of molecules into primitive living cells could take place. Thus, the earliest phase of biological evolution takes place not in space or on Earth but inside comets. In its first billion years, the planets of our solar system suffered from severe bombardment by comets. Some planetary scientists believe it is probable that incoming comets transported water to Earth. Hoyle and Wickramasinghe took this idea a stage further, proposing that comets introduced microorganisms to Earth about 4 billion years ago.[23]

Their boldest move was the proposition that "life in space" should have progressed as far as bacteria. The most prevalent life form on Earth today is the bacterium. Bacteria are an example of life based on the cell, which is a package containing the genetic material. Under ideal conditions of nutrition, bacteria reproduce stupendously fast and undergo rapid evolution. The primitive Earth had almost no oxygen, so the first bacteria were adapted to flourish without oxygen. The genes that encoded for metabolism in the oxygen-free world 4 billion years ago have been preserved in evolution all the way from bacteria to humans. Molecular biologists working on the origin of life consider that accounting for the steps from simple molecules to bacteria is a much more difficult task than explaining the steps from bacteria to humans. That is one of the reasons why Hoyle-Wickramasinghe panspermia has never been taken seriously by molecular biologists. The astronomical community also viewed the hypothesis with skepticism bordering on scorn. Biologists were also unimpressed by Hoyle and Wickramasinghe's rejection of Darwinian evolution, which neither accepted.

What really upset the biologists, though, were Hoyle's excursions into the causes of disease. If comets could bring bacteria, could they not also scatter viruses, a simpler form of life? Hoyle and Wickramasinghe asserted that the root cause of a major epidemic or pandemic is the seeding of the atmosphere with viruses expelled from a passing comet. Of course, they could not get a hypothesis as outlandish as this accepted for publication in a scientific journal; instead, they wrote a popular book, *Diseases from Space*, which gives their argument and the supporting research.[24] They used real data, from the attendance registers in schools, to investigate the 1978 influenza epidemic in England and Wales, to which they attributed a cosmic origin. The sudden appearance of completely new pathogens, such as Legionnaires' disease in 1976, only added to their conviction that nasty bugs came from space.

In 1985, in an abrupt change of direction, Hoyle and Wickramasinghe allowed themselves to be dragged into a ridiculous argument about a dead bird. One of the prize possessions of Britain's Natural History Museum is its fossil of *Archaeopteryx lithographica*, one of the most important fossils ever discovered. It is an outstanding example of a transition evolutionary form, since the fossil has some features inherited from reptiles, and others, such as feathers, found only in birds. Hoyle and Wickramasinghe wrote an article in the *British Journal of Photography* claiming that the museum's *Archaeopteryx* is a fake.[25] An ordinary dinosaur fossil, they suggested, had been treated with a paste of powdered limestone, into which bird feathers have been pressed in order to create the illusion of an extraordinary, dinosaur-like bird.

The paleontologists at the museum were absolutely outraged when they learned of the contents of this paper. They had every right to feel insulted: The claim had not gone to a peer-reviewed journal, and the bold assertions of a man as famous as Fred Hoyle could damage the museum, which depended on the public purse, and therefore dreaded controversy.

The Natural History Museum conducted an extensive study on its *Archaeopteryx* specimen and refuted all of Hoyle and Wickramasinghe's claims. The team investigating the claims found that microcracks in the limestone fossil extended into and beyond the feathered areas in an unbroken line, indicating that no material had been added later to the original rock. The team also found feather imprints underneath the fossil bones, which would have been impossible for a hoaxer to accomplish. Hoyle did not accept the museum's response, and together with Wickramasinghe he wrote a book on the topic (maybe Hoyle needed the royalty income) in 1987. The same year the museum mounted an impressive exhibition of the fossil, with the title *Archaeopteryx—Is It a Fake?* They invited the public to view all the evidence and draw their own conclusion. This excellent exhibition, which was heavily promoted by the museum, probably damaged Hoyle's scientific reputation. A different aspect of this conflict would not have pleased Hoyle at all: In the United States, he is still regarded as a hero by the creationists because, in their eyes, he exposed one of the best pieces of fossil evidence for evolution as a fake.

Some old rocks of a different kind had attracted Fred's attention in the 1960s and 1970s: Stonehenge, which he first visited in the 1950s. The traveler approaching Stonehenge from the west reaches the crest of a hill, from which the truly massive group of stones is visible to the right. Today, it is unfortunate that numerous tourists and their cars diminish the experience of looking in wonder at this ancient structure, constructed in three stages from about 3,000 B.C. to 1,800 B.C. In 1963 and 1964, Gerald Hawkins, a British astronomer working at Boston University, published two papers making the remarkable claim that Stonehenge is a neolithic computer-cum-observatory for predicting eclipses of the Sun and Moon.[26] He had made a thorough survey of the site, plotted every stone and pit, punched coordinates onto cards and fed them, and astronomical data, into an IBM 704. Almost 3 years later, Hawkins produced a popular book, *Stonehenge Decoded*, setting out his ideas in nontechnical language. That book became a best-seller and sparked a fierce controversy between the astronomer with no training in archeology, and the archeologists with no knowledge of astronomy.[27]

Glyn Daniel, a fellow of St. John's College and an expert on the archeology of the Stone Age, had a sideline in common with Hoyle: writing detective and mystery fiction set in Cambridge. (Glyn's novels found a publisher and are still in print, something Fred never achieved.) One evening, over a High Table dinner at St. John's, Daniel invited Hoyle to check the astronomy in *Stonehenge Decoded*. He explained the reason for the furious row between Hawkins and the archeologists: "That man is seriously astray in his archaeological assertions, so how can we trust the astronomy?" Fred, who was about to take a hill-walking holiday in Scotland, said he would look into the matter during his vacation. Mindful of Scotland's frequent rain, he took plenty of reading material: *Stonehenge Decoded*, logarithm tables, a textbook on spherical astronomy, and a couple of pads of paper.[28]

As he read the book, Hoyle soon understood the fury of the archeologists, who evidently disliked its popular style. However, the main claim, that Stonehenge is an astronomical observatory, "seemed to me to have a genuine ring of truth about it." Hoyle found the argument so convincing that he set to work for a few hours on the arithmetic. He was able to confirm all of Hawkins's astronomical claims.

So what exactly had Hawkins done, and why did Hoyle take so much interest? According to Hawkins, Stonehenge had been constructed with a series of sight lines that marked out on the distant horizon the critical rising and setting points of the Sun and Moon. Antiquarians had long known the most famous of these alignments: The sight line from the center of the monument to the outlying Heelstone lies close to the sunrise on midsummer's day. Hawkins had identified the four sight lines for midsummer and midwinter sunrise and sunset, which would not have been particularly challenging to construct in the Stone Age. For the Moon, it was a different matter entirely because the motion of the Moon in our sky is vastly more complicated than the annual motion of the Sun. The rising and setting points of the Moon vary through an 18.61-year cycle. Hawkins believed he had identified sight lines to mark the most southerly and northerly points of moonrise and moonset. When Hoyle checked this work, and carefully plotted the lines, he noticed a truly amazing symmetry: Sight lines intersected exactly at right angles, and some lines even served a dual purpose, marking a critical moonrise in one direction and a moonset in the other. This symmetry is broken if one imagines a Stonehenge built only tens of miles to the north or south of the actual location. Only at the latitude at which Stonehenge was built does the rectilinear pattern yield right-angle intersections and dual sight lines. Hoyle "found it impossible to dismiss this property as a mere coincidence." He began seriously to accept the view that Stonehenge had been an astronomical observatory.

In the astronomical interpretations, much significance is attached to a set of 56 holes arranged almost symmetrically in a circle that surrounds the main group of stones. These holes predate the erection of the stones. They are named after the seventeenth-century antiquarian John Aubrey who mentions them in a manuscript now in the Bodleian Library, Oxford. Hawkins had proposed an observing method for predicting eclipses of the Sun and Moon that involved using the Aubrey holes as a counter. The coincidence that the number 56 is almost the same as the length of three 18.61-year lunar cycles greatly impressed Hawkins. He imagined that, following an eclipse at Stonehenge, a marker stone could be moved by three Aubrey holes a year, and when it returned to its starting position, 18.67 years later (i.e., 56/3 years later), an eclipse might be seen.

Hoyle did not like Hawkins's hypothesis at all: For him, it was difficult to see how Stone Age humans would have calibrated the system and, in any case, it predicted only a small fraction of all eclipses. Hoyle discovered a superior system that required careful observation of moonrises. He treated the 56 Aubrey holes as positional markers on a giant orrery, which kept track of the Sun and Moon in their orbits, as well as major axis of the Moon's orbit. Hoyle's model is breathtakingly elegant to anyone who *already understands* spherical astronomy. Furthermore, it predicts all eclipses, with no false alarms. Only a small fraction of these, however, would actually be visible from Stonehenge.[29]

Of course, the archeologists dismissed Hoyle's proposals as nonsense and undisciplined speculation because they appeared to attribute great intellect to Stone Age people. He responded thusly:

> It is not a speculation to assert that we ourselves could use Stonehenge to make eclipse predictions. We could certainly do so without making any substantive changes to the layout. While this does not prove that Stone Age man did in fact use Stonehenge for making eclipse predictions, the measure of coincidence otherwise implied would be quite fantastic.[30]

Hoyle developed his ideas on Stonehenge into a popular lecture, which he gave several times, often earning a useful fee. In the Cumbria years, when he needed to earn his way as a writer, he produced *On Stonehenge*. Published in 1977, this book sold particularly well in the United States. His popular fiction continued to flow at a steady rate, now jointly authored with his son Geoffrey. Fred provided the scientific framework, to which Geoffrey added realistic characters and events. Their output included *The Inferno* (1973), *Into Deepest Space* (1974), *The Incandescent Ones* (1977), and *The Westminster Disaster* (1978). The world of astronomy forms the background for *The Inferno*, a fast-paced novel set partly at the Anglo-Australian Telescope. Its narrative has a strong autobiographical streak: There is even a blazing row about who can use the British "time" on the telescope!

In 1982, Ladybird, a major publisher of children's books, promoted a series of four short science fiction titles. The publisher's blurb for one of them, *The Frozen Planet of Azuron*, can serve to convey the spirit of the series:

> As the Earth's winter gets colder and colder, Professor Gamma realises that a powerful villain must be at work. He sets off for a distant planet to find the culprit, knowing that it will be a very dangerous task. . . .

The ice age theme led to a factual account, his book *Ice*, in 1981: "How the next ice age will come—and how to prevent it." From reading this book (which is intended for a college-educated readership), one gets the impression that Hoyle cared very deeply about the possibility of a new ice age, the consequences of which he examines in considerable detail. As to the causative agent, he imagined that global climate change, such as an ice age, could be triggered by "the strike of a giant meteorite, which could happen at any time." The book's final paragraph shows that he despaired at his inability to persuade society to take his claims more seriously:

> The risk of the next ice age is not just the biggest of the risks we run. It is a risk that would hopelessly compromise the future. Besides wiping out a considerable fraction of those now alive, it would leave a wan, grey future from which the survivors and their descendents could do nothing to escape. It would be a condition that might last 50,000 years or more, a future in which the prospects for mankind would be much less favorable than they are today. This is why our modern generation must take action to avoid catastrophe, an ultimate catastrophe besides which the problems that concern people, media and governments from day to day are quite trivial.

Every winter, Barbara and Fred had a reminder of former ice ages. Their cottage, situated at an elevation of 1,400 feet, became snowed in by raging blizzards. By 1990, the Lake District winters had become too much for the aging couple. In a quest for better weather, they moved into a modern two-bedroom apartment in the south coast town of

Bournemouth, with views over the sea and 7 miles of golden beaches. This beautiful resort also offered numerous hotels in which visiting professional colleagues could find accommodation.

The professional relationship that lasted until the end was a triangular affair involving Fred Hoyle, Geoffrey Burbidge, and Jayant Narlikar. Narlikar visited about once a year until the late 1980s, while Burbidge had a lengthy telephone call from La Jolla to Bournemouth about once a week. Sometimes Geoffrey and Margaret were in Cambridge for a summer conference, and that always gave them an opportunity to meet up with Fred to do some science.

The trio worked on what had become for them familiar topics: the nature of quasar redshifts, matter creation, and alternative cosmologies. Geoffrey Burbidge stoutly maintained the view that quasar redshifts are not indicative of immense cosmological distances because quasars are local objects ejected at very high velocities from the nuclei of galaxies. At conferences, he would say, "Our general conclusion is that quasi-stellar objects and galaxies are clustered together in space although they sometimes have very different redshifts. We believe that the quasi-stellar objects are ejected from galaxies."[31] The three convinced themselves that violent events in the nuclei of galaxies are manifestations of matter creation. Their basic concept was that condensed objects are generated in the nuclei of galaxies as a result of mass creation and are then ejected periodically, accompanied by hot gas and accelerated charged particles. However, their arguments failed to win any support outside their own small group.

In the 1990s, the three friends produced a major new cosmological model, which they called the quasi-steady-state theory. This borrowed some of the ideas on matter creation from the original steady-state model of 1948, challenged the apparent support for the big bang cosmology from observations of the microwave background, and gave a unified treatment of creation in the centers of active galaxies. The theory is remarkably imaginative, to such an extent that it is hard to avoid the conclusion that they had by now become their own worst enemies, in the sense that anything they wrote was quickly dismissed as the work of cranks.

The quasi-steady-state model is based on the idea that the universe has experienced an enormous number of little bangs, each involving about 10^{16} solar masses of material, spread through space and time. Each little bang cooks up the light elements (helium, lithium, beryllium, and boron) and its heat energy becomes the microwave background. The little bangs are associated with the creation of matter, in the form of exotic Planck particles,[32] each of which contains about a millionth of a gram of matter. Although this appears to be a small mass, it is absolutely enormous to a particle physicist because it is equivalent to the mass of a million million million protons. Hoyle and his two colleagues dreamed up a clever scheme for creating the new particles without violating the laws of physics: The positive energy of the Planck particle is balanced by negative energy in the creation field. Such Planck particles are wildly unstable, decaying within 10^{-43} seconds of creation into a fireball of energetic particles that undergo nuclear reactions to make the light elements. The creation itself takes place only in strong gravitational fields associated with dense concentrations of matter.

This model fits the observations best if the time that elapses between major creation events is about equal to the so-called "age" of the big bang universe, and they are interspersed with long periods of weak creation. Such a universe is infinite in both space and time. At present, we are living in a part of the universe with weak creation, which is associated with the violent activity at the centers of some galaxies and the expulsion of massive objects. Within the big bang picture, the observed universe has a large amount of "missing" mass—invisible matter, the presence of which is inferred from gravitational fields. In the quasi-steady-state universe, the missing mass is assigned to compact massive objects, "waiting like buds to burst out in a new generation of creation events."[33]

Their paper in *Astrophysical Journal*, which is where they announced the new model, pleaded with the astronomical community to take their ideas seriously:

> This paper is not intended to give a finished view of cosmology. It is intended rather to open the door to a new view which is at present blocked by a fixation with big bang cosmology. We believe that the alternative scheme described here has sufficient robustness to withstand observational tests.

In September 1992, Hoyle gave a short preview of the new theory at an international symposium on observational cosmology. Speaking in Italy, he said:

> Our aim is not to convince people of a particular cosmological model. Our purpose is to convince you of the logical necessity for considering the physical problem of the creation of matter. When this is done along quite standard lines, hitherto unexpected possibilities, like the properties of the Planck particles, emerge. The need in cosmology is to open doors, not to close them, as the present organisation of science tends to do.

Informally, the quasi-steady-state theory has been dismissed as a hypothesis promoted by formerly distinguished theorists who sadly have lost their way. However, there is evidence that some professionals have given it serious thought: The *Astrophysical Journal* paper has been quite widely cited by cosmologists who are not afraid to work on alternative cosmology. Hoyle's plea to work hard on matter creation has certainly been addressed because a philosophical weakness of the big bang is the creation of everything out of nothing.

In the second half of November 1997, Fred drove up to west Yorkshire, back to his childhood home in Gilstead, where his sister Joan has spent her entire life. He wanted to spend a couple of weeks driving and walking around the district. On Monday, November 24, early in the afternoon, Fred announced to Joan that he wanted to take a moorland walk. It had been raining for 3 days, so he had been cooped up indoors, working on a draft of a cosmology book. However, Joan thought it already rather late in the day, 2:30 p.m., with the light fading fast. Fred nevertheless stepped out for some fresh air, without giving a destination.

He in fact headed for Eldwick, on a route that took him across the upper slopes of a steep ravine called Shipley Glen. A wooded path, supposedly an easy ramble, runs 400 feet above the stream at the bottom of the glen. Fred, walking in trainers, not boots, took this upper path with its thin covering of leaves. Those leaves concealed a landslip on the path at the steepest part of the glen. Fred lost his footing at this place, falling into a 10-foot drop. He rolled over, hit a tree, and then smashed his head on a rock. Fred lay unconscious at first, but then recovered from the concussion sufficiently to be able to assess the situation. He did not have a mobile phone, twilight was closing in fast, there were no people walking dogs to whom he could cry for help, and it would be too dangerous to climb back up the landslip.

To his right, he had a small stream, swollen by 3 days of rain, which plunged straight down the ravine. On the other side of this, Fred could see that the going would be relatively easy. He tried nimbly to cross the stream, but slipped head over heels. He plunged 300 feet down a very steep slope, through trees and bushes, ending with a sheer pitch into the stream bed, where he took the final impact on his right shoulder. His glasses were gone and his wristwatch had been torn off.

He quickly returned to consciousness. He was face down in the stream. With a great struggle, he was on his feet but, lacking the strength to climb out of the rift, he slipped backward, with his back propped by two boulders, which kept his head above the rising water level. By now, there was little daylight.

By 4:30 p.m., Joan called the police. Initially, they did not take the call too seriously—there could be a mundane explanation for the delay. But by 6:00 p.m., she had persuaded an official in charge of rescues that this absence was out of character for her 82-year-old brother. The police called out the Calder Valley Mountain Rescue Team, a highly trained volunteer rescue squad. By 10:00 p.m., some 25 volunteers, including two doctors, drawn from all over Lancashire and Yorkshire had arrived in Bingley. Unfortunately, no one knew where to start searching. They lost the first 2 hours on a wild goose chase after a man in a pub claimed to have seen Fred staggering all over the road at a location far from the site of the accident.

Shortly after midnight the team leader held an urgent conference with Joan. When he realized how close the house in Gilstead was to Shipley Glen, he ordered the searchers to concentrate there. The team arrived at the northern end of the glen and started to work downstream. This was difficult in the darkness, so they brought in their dogs, a crucial step as it turned out.

At 3:30 a.m., one of the dogs found Fred, some 12 hours after he had slipped. The two doctors quickly determined that Fred's legs were rigid and his body temperature had dropped to 73°F (23°C), the team's lowest record to date. The doctors thought he was dead until he suddenly grasped one of them by the arm. They immediately gave him oxygen. They lashed him to a stretcher and then struggled, in some desperation, up the eastern side of the glen. In the lights of the emergency vehicles, they could see that the whole left side of his body appeared damaged from head to toe, which suggested they should go to the major hospital in Leeds. That would have taken an hour, which the medical team thought too long for an elderly man whom they thought was dying. Instead, the ambulance paramedics took him to the accident service in a smaller hospital in nearby Bradford.

Geoff Hoyle and Barbara flew from Bournemouth to Bradford early the next morning, while his daughter Elizabeth took the train from London. By early afternoon the medics had raised his body temperature to 86°F (30°C), at which point he regained consciousness. He recognized his family immediately and gave them a monosyllabic account. He could not, and never did, remember precisely what had happened; the account given here is reconstructed from Geoffrey Hoyle's later search of the scene and the report of the rescuers.[34]

In the first few days in hospital, he had pneumonia as well as a serious kidney problem caused by hypothermia. His mental state became very confused—he thought nurses giving oxygen were trying to gas him. And he had a badly smashed shoulder. On December 4, the family arranged for him to be moved by private ambulance to the hospital in Bournemouth, which released him on December 15, with his shoulder in a restraining harness, to recuperate at home. In mid-January an orthopedic surgeon managed to stabilize the shoulder without surgery. Fred was comfortable enough to use his right hand soon after this: His manuscript letter of record to Don Clayton is dated February 10.

Did he fall or was he pushed? He had £200 in banknotes in his wallet when he stepped out on November 24. When he arrived at Bradford Royal Infirmary early on November 25, the wallet had been stripped of cash but nothing else taken. At the scene of the fall the rescue team had found a discarded balaclava helmet, a disguise used by muggers, within 2 yards of the body.

The accident affected Hoyle mentally. As he recovered, people asked him what had happened, but he never experienced a returning memory. Two years after the accident he wrote:

> What hasn't returned to normal is the inside of my head. I can still feel faintly the places where I was hit, the place where the police thought a mugger had got me. I know definitely that some things are for the worse. My capacity for mental arithmetic has worsened. I cannot multiply 57 by 32. I can get as far as the first multiplication, $57 \times 3 = 171$, but my ability to hold this intermediate result in memory is gone. At the time of the accident I was writing a book on cosmology with Geoffrey Burbidge and Jayant Narlikar. Luckily my part of the book was largely finished, so that my role in this enterprise has been limited mainly to proof-reading, which hasn't been too much of a test.[35]

One evening late in 1998, Geoffrey Burbidge phoned me at home with an interesting proposition. I had first met him a quarter of a century earlier, at a summer school in Sicily for graduate students and young professionals. Subsequently, we kept in touch at meetings of the American Astronomical Society, at which he, Margaret, and I would sometimes dine together. Geoff knew that I was the science director of Cambridge University Press, where I had responsibility for managing the astronomy list. He informed me that Hoyle, Narlikar, and he had completed a major book on cosmology, pulling together all their work for the past 30 years or so. "Simon, would you like to look it over for CUP?" he said. My first thought, literally, was that I would never persuade my senior colleagues at the press to accept a book by these gadflies. The second thought was a sense of excitement at the chance of publishing a controversial book. I asked Geoff to send the typescript.

Three days later a FedEx package arrived from La Jolla. I opened the box and removed the typescript. The title page read: *A Different Approach to Cosmology—from a Static Universe Through the Big Bang Towards Reality.* As soon as I sifted through the pages, I began to warm to the book because the authors had taken an historical approach. They handled the successive unfolding of the intellectual puzzles with great skill, and then in the last third of the book gave a unified account of their own ideas. It was good to see their corpus in a single work rather than as a motley stack of photocopies of their papers. I decided I wanted to publish the book, but first I needed academic support.

I sent the typescript to four reviewers, all of them distinguished professors: Two were in the United States and two in the UK. When I received the reports, I saw that I had a draw: 2–2. Two reviewers were extremely negative, giving the advice that publication would do irreparable damage to the reputation of Cambridge University Press. One reviewer colorfully commented, "This book is full of black pots calling black kettles black." And the other two reviewers offered equivocal advice: The science is almost certainly wrong; no one takes these ideas seriously, but for the sake of future historians of cosmology, it will do no harm for the press "to let them have their say." I decided to take a chance: I would use the future historians and a "for the record" argument to persuade the press to offer publication. My senior colleagues in the press reacted with skeptical excitement: I had a reputation for making outlandish proposals. But before we could offer a contract, we needed to take the matter to a university committee, the board of trustees of the press, collectively known as the Syndicate.

And so, one Friday afternoon in the Lent term 1998, I cycled from the modern corporate offices of the press to the historic Pitt Building in the center of Cambridge. Tourists sometimes misread its architecture as an ecclesiastical structure. The Syndics and officers, clad in academic gowns, were meeting to discuss recommendations to offer publishing agreements for some five dozen titles. Mostly, these were accepted on the nod with no discussion. But I had been summoned to present the case for offering publication of *A Different Approach to Cosmology*. After my presentation, a lengthy discussion ensued and, for a moment, it looked as though Fred Hoyle was going to be crushed yet again by a university committee at which he had no voice. One Syndic who had vivid memories of the 1949 radio broadcasts now came to the rescue of my situation. She felt that the university should "let Fred have his say at last," and she agreed with my point on the importance of the book for future historians. After listening to her comments, the Syndicate agreed to offer a contract.[36]

Geoff Burbidge took responsibility for the final version of the book and was the point of contact with Cambridge University Press. He, Fred, and I had a couple of summer lunches at the Plough in Coton, sitting out in the Sun on both occasions. We had ploughman's lunches and half-pint beers. Fred really enjoyed talking about the old times and cosmological models. He was greatly amused by the photograph that appears on page 188 of the book. This shows a gaggle of approximately 300 white geese waddling across a field. The farmyard geese are strung out in a line, tightly packed, all looking in the same direction and head-ing for the same objective. Geoff supplied this caption:

> This is our view of the conformist approach to the standard (hot big bang) cosmology. We have resisted the temptation to name some of the leading geese.

On publication, the book received very good reviews in the scien-tific press. Of course it did not find any new recruits for alternative cosmology, but the reviewers were pleased to see this rich theory written up in single source. The press sold more than 3,500 copies, an absolutely staggering number for an academic monograph where a sale of more than 500 copies is considered on target.[37]

Geoffrey Burbidge stayed in touch with Fred right to the end, usu-ally through a weekly phone call. Fred's health deteriorated slowly but not to the point that he was forced to give up intellectual activity. He returned to old problems: the origin of angular momentum, nuclear cosmochronology, the age of Earth, and the origin of the Earth and Moon. He produced the typescript of a last book, *A Different Approach to the Age of the Earth*. This lacked the sparkle and insight of his earlier books, in consequence of which I was unable to recommend Cam-bridge to offer publication.

Fred Hoyle lived into the twenty-first century, celebrating his 86th birthday on June 24, 2001. Six weeks later he had the last of a series of strokes and died on August 20, 2001.

Bernard Lovell wrote the obituary for *The Guardian*, which he opened with the comment: "Fred Hoyle will be remembered as one of the most distinguished and controversial scientists of the 20th century."[38] Chandra Wickramasinghe writing in *The Independent* highlighted Fred's popularization of science:

> Fred Hoyle will also be remembered as one of the greatest popularisers of science in the 20th century, following in the distinguished traditions of H. G. Wells, James Jeans and Arthur Eddington. He had a rare gift of explaining complex scientific concepts in the simplest of terms, and in so doing he never failed to captivate huge audiences on radio, TV, public lectures as well as through his popular books.[39]

In London *The Times* proclaimed:

> Between 1945 and 1970, the range and significance of Fred Hoyle's contributions to astrophysics and cosmology probably surpassed those of any other scientist in the world. He was internationally acclaimed for his original work on stars, galaxies, gravity and the origin of atoms.[40]

Walter Sullivan, in a news item for *The New York Times,* included a quote from John Faulkner:

> Dr. John Faulkner, of the Lick Observatory in California, said that during the "magical six years" after the establishment of the Institute of Theoretical Astronomy at Cambridge in 1966, it became "an obligatory Mecca" for young American astronomers, many of whom felt the Institute "fostered their best work."[41]

Hoyle's good friend John Maddox authored the obituary for *Nature.* His summary of the man described him as "a kind of Leonardo," who made "monumental contributions to astronomy and cosmology, and was a brilliant popularizer." Maddox criticized him for putting "his name to much rubbish," singling out the campaign on *Archaeopteryx* as bordering on the ridiculous.[42]

Physics Today asked Sir Martin Rees, the astronomer royal, for an obituary. Rees produced a masterly summary of Fred's life in physics, concluding thusly:

> Hoyle's enduring insights into stars, nucleosynthesis, and the large-scale universe rank among the greatest achievements of 20th-century astrophysics. Moreover, his theories were unfailingly stimulating, even when they proved transient. He will be remembered with fond gratitude not only by colleagues and students, but by a much wider community who knew him through his talks and writings.[43]

Fred's colleagues organized two memorable celebratory conferences, in Cambridge and Cardiff. The Cambridge meeting, which lasted just 1 day, was on Tuesday, April 16, 2002. It started in the new lecture room of the Institute of Astronomy, which had been added to the former IoTA building. Sir Hermann Bondi spoke on the accretion work; Wal Sargent gave a summary of "Fred's major contributions in the context of astrophysics and cosmology today"; George Efstathiou reviewed galaxy formation, Philip Solomon star formation, and David Arnett nucleosynthesis. In the afternoon, we transferred to St. John's

College for talks by Chandra Wickramasinghe, John Barrow, Malcolm Longair, Jayant Narlikar, John Faulkner, and Margaret Burbidge. A reception and magnificent banquet followed the scientific papers. At this dinner, Geoffrey Burbidge gave the speech and proposed a toast "To Fred Hoyle." In his address he reminded the audience that " the fact that Fred's view was often a minority view probably means that he was far ahead of his time." He concluded:

> In 60 years Fred had a huge number of ideas. Like all very great scientists he was not afraid to be wrong. In my view, he was much more often right than wrong, because he was always guided by the evidence (all of it), and by an excellent intuition. In the latter part of his career the ideas diverged a long way from what most people want to believe, particularly in cosmology and high-energy astrophysics.

The Cardiff meeting lasted 2 days, June 25–26, 2002, and its proceedings were quickly published.[44] The meeting commenced with personal reminiscences, including a remarkable contribution from Cyril Domb describing the war work for naval radar. Unlike the Cambridge meeting, the Cardiff gathering gave significant coverage to Hoyle's work in astrobiology and panspermia. John Faulkner gave a witty and memorable after-dinner speech with the title "Remembering Fred Hoyle." Here is Faulkner's final toast:

> [This] brings me to the end of my reminiscences of Fred Hoyle, a man always so much larger than life. And now, as we think about Fred, and what knowing him meant to each of us, let me ask you all to take your glasses, to raise them, and to drink a toast to the memory of Fred Hoyle, one of the most creative, most inspiring, most gifted and accomplished of giants in the history of science, a man who truly dominated the astrophysics of the last half of the twentieth century. To Fred Hoyle!

We can all drink to that!

Notes

My principal aims are to acknowledge the sources and to give the contexts for my own opinions. In terms of professional training, I am an astrophysicist, not a historian of science. Therefore my Notes should be received as pointers for further scholarship rather than as a complete guide to the literature and archives.

References to St. John's College Archives are to the personal papers of Fred Hoyle in the Special Collections at St. John's College, Cambridge. Where I have used 23/1, for example, 23 is the number of the box and 1 is the number of folder inside that box.

CHAPTER 1

1. Hoyle (1972).
2. Hoyle (1973a).
3. Letter to the Editor. *The Times.* January 4, 1972.
4. Hoyle (1994, p. 375).
5. Burbidge (2003, p. 215).
6. And did the Countenance Divine
   Shine down upon those clouded hills;
   And was Jerusalem builded here
   Among those dark Satanic mills?
   —William Blake (1757–1827)

7. In 1932, when writing to Emmanuel College Cambridge, Fred Hoyle was still using "34 Gilstead, Bingley" as his postal address. Emmanuel College Archives.
8. Hoyle (1986, p. 10).
9. Burbidge (2003, p. 215).
10. Priestley, J. B. 1933. *English Journey*. London: Heinemann, p. 166.
11. Hoyle (1994, p. 42).
12. Hoyle, F. 1982. The universe: Past and present reflections. *Annual Review of Astronomy and Astrophysics* 20:1–36.
13. Hoyle (1986, p. 71).
14. Joan Hoyle, interview of by author, June 12, 2004.
15. Hoyle (1968, p. 53; 1986, p. 90).
16. Personal effects of Sir Fred Hoyle, St. John's College Archives.
17. Burbidge (2003, p. 216).
18. Personal effects of Sir Fred Hoyle, St. John's College Archives. The collection also includes his steam engine and chessboard.
19. Joan Hoyle, interview by author, June 12, 2004.
20. The astronomy articles are marked in pencil. St. John's College Archives.
21. Personal effects of Sir Fred Hoyle, St. John's College Archives. A silver queen was awarded in 1932 and a pawn in 1933.
22. Letter, Alan Smailes to Dr. Peter Giles, Master, December 3, 1932. Emmanuel College Archives.
23. Hoyle's personal student file. Emmanuel College Archives.
24. Hoyle (1986, p. 15).
25. Hoyle's tutorial correspondence file. Emmanuel College Archives.
26. Hoyle (1994, plate 8).

CHAPTER 2

1. One magnificent chimney is preserved at the Damart factory.
2. The titles president, principal, provost, warden, and mistress are also in use in Oxford and Cambridge.
3. The system at Cambridge has changed since the 1930s. When Hoyle was admitted, tutors were centrally involved in college teaching, but this is now arranged by directors of studies.
4. Today, admissions to Cambridge are running at over 5,000 annually, a number that is equal to the entire student body in the 1930s.
5. Here, and in what immediately follows on Hoyle as an undergraduate, I used *Emmanuel College Magazine*, volumes 29 and 30.
6. See Stubbings (1995).
7. A prize of about $150 today.
8. *Emmanuel College Magazine*, volume 29.
9. I am indebted to Professor Don Clayton of Clemson University for providing me with a copy of a personal letter dated June 11, 1997, from Hoyle to Clayton. In this letter Hoyle describes his first climb in Scotland in detail. My account is taken directly from this letter.

10. The B.A. degree was awarded after 3 years of study. Hoyle had skipped the second year entirely, and so he could not proceed to the degree at Part II because he had only been in residence at the university for 2 years.

11. Quoted by Knox and Noakes (2003, p. 393).

12. Manuscript notes for a lecture to the Cambridge Society, St. John's College Archives, 23/1.

13. Personal recollection of John Faulkner, April 2003.

14. Winning a Blue continues to be the highest sporting award at Oxford and Cambridge.

15. Hoyle (1994, p. 98).

16. See Young (1899).

17. St. John's College Archives contain 10 Ordnance Survey 1-inch-to-the-mile maps with the Munros marked in red ink. He was in his fifties when he started climbing them in earnest.

18. The closing date for applications for studentships from the Department of Scientific and Industrial Research was April.

19. Peierls (1984).

20. Rutherford (1904. *Radioactivity and Radioactive Substances.* Cambridge, UK: Cambridge University Press. Chapter 1–4.

21. Peierls (1985, p. 90).

22. Wilson (1983, chap. 16).

23. Hendry (1984, p. 104).

24. Hoyle (1994, p. 124).

25. Fermi (1934, p. 161).

26. The article was by Bethe and Bacher (1936). Unfortunately, the actual copy of the journal used by Hoyle has been missing from the Rayleigh Library at the Cavendish for many years.

27. Hoyle (1994, p. 122).

28. Peierls (1985, p. 135).

29. Hoyle (1937a).

30. Hoyle (1937b).

31. Hoyle (1968, p. 75).

32. Bethe, H. A., F. Hoyle, and R. Peierls. 1939. Interpretation of beta-disintegration data. *Nature* 143:200.

33. Leedham-Green (1996, p. 195).

34. Kragh (1990, p. 195).

35. Hoyle, F. 1939. Quantum electrodynamics I. *Proceedings of the Cambridge Philosophical Society* 35:419; Hoyle, F. 1939. Quantum electrodymanics II. *Proceedings of the Cambridge Philosophical Society* 35:438.

36. Lyttleton (1936).

37. Lyttleton's 1933 lecture notes. St. John's College Archives, 24/6.

38. Hoyle (1994, p, 131).

39. Personal recollection of author. Lyttleton proposed courses to the faculty of mathematics that were deliberately so obscure no student elected to attend.

40. F. Hoyle, manuscript notes, St. John's College Archives, 44/4.

41. Hoyle, F. 1968. *Encounter with the Future.* New York: Simon and Schuster, p. 75.
42. Burbidge (2003, p. 241).

CHAPTER 3

1. Transcript of tape-recorded interview with Hoyle made by Alan Lightman, August 15, 1989, p. 8. American Institute of Physics.
2. Dirac (1931, p. 133).
3. Hoyle M.S., personal comments on the history of nuclear astrophysics. St. John's College Archives 23/1.
4. Eddington (1935, p. 168).
5. G. Burbidge (2003, p. 216).
6. Hoyle and Lyttleton (1939a, p. 405).
7. Jeffreys (1929).
8. Citation data were analyzed, at author's request, by Christopher King, Institute for Scientific Information, Philadelphia, 2003.
9. Hoyle and Lyttleton (1939b).
10. van de Hulst (1945).
11. Ewen and Purcell (1951).
12. Atkinson (1940a, p. 500; 1940b, p. 314).
13. Atkinson (1940a, p. 5007).
14. Letters, R. Atkinson to R. A. Lyttleton, St. John's College Archives 87/1.
15. Hoyle and Lyttleton (1940).
16. Eddington (1937).
17. Adams (1941).
18. Carruthers (1970).
19. Hoyle and Lyttleton (1941).
20. Thackeray (1941).
21. Bondi (1990, chap. 7).
22. Bondi and Hoyle (1944).
23. Bondi et al. (1947).
24. Hoyle (1949b).
25. McCrea (1953).
26. Dodd (1952).
27. Mestel (1954).
28. Lyttleton (1953).
29. Davidson and Ostriker (1973).
30. Pfahl and Rappaport (2001).
31. F. H. Milne Lecture, Oxford 1978. St. John's College Archives, MS 23/1.

CHAPTER 4

1. Howse (1993, p. 45).
2. Treasury to Admiralty, October 9, 1941. UK National Archives, ADM 116/5475.
3. Director of Signal Department, Minutes of December 14, 1941. UK National Archives, ADM 116/5475.

4. Recollection of S. T. Wright written to Cmdr. Derek Howse, February 12, 1989. Churchill College Archives Centre. Naval Radar Trust.

5. Hoyle (1994, p. 164).

6. Hoyle (1994, p. 174).

7. *Radio Times*, October 3, 1940.

8. Letter, Lyttleton to Hoyle, undated, but from the context probably October 1940. St. John's College Archives.

9. The BBC Home Service regularly had 15-minute talks at 9:20 p.m., given by ministers, miliary men, commentators such as Alastair Cooke, and academics. Lyttleton may have missed the talk because the *Radio Times* did not always publish the speaker's name. The talk in question was published in *The Listener* on January 2, 1941, pp. 23–24.

10. Redshift is an increase in the wavelength of electromagnetic radiation—light, for example—between its emission and its reception. Three separate sources of galaxy redshifts are now distinguished: the cosmological redshift (due to the expansion of the universe), the Doppler effect (motion other than the expansion of the universe), and gravitational redshift.

11. Norman Vidler, The Story of ASWE. Unpublished notes written in 1969, recalling his experiences at the Signal School in the 1940s. I am grateful to Dr. Leigh Bailey, University of Vienna, for finding this piece of emphemera.

12. Admiralty Signal Establishment Monograph M382, "Notes on the Vertical Polar Diagrams." No author was given, but Hoyle's recollection (St. John's College Archives 1/3) is that he wrote this report.

13. Papers of the Naval Radar Trust, Archives Centre, Churchill College, Cambridge.

14. Interview recorded by author, March 2003.

15. I am indebted to my colleague Leigh Bailey for this cameo of the local pub. Bailey's father worked with Hoyle at the Signal Establishment. His family papers included an autobiographical sketch of the wartime period.

16. Churchill to Pound, September 22, 1939. National Archives, ADM 205/2.

17. Hoyle (1994, p. 188).

18. Interview, H. Bondi by author, March 2003.

19. A large German radar dish about 7.4 meters (24 feet) in diameter. After the war, captured Würzburg antennas played an important part in the birth of radio astronomy in England and the Netherlands.

20. Interview, H. Bondi by author, March 2003.

21. I am indebted to Dr. Stirling Colgate, Los Alamos National Laboratory, for this recollection, January 6, 2004.

22. Quoted in G.Burbidge (2003, p. 218). In 2003, Gold wrote to Burbidge with a number of his recollections, which are recorded by this source.

23. Bondi (1990a, p. 49).

24. The building is still there. In January 2004, planning permission was obtained to improve the facilities, at a cost of £9million.

25. This railway is now a magnet for tourists and has helped with the economic regeneration of northern Snowdonia following the collapse of the slate industry.

26. Interview, H. Bondi by author, March 2003.

27. Speech given by Hermann Bondi at Churchill College, Cambridge, July 21, 1990.

28. Bondi (1990a, p. 44).
29. The transcript of the talk can be found in *The Observatory* 36:324–329, 1913.
30. Russell (1929).
31. The Vogt-Russell theorem.
32. Hoyle (1982b, p. 1).
33. E-mail, Donald Osterbrock to Mitton, October 21, 2003.
34. Hoyle (1994, p. 226)
35. *Ibid.*
36. Osterbrock (2001, p. 101).
37. Attributed by Osterbrock to Albert Whitford.
38. Osterbrock (2001, p. 102).
39. Baade (1944a,b).
40. Hoyle (1994, p. 198).

CHAPTER 5

1. For example, the square of the ratio of the strengths of the electromagnetic and gravitational force is equal to the number of particles, 10^{80}, in the Eddington-Lemaître universe.
2. Ferris (1979, p. 113). Also see interview, Thomas Gold by Spencer Weart, recorded April 1, 1978. American Institute of Physics Oral History Archives.
3. Interview, Hermann Bondi by author, March 2003.
4. Roy Garstang's Cambridge notebooks and other records are on deposit at the American Institute of Physics Library. This quote is from an essay he wrote in 1996.
5. Bondi (1990, p. 68).
6. Bondi (1990, p. 62).
7. Transcript of tape-recorded interview with Hoyle made by Alan Lightman, August 15, 1989, p. 11. American Institute of Physics. Oral History Archives.
8. Bondi (1948b).
9. Gamow (1942).
10. Gamow (1946).
11. Alpher (1990, p. 134).
12. Alpher et al. (1948).
13. Transcript of tape-recorded interview with Hoyle made by Alan Lightman, August 15, 1989, p. 7. American Institute of Physics. Oral History Archives.
14. Robertson (1933).
15. Hoyle (1982, p. 11).
16. Transcript of interview, Thomas Gold by Spencer Weart, recorded April 1, 1978, p. 33. American Institute of Physics. Oral History Archives.
17. Bondi, H. 1990. P. 191 in *Modern Cosmology in Retrospect*, B. Bertotti, R. Balbinot, S. Bergia, and A. Messina, eds. Cambridge, UK: Cambridge University Press.
18. Transcript of interview, Thomas Gold by Spencer Weart, recorded April 1, 1978, p. 39. American Institute of Physics. Oral History Archives.
19. Weart 34. [*Ibid.*, p. 34]

20. Ferris (1979, p. 114).

21. Jeans (1928, p. 352).

22. Dirac (1937).

23. Bondi and Gold (1948).

24. Hoyle (1948).

25. Hoyle (1949a, p. 368 n).

26. *Ibid.*, 109:365.

27. Weart, 7. Transcript of interview, Thomas Gold by Spencer Weart, recorded April 1, 1978, p. 7. American Institute of Physics. Oral History Archives.

28. Bondi (1952, p. 155).

29. The Astrophysics Data System (ADS) had recorded about 60 citations to each paper by late 2003. ADS citations analysis is restricted to citations by papers in astronomy and astrophysics.

30. Data supplied by the Institute for Scientific Information, Philadelphia. The paper had more than 200 citations by early 2003, searched across the entire physical sciences literature.

31. Bondi (1948a).

32. An informal report on this General Assembly was published in *The Observatory* 68:161, 1948.

33. Bondi (1990b, p. 193).

34. Hoyle (1994, p. 252).

35. Proceedings published in *The Observatory* 68:209, 1948.

36. Transcript of interview, Thomas Gold by Spencer Weart, recorded April 1, 1978, p. 44. American Institute of Physics. Oral History Archives.

37. *The Observatory* 68:218, 1948.

38. Proceedings published in *The Observatory* 68:218, 1948.

39. The events that follow, together with scripts, documents, memos, and letters, are all extracted from the archived scripts and contributor files made available to S.M. at the BBC Written Archives Centre, Caversham Park, Reading, Berkshire. The files are arranged by source department.

40. Bondi et al. (1947).

41. Hoyle (1949b).

42. This hotel is now in the super deluxe class. When Hoyle stayed, it was one of the larger hotels within easy walking distance of the BBC.

43. The minutes of the meeting are published in *The Observatory* 68:41, 1948.

44. Tayler (1987, p. 144).

45. The BBC Written Archives Centre holds a microfilmed "Speaker's Copy" of the script. The script has several last-minute changes in Hoyle's handwriting, mainly aimed at achieving a more colloquial style.

46. Hoyle (1994, p. 253).

47. *Radio Times*, March 28, 1949.

48. In the spring of 1949, Laslett commissioned six talks on Christianity and the history of mankind from Professor Herbert Butterfield. Meanwhile, Clow commissioned Professor Nevill Mott, FRS (1977 Nobel Prize in physics), Professor Alexander Todd, FRS (1955 Nobel Prize in chemistry), and Professor Eric Ashby for "New Frontiers in Science."

49. The Doppler effect is the change in the observed frequency of sound or electro-magnetic waves when the source of the waves and the observer are moving apart or toward each other.

50. Alpher (1990, p. 135).

51. *The Listener* 41:567, 1949.

52. This is the meeting of the RAS held the previous October in Edinburgh. The writer either attended the meeting or had read the record in *The Observatory*, which all fellows received free of charge.

53. Dingle (1968).

54. Carpenter (1996, p. 110).

55. *The Listener* 42:103, July 1, 1949.

56. The correspondence file for Hoyle at the BBC Written Archives Centre shows that on November 21, 1949, Mary Somerville (assistant controller talks) proposed using Laslett to produce five programs to be given by Fred Hoyle. As early as November 25, 1949, Laslett, writing from St. John's College, had the titles of five talks from Hoyle. Hoyle (1996, p. 253) states that Laslett did not approach him until mid-January 1950. This is incorrect. However, the draft of the first talk was not available until January 10, 1950.

57. Mary Somerville (1897–1963) was educated at Somerville College, Oxford. She joined the BBC in 1925 and retired as controller of talks in 1955. Memos at the BBC Written Archives Centre show that she played a major diplomatic role in handling the controversy engendered by the 1950 talks.

58. The Reith Lectures were inaugurated in 1948 by the BBC to mark the historic contribution made to public service broadcasting by Sir John (later Lord) Reith, the corporation's first director-general. The 1950 lectures are reproduced in *The Listener*, volume 44.

59. *Radio Times*, January 20, 1950, p. 13. Although the text is unsigned, it is very much in the style used by Mary Somerville.

60. *The Listener*, 43:227, 271, 321, 375, 419, 1950.

61. Hoyle (1950).

62. Hoyle (1996, p. 254).

63. *Cambridge University Reporter*, March 9, 1949. Salaries of university officers are given on p. 950.

64. I am indebted to Bertie Bellis, St. John's College, 1948–1950, and former headteacher at the Leys School Cambridge, for this personal recollection, January 1, 2003.

65. Hoyle (1996, p. 254).

66. Hoyle (1950, p. 95). Test cricket had resumed in 1946. At the date of his broadcast, Australia had won all seven test matches played in the postwar period.

67. *The Listener*, 43:304, 345, 395, 433, 477, 520, 565, 611, 1950.

68. This is fully documented in the Hoyle papers at the BBC Written Archives Centre.

69. Lovell commented: "We are living through one of those rare epochs which yield a succession of the most unexpected cosmological discoveries." *The Listener*, 44:11, 1950.

70. I am indebted to Erin O'Neill of the BBC Written Archives Centre for these data. The rating is the percentage of adults who listened to the final broadcast.
71. *The Listener* 44:456, 1950, has the full text.
72. *The Listener* 44:496, 1950, has the full text.
73. *The Listener* 44:547, 594, 647, 693, 743, 798, 1950.
74. Dingle wrote: "It is his incredible inability to conceive that there were astronomers before he was vouchsafed to us, and there will be others, that is partly responsible for his ideas being treated with so much less respect.... Mr. Hoyle is having an abnormally protracted youth. This discussion should assist his progress to maturity."
75. In 1909, he made the first authenticated powered flight in Britain. The same year he won a £1,000 prize from the *Daily Mail* for being the first to fly a circular mile.
76. Cohen (1951).
77. *New Yorker*, April 5, 1951, p. 117.
78. Payne-Gaposchkin (1951).
79. I am indebted to Helge Kragh's monograph *Cosmology and Controversy* (1996, chap. 4) for his account of the reception of the theories by the professional community.
80. Edmondson (1951). I knew Edmondson, a kindly man, who married the daughter of H. N. Russell. When I read his review in the Cambridge University Library, I felt that he must have experienced very strong revulsion to Hoyle's message and conduct.
81. Williamson (1951).
82. O'Connell (1953).
83. Bondi, (1952).
84. Gamow (1952).
85. Gamow (1965).
86. Dingle (1953).

CHAPTER 6

1. I am indebted to Peter Hingley, librarian of the RAS, for showing me F.H.'s application form as well as the admission form.
2. These teas were still featured as late as 1968–1969, when I was a research student at the Cavendish Laboratory. The energetic questioner in those days was often young Martin Rees, who succeeded Hoyle as Plumian professor and became astronomer royal.
3. Hoyle (1994, pp. 152–153).
4. "Burning" is often used as a concise term for nuclear fusion processes, but it is totally different from the chemical burning that creates an ordinary flame.
5. Eddington died in 1944. The vacancy was filled in 1946 by the geophysicist Harold Jeffreys. Hoyle succeeded him in 1958.
6. Astronomers use kelvin (abbreviated K) as their unit for temperature. One kelvin is equivalent to 1 degree Celsius, and temperatures are measured from absolute zero, which is approximately –273°C.

7. Eddington (1924).

8. Hoyle (1994, p. 153).

9. Hoyle and Lyttleton (1942a). Lyttleton uses an old-fashioned spelling of the verb "to show," namely, "shew." Also, he handled all the correspondence with the RAS.

10. Pencil manuscript dated April 23, 1942. St. John's College Archives 78/1.

11. Hoyle and Lyttleton (1942b).

12. Some earlier work by E. J. Opik, published in Estonia, was then unknown. I do not have the original paper.

13. See Sadler (1987).

14. *The Observatory* 64:309, 1942.

15. Typed letter, Lyttleton to Hoyle, dated September 5, 1942. St. John's College Archives 78/1.

16. Hoyle and Lyttleton (1943).

17. Hoyle (1994, p. 153).

18. Unsöld (1929).

19. McCrea (1929).

20. Russell (1929).

21. Strömgren (1983).

22. Dunham (1939).

23. Hoyle (1946).

24. Hoyle (1994, p. 263).

25. Hoyle (1952).

26. This section of old Route 66 is now Interstate 40.

27. Hoyle has left no record of visiting the observatory on the way out, but he did visit on his return journey.

28. Hoyle (1994, p. 264).

29. St. John's College Archives 86/10.

30. An early version of this lecture was published the previous year, in *The New York Times Magazine*, June 1, 1952.

31. The house is at 1340 Woodstock Road, San Marino, Los Angeles County, California. It received designation as a National Historic Landmark (for astronomy) on December 8, 1976. I am indebted to Gale E. Christianson, Hubble's biographer, for the information on Hubble's fascination with all things English.

32. Grace Hubble's airmail letter is dated February 1, 1958, and postmarked Pasadena. St. John's College Archives 86/8.

33. Christianson (1997, p. 363). Grace kept up her circle of intimate friends for a few years after Edwin's death.

34. Osterbrock (2001, p. 142).

35. Kragh (1996).

36. Hoyle (1955, p. 308).

37. Hoyle (1994, p. 277).

38. Tombaugh discovered the planet on February 18, on two plates taken on January 23 and 29 with the 13-inch Cooke refractor at the Lowell Observatory. Letter, Tombaugh to author, January 23, 1986.

39. These descriptions of the observatory are adapted from Christianson (1997, pp. 90–101).
40. Hoyle (1994, p. 279).
41. Hoyle and Schwarzschild (1955).
42. Wickramasinghe et al. (2003, p. 50).
43. W. Baade, A. Blaauw, D. Chalonge, W. A. Fowler, O. Heckmann, G. Herbig, F. Hoyle, G. Lemaître, B. Lindblad, W. W. Morgan, J. J. Nassau, J. Oort, E. Salpeter, A. Sandage, M Schwarzschild, L. Spitzer, B. Stromgren, and A.D. Thackeray, were the invitees. The Vatican's own small staff of astronomers participated as well.
44. Pontificiae Academia Scientiarvm (1958).
45. Hoyle (1957, p. 37), Hoyle (1962, chap. 4).
46. Interview, Joyce Wheeler by author, March 19, 2003.
47. See Hoyle (1994, p. 293), for example.
48. Haselgrove, C. B., and F. Hoyle. 1956. A mathematical discussion of the problem of stellar evolution, with reference to the use of an automatic digital computer. *Monthly Notices of the Royal Astronomical Society* 116:515.
49. Haselgrove and Hoyle (1956).
50. Haselgrove and Hoyle (1959).

CHAPTER 7

1. As a research student in the radio astronomy group at the Cavendish Laboratory from 1968 to 1971, I worked on a daily basis with Martin Ryle.
2. Lovell (1985).
3. Smith (1987).
4. Ryle et al. (1971, p. 11).
5. The battleships were the *Scharnhorst* and *Gneisenau,* accompanied by the heavy cruiser *Prinz Eugen.*
6. Hey (1971, p. 91).
7. In 1937, Grote Reber mapped the Milky Way and detected two sources of enhanced emission in Cygnus and Cassiopeia, but these were steady sources, not outbursts.
8. It is difficult to describe the principle of an interferometer using words alone. The Ryle-Vonberg interferometer exploited the tiny differences in the radio signals from the same source detected at the separated antennas. The "wave fronts" from a cosmic source do not hit the antennas at precisely the same instant. When the two wave motions are brought together, they are said "to interfere," producing a stronger or weaker signal according to the extent to which they reinforce or cancel each other out. By capturing the differences in the signals, they could model (crudely) the structure and position of the source. In the first interferometer, one of the two antennas was on the university rugby ground at Grange Road, Cambridge.
9. Ryle (1948).
10. I am indebted to Sir Hermann Bondi, who was present, for sharing this recollection, March 11, 2004.

11. Letter, Ryle to Hoyle, November 26, 1949. Churchill College Archives, Ryle Correspondence.

12. Ryle (1950).

13. Ryle (1952). This is known as a phase-switching interferometer.

14. The citations are today posted in the Archives section of the Royal Society Web pages. Ryle's citation reads: "Leader of a research team which has obtained results of the greatest importance in radio astronomy. Distinguished for his work both on the theoretical and experimental sides. By ingenious design of simple aerial arrangements and by using phase-switch methods has obtained very high resolving power with his Michelson-type interferometer for locating radio stars. Fertile in new ideas both for elegant and simple experimental devices, and for new theories of the origin of the radiations, he has obtained world-wide recognition as a leading expert in this interesting and important branch of science. In charge of group responsible for design of radar jamming equipment for Bomber Command during the last war."

15. Hoyle et al. (2001, p. 47).

16. Kragh (1996, p. 308).

17. Hoyle, unpublished manuscript. St. John's College Archives, 37/4.

18. Hoyle, unpublished manuscript, On the nature of strong radio sources. St. John's College Archives 37/4.

19. Edge and Mulkay (1976, p. 97).

20. Lovell (1951).

21. *The Observatory* 71:209, 1951.

22. The Smith-Baade correspondence is reproduced extensively by Edge and Mulkay (1976, pp. 101–111).

23. Osterbrock (2001, p. 154).

24. Mitton and Mitton (1972).

25. Phone interview, G. Burbidge by author, March 10, 2004.

26. Edge and Mulkay (1976, p. 139).

27. Technically, the slope is −1.5, but expressing it thusly makes the argument more difficult to grasp.

28. St. John's College Archives 37/4.

29. Peter Eggleton and John Faulkner, for example.

30. *New York Times*, May 25, 1955, p. 22.

31. *Manchester Guardian*, July 15, 1955, p. 6.

32. Ryle (1955).

33. *The Observatory* 75:102, 1955.

34. The was the Fourth IAU Symposium. The papers are published in van de Hulst's (1957) *Radio Astronomy*.

35. Pawsey (1957).

36. Mills and Slee (1957).

37. Pawsey (1957).

38. Transcript of interview, Ryle by Woodruff Sullivan, August 19, 1976. Churchill College Archive Centre.

39. The quotation is taken from an unpublished autobiographical fragment. St. John's College Archives 1/3.

40. Ryle (1958).
41. Quoted by Edge and Mulkay (1976, p. 164).
42. Scheuer (1958).
43. Hill and Mills (1962).
44. *The Observatory* 81:45, 1961.
45. Hoyle (1994, p. 407).
46. E-mail, F. G. Smith to author, February 16, 2003.
47. Manuscript. St. John's College Archives 1/3.
48. Bernstein (1984, p. 207).
49. Public Broadcasting System. People and Discoveries. Available at http://www.pbs.org/wgbh/aso/databank/entries/dp65co.html. December 9, 2004.
50. Roll and Wilkinson (1966).
51. Penzias and Wilson (1965).
52. Malcolm Longair, lecture at St. John's College, 3:30 p.m., April 16, 2002.
53. *The Observatory* 88:185, 1968.
54. Personal e-mail, Tritton to author, February 23, 2003.
55. Hoyle et al. (2001).
56. Guth (1981).
57. Perlmutter et al. (1999).
58. Longair (2004).

CHAPTER 8

1. Dalton (1808).
2. Mendeleyev (1891).
3. *Encyclopedia Britannica*, 1911 edition.
4. Eddington (1927, p. 100).
5. Eddington (1935, p. 146).
6. A. S. Eddington, New Pathways in Science, Cambridge University Press (Cambridge), 1935, 146.
7. G. Burbidge (2003, p. 223).
8. Alpher and Herman (1950).
9. G. Burbidge (2003, p. 223).
10. F. Hoyle, unpublished manuscript essay, personal comments on the history of nuclear astrophysics. St. John's College Archives 23/1.
11. Hoyle (1982b, p. 2).
12. F. Hoyle, manuscript notebook, Synthesis of Elements by Stars. St. John's College Archives 86/5.
13. Mitton (1979).
14. Hoyle (1946).
15. E. M. Burbidge, Modern alchemy: Fred Hoyle and element building by neutron capture. Lecture at St. John's College, Cambridge, April 16, 2002.
16. The star is Alpha-2 Canes Venaticorum (α^2CVn), a sixth-magnitude companion to the brightest star in the constellation Canes Venatici. The paper is Burbidge and Burbidge (1954).
17. Burbidge and Burbidge (1955).

18. I am indebted to Ward Whaling who provided many of the details given in my account, as well as drawing my attention to errors in other publications.
19. Bethe (1939).
20. F. Hoyle, unpublished manuscript essay, personal comments on the history of nuclear astrophysics. St. John's College Archives 23/1.
21. Salpeter (1952).
22. Transcript of interview, Willy Fowler by Charles Weiner, February 5, 1973. American Institute of Physics. Oral History Archives.
23. E-mail, Ward Whaling to Ray Spear, Australia, June 4, 2001.
24. Chown (1999). Chapter 13 has much interpolation.
25. E-mail, Ward Whaling to author, April 26, 2004.
26. Spear (2002). My account is based on this thoroughly researched paper in which Professor Spear corrects numerous embellishments that have been added to this story over the years.
27. Transcript of interview, Willy Fowler by Charles Weiner, February 5, 1973. American Institute of Physics. Oral History Archives.
28. Phone conversation, G. Burbidge with author, April 20, 2004.
29. Hoyle (1954).
30. Subsequent research has shown that the source of iron-56 is the radioactive decay of nickel-56.
31. Merrill (1952).
32. F. Hoyle, unpublished manuscript essay, personal comments on the history of nuclear astrophysics. St. John's College Archives 23/1.
33. Cameron (1999).
34. Cameron (1955).
35. Personal e-mail, G. Hoyle to author, April 21, 2004.
36. Burbidge et al. (1957).
37. Suess and Urey (1956).
38. Osterbrock (2001, p. 69). The galaxy was IC4182.
39. St. John's College Archives 86/10.
40. Hoyle et al. (1956).
41. Hoyle (1994, p. 301).
42. E-mail, G. Hoyle to author, April 23, 2004.
43. E-mail, A. Cameron to author, April 24, 2004.
44. Hoyle and Fowler (1960).
45. Hoyle (1959).

CHAPTER 9

1. Hoyle (1994, p. 304).
2. O'Brien (2003, p. 108).
3. St. John's College Archives box 101.
4. Clayton (1999).
5. Hoyle and Fowler (1960).
6. Hoyle (1955, p. 214).
7. Interview, Peter Eggleton by author, January 2003.

8. Transcript of interview, W. A. Fowler by Charles Weiner, February 5, 1973, p. 73. American Institute of Physics. Oral History Archives.
9. The official record of the Sixth Herstmonceux Conference is in *The Observatory* 82:143, 1962.
10. Fowler and Hoyle (1964).
11. Clayton (2003a, p. 236).
12. E-mail, John Falkner to author, July 7, 2004.
13. Mitton and Mitton (1972).
14. Mitton (1976, p. 133).
15. G. Burbidge (1959).
16. Transcript of interview, Willy Fowler by Charles Weiner, May 30, 1974, p. 81. American Institute of Physics. Oral History Archives.
17. G. Burbidge (1961).
18. Hoyle and Fowler (1963b).
19. Hoyle (1981b, p. 14).
20. Hoyle and Fowler (1963a).
21. Schmidt (1963).
22. Hoyle (1982b, p. 7).
23. Hoyle (1994, p. 326).
24. Schmidt (1990, p. 350).
25. The original letter is in the BBC Written Archives Centre.
26. In 2004, Julie Christie played a small part in the movie *Harry Potter and the Prisoner of Azkaban*. Fred Hoyle would have approved of the very clever scenes in this movie that involve time travel, the theme of his science fiction novel *October the First Is Too Late*.
27. J. Faulkner, after-dinner speech delivered June 26, 2002, in Cardiff. The text is reproduced in Wickramasinghe et al. (2003, p. 313).
28. Quoted in White and Gribbin (1992, p. 58). Gribbin had been a research student of Hoyle.
29. *Nature* 184:686, Applied mathematics and theoretical physics at Cambridge, 1959.
30. Interview, John Faulkner by J. Mitton, March 11, 2003. I am indebted to Jacqueline Mitton for conducting this interview at the National Astronomy Meeting in Dublin, Ireland.
31. Hoyle (1994, pp. 334–335).
32. Typed letter, L. Mestel to colleagues, May 6, 1964.
33. Pencil manuscript. St. John's College Archives 8/1.
34. Typed statement of Batchelor handed to Mestel on May 7, 1964.
35. Manuscript letter, J. Polkinghorne to L. Mestel, May 26, 1964.
36. Hoyle (1994, p. 335).
37. MS draft, signed F. Hoyle, private papers of L. Mestel.
38. Atticus column, *Sunday Times*, July 19, 1964. The manuscript is at St. John's College Archives 8/1.
39. Hoyle (1994, p. 332).
40. I am indebted to Professor Leon Mestel who kindly gave me access to his collection of university papers that he assembled during the events of the 1960s and 1970s.

41. Hoyle (1994, p. 333).
42. The management board of the Observatories.
43. St. John's College Archives 7/7, 8/2, and 8/3.
44. St. John's College Archives 1/2, p. 8.6.
45. G. Burbidge (2003, p. 238).
46. St. John's College Archives 8/3.
47. Hoyle (1994, p. 345).
48. International Computers and Tabulators, ICT. Westwater had a distinguished navy career after the war.
49. I am indebted to John Faulkner for e-mails on June 24, 2004, recounting his interactions with Hoyle.
50. E-mail, John Faulkner to author, July 7, 2004.
51. E-mail, John Faulkner to author, July 26, 2004.
52. Interview, J. Faulkner by J. Mitton, April 11, 2003.
53. J. Faulkner, after-dinner speech delivered June 26, 2002, in Cardiff; reproduced in Wickramasinghe et al. (2003, p. 313).
54. E-mail, Don Clayton to author, May 12, 2004.
55. Interview, J. Faulkner by J. Mitton, April 11, 2003.

CHAPTER 10

1. Hoyle (1994, p. 346).
2. Hoyle (1994, p. 347).
3. The latest edition, 1997, is *Munro's Tables and Other Tables of Lower Hills*, revised by Derek A. Bearhop and published by the Scottish Mountaineering Club. I am indebted to a member of the club, David Purchase, FRAS, for assistance with this section. Letter from D. Purchase to S. Mitton, June 20, 2004.
4. Clayton (1975, p. 164).
5. Hoyle (1994, p. 358).
6. This account is taken from Gascoigne et al. (1990).
7. M. Burbidge (2003, p. 24).
8. Lovell (1991, p. 2).
9. Hoyle (1994, p. 369).
10. Hoyle (1994, pp. 369–370).
11. Lovell (1991) is the source of these events and the supporting committee papers.
12. G. Burbudge (2003, p. 238).
13. Personal recollection of author.
14. Here and in what follows, I am indebted to Professor Don Clayton for reminiscences and personal correspondence, as well as e-mail to me dated April 27, 2004. The events are partly recounted in his autobiography (Clayton, 1975).
15. Clayton (1996).
16. J. Faulkner, after-dinner speech delivered June 26, 2002, in Cardiff. The text is reproduced in *Fred Hoyle's Universe* (Wickramasingh et al., 2003, p. 325).
17. Tony Hewish, in a private conversation with S. Mitton, said he endured a short period of sleepless nights while they made the observations to exclude the extraterrestrial civilization hypothesis.

18. Pilkington et al. (1968).
19. E-mail, Jocelyn Bell Burnell to author, January 5, 2004.
20. Hewish et al. (1968).
21. Ostriker (1968).   .
22. Gold (1993).
23. Gold (1968).
24. Hoyle et al. (1966).
25. Hoyle and Burbidge (1966).
26. Hoyle (1981, p. 24).
27. Hoyle (1981, p. 22).
28. Personal recollection of author, 1968–1971.
29. Hoyle (1981, p. 40).
30. Hoyle (1981, p. 43).
31. Arp (2003, p. 174).
32. Hoyle and Narlikar (1962).

CHAPTER 11

1. Hoyle (1994, p. 373).
2. University Library, University Archives, folder UA/OBSY J.3 (xviii), inspected by author in March 2003. When the University Library learned of the preparation of this biography, this folder was subsequently closed to readers until 2053.
3. Report of the Council of the School of the Physical Sciences to the General Board, dated February 24, 1971, Recommendation VIII.
4. Hoyle (1994, p. 373).
5. Cambridge University Reporter, 1971–1972, p. 78.
6. Letter, B. Lovell to author, May 20, 2004. I am indebted to Sir Bernard Lovell who has felt able to share his memories of the board meetings with me since more than 30 years have elapsed.
7. Hoyle (1994, p. 374). Historians (and conspiracy theorists) of Cambridge politics should note that Hoyle wrongly has (Sir) Brian Pippard on both boards. This error invalidates his arithmetic. The electors appointed to September 30, 1971, were P. Hall, M. Ryle, H. Massey, N. Mott, W. Hodge, B. Lovell, R. Woolley, and F. Hoyle. From October 1, 1971, F. G. Smith and B. Pippard replaced B. Lovell and P. Hall. Hoyle's version of the politics assumes that Lovell and Hall would have supported him, but that Smith and Pippard would not.
8. Hoyle (1994, p. 375).
9. I am grateful to Brian Pippard for this information on the electors. Private conversation, September 3, 2004.
10. The difficulties are described at great length and in detail by Gascoigne et al. (1990, pp. 125–202).
11. Letter, Fred Hoyle to Barbara Hoyle, February 14, 1972. St. John's College Archives 24/3.
12. Letter, Fred Hoyle to the Vice Chancellor, February 14, 1972. St. John's College Archives 24/3.
13. Public announcement signed by Fred Hoyle, February 14, 1972. St. John's College Archives 24/3.

14. Letter, Fred Hoyle to Barbara Hoyle, February 14, 1972. St. John's College Archives 24/3.
15. Public announcement signed by Fred Hoyle, February 14, 1972. St. John's College Archives 24/3.
16. Hoyle (1994, p. 375).
17. *Ibid.*
18. Letter, D. Lynden-Bell to F. Hoyle, February 14, 1972. St. John's College Archives 24/3.
19. St. John's College Archives 1/3, 8.13, and 8.14.
20. I am grateful to the BBC Written Archives Centre for a photocopy of its contributors' cuttings file on F. Hoyle.
21. Letter, Margaret Thatcher to Fred Hoyle, June 29, 1972. St. John's College Archives 24/3.
22. St. John's College Archives 45/5.
23. *Ibid.*

CHAPTER 12

1. Letter, Sir Bernard Lovell to author, May 20, 2004.
2. Hoyle (1973b).
3. The account that follows draws heavily on Gascoigne et al. (1990, pp. 130–165) and Hoyle (1994, pp. 384–391).
4. Hoyle (1982c, p. 23).
5. Willy Fowler said this to Don Clayton. E-mail, Clayton to author, August 12, 2004.
6. Clayton and Hoyle (1974).
7. Clayton (2003, p. 115).
8. E-mail, D. Clayton to author, August 10, 2004.
9. E-mail, Clayton to author, August 16, 2004.
10. *The Times,* London, March 22, 1975.
11. E-mail, Clayton to author, August 16, 2004.
12. I am grateful to Robert Temple, who conducted the interview, for this recollection. I am also grateful to Don Clayton who told me of the impact of the affair on Hoyle; e-mail, Clayton to author, August 16, 2004.
13. Clayton and Hoyle (1976).
14. Hoyle and Wickramasinghe (1962).
15. Clayton (2003b, p. 358).
16. Amari et al. (2001).
17. The account of the Venice meeting is drawn from letters and papers in St. John's College Archives 22/15 and 36/4.
18. Manuscript notes for speech. St. John's College Archives, 22/15.
19. Transcript of interview broadcast on July 30, 1975. St. John's College Archives, 36/4.
20. E-mail, J. Narlikar to author, August 13, 2004.
21. Hoyle and Narlikar (1980).
22. Hoyle (1994, p. 395).

23. The theory is described in detail in *Lifecloud* (Hoyle and Wickramasinghe, 1978). The key papers are reprinted in *Astronomical Origins of Life* (Hoyle and Wickramasinghe, 2000).
24. Hoyle and Wickramasinghe (1979).
25. Hoyle et al. (1985).
26. Hawkins (1963, 1964).
27. Hawkins (1965).
28. Hoyle (1972, p. 25).
29. Hoyle (1966).
30. Hoyle (1977, p. 100).
31. Hoyle et al. (2001, p. 141).
32. Three important fundamental constants—the gravitational constant, the Planck constant, and the velocity of light—can be combined to yield a length known as the Planck length. It is about 10^{-35} meters. A Planck particle has the mass required for the Planck length to equal the radius of a black hole of that mass. The importance to Hoyle, Burbidge, and Narlikar was that Planck particles were (and still are) widely discussed as examples of exotic matter.
33. Hoyle et al. (1993, p. 451).
34. I am indebted to Don Clayton and Jayant Narlikar for a copy of Fred's own account of the misfortune, which he wrote on February 10, 1998, after his recovery, to Don Clayton. There is a further account in the unpublished typescript of Fred's last book (*A Different Approach to the Age of the Earth*), the first two chapters of which describe the accident and its aftermath.
35. From the second chapter of the above-mentioned typescript.
36. The supportive Syndic was the principal of Newnham College, the distinguished philosopher Baroness Onora O'Neill.
37. E-mail, Jacqueline Garget (Cambridge University Press editor) to author, August 23, 2004.
38. Lovell (2001).
39. Wickramasinghe (2001).
40. *The Times* (London), August 22, 2001.
41. Sullivan (2001).
42. Maddox (2001).
43. Rees (2001).
44. Wickramasinghe et al. (2003).

Bibliography

Adams, W. S. 1941. Some results with the Coudé spectrograph of the Mount Wilson Observatory. *Astrophysical Journal* 93:11.

Alpher, R. 1990. Early work on big-bang cosmology and the cosmic blackbody radiation. In *Modern Cosmology in Retrospect*, B. Bertotti, R. Balbinot, S. Bergia, and A. Messina, eds. Cambridge, UK: Cambridge University Press.

Alpher, R. A., and R. C. Herman. 1950. Theory of origin and relative abundance distribution of the elements. *Reviews of Modern Physics* 22:153.

Alpher, R. A., H. Bethe, and G. Gamow. 1948. The origin of the chemical elements. *Physical Review* 73:803.

Amari, S., X. Gao, L. R. Nittler, E. Zinner, J. José, M. Hernanz, and R. S. Lewis. 2001. Presolar grains from novae. *Astrophysical Journal* 551:1065.

Arp, H. 2003. Research with Fred. In *Fred Hoyle's Universe*, C. Wickramasinghe, G. Burbidge, and J. Narlikar, eds. Dordrecht, The Netherlands: Kluwer Academic.

Atkinson, R. d'E. 1940a. Accretion and stellar energy. *Monthly Notices of the Royal Astronomical Society* 100:500.

Atkinson, R. d'E. 1940b. On the capture of interstellar matter by stars. *Proceedings of the Cambridge Philosophical Society* 36:314.

Baade, W. 1944a. The resolution of Messier 32, NGC 205, and the central region Andromeda Nebula. *Astrophysical Journal* 100:137.

Baade, W. 1944b. NGC 147 and NGC 185, two new members of the local group of galaxies. *Astrophysical Journal* 100:147.

Bernstein, J. 1984. *Three Degrees Above Zero*. New York: Scribner.

Bethe, H. A. 1939. Energy production in stars. *Physical Review* 55:434.

Bethe, H. A., and R. F. Bacher. 1936. Stationary states of nuclei. *Reviews of Modern Physics* 8:82–229.

Bethe, H. A., F. Hoyle, and R. Peierls. 1939. Interpretation of beta-disintegration data. *Nature* 143:200.

Bondi, H. 1948a. Observation and theory in cosmology. *The Observatory* 68:111.

Bondi, H. 1948b. Review of cosmology. *Monthly Notices of the Royal Astronomical Society* 108:104.

Bondi, H. 1952. *Cosmology* (Cambridge Monographs on Physics). Cambridge, UK: Cambridge University Press.

Bondi, H. 1990a. *Science, Churchill and Me.* Oxford: Pergamon Press.

Bondi, H. 1990b. The cosmological scene 1945–1952. In *Modern Cosmology in Retrospect*, B. Bertotti, R. Balbinot, S. Bergia, and A. Messina, eds. Cambridge, UK: Cambridge University Press.

Bondi, H., and T. Gold. 1948. The steady-state theory of the expanding universe. *Monthly Notices of the Royal Astronomical Society* 108:252.

Bondi, H., and F. Hoyle. 1944. On the mechanism of accretion by stars. *Monthly Notices of the Royal Astronomical Society* 104:273.

Bondi, H., F. Hoyle, and R. A. Lyttleton. 1947. On the structure of the solar corona and chromosphere. *Monthly Notices of the Royal Astronomical Society* 107:184.

Burbidge, G. 1959. Estimates of the total energy in particles and magnetic field in non-thermal radio sources. *Astrophysical Journal* 129:849.

Burbidge, G. 1961. Galactic explosions as sources of radio emission. *Nature* 190:1053.

Burbidge, G. 2003. Sir Fred Hoyle. *Biographical Memoirs of Fellows of the Royal Society*, 49:215.

Burbidge, M. 2003. Fred Hoyle and the Anglo-Australian Telescope. In *Fred Hoyle's Universe*, C. Wickramasinghe, G. Burbidge, and J. Narlikar, eds. Dordrecht, The Netherlands: Kluwer Academic.

Burbidge, G. R., and E. M. Burbidge. 1954. A spectrophotometric study of $_-^2$Cvn. *Astrophysical Journal* 59:318.

Burbidge, G. R., and E. M. Burbidge. 1955. An analysis of the magnetic variable $_-^2$ Canum, Venaticorum. *Astrophysical Journal Supplement Series* 1:431.

Burbidge, E. M., G. R. Burbidge, W. A. Fowler, and F. Hoyle. 1957. Synthesis of the elements in stars. *Reviews of Modern Physics* 29:547.

Cameron, A. G. W. 1955. Origin of anomalous abundances of the elements in giant stars. *Astrophysical Journal* 121:144.

Cameron, A. G. W. 1999. Adventures in cosmogony. *Annual Review of Astronomy and Astrophysics* 37:1.

Carpenter, H. 1996. *The Envy of the World: Fifty Years of the BBC Third Programme and Radio 3, 1946–1996*, London: Weidenfeld and Nicolson.

Carruthers, G. R. 1970. Rocket observation of interstellar molecular hydrogen. *Astrophysical Journal* 161:L81.

Chown, M. 1999. *The Magic Furnace.* London: Jonathan Cape.

Christianson, G. E. 1997. *Edwin Hubble.* Bristol: Institute of Physics Publishing.

Clayton, D. 1975. *The Dark Night Sky.* New York: Quadrangle/New York Times Books.

Clayton, D. D. 1996. Obituary of William Alfred Fowler. *Proceedings of the Astronomical Society Pacific* 108:1.

Clayton, D. D. 1999. Radiogenic iron. *Meteoritics and Planetary Science* 34:A145.

Clayton, D. 2003a. *Isotopes in the Cosmos.* Cambridge, UK: Cambridge University Press.

Clayton, D. D. 2003b. Novae as thermonuclear laboratories. In *Fred Hoyle's Universe*, C. Wickramasinghe, G. Burbidge, and J. Narlikar, eds. Dordrecht, The Netherlands: Kluwer Academic.

Clayton, D. D., and F. Hoyle. 1974. Gamma-ray lines from novae. *Astrophysical Journal* 187:L101.

Clayton, D. D., and F. Hoyle. 1976. Grains of anomalous isotopic composition from novae. *Astrophysical Journal* 203:490.

Cohen, R. C. 1951. The nature of the universe. *Christian Science Monitor*. April 3, p. 9.

Dalton, J. 1808. *A New System of Chemical Philosophy*. London: John Weale.

Davidson, K., and J. P. Ostriker. 1973. Neutron-star accretion. *Astrophysical Journal* 179:591.

Dingle, H. 1953. Science and modern cosmology. *Monthly Notices of the Royal Astronomical Society* 113:393.

Dingle, H. 1968. The case against the special theory of relativity. *Nature* 217:19.

Dirac, P. A. M. 1931. Quantised singularities in the electromagnetic field. *Proceedings of the Royal Society A*, 133:60.

Dirac, P. A. M. 1937. The cosmological constants. *Nature* 139:323.

Dodd, K. N. 1952. The unsteady accretion problem. *Monthly Notices of the Royal Astronomical Society* 112:374.

Dunham, T. 1939. The material of interstellar space. *Proceedings of the American Philosophical Society* 81:277.

Eddington, A. S. 1924. On the relation between the masses and luminosities of the stars. *Monthly Notices of the Royal Astronomical Society* 84:308.

Eddington, A. S. 1926. *The Internal Constitution of the Stars*. Cambridge: Cambridge University Press.

Eddington, A. S. 1927. *Stars and Atoms*. London: Oxford University Press.

Eddington, A. S. 1935. *New Pathways in Science*. Cambridge, UK: Cambridge University Press.

Eddington, A. S. 1937. Interstellar matter. *The Observatory* 60:99.

Edge, D. O., and M. J. Mulkay. 1976. *Astronomy Transformed*. New York: John Wiley.

Ewen, H. I., and E. M. Purcell. 1951. Observation of a line in the galactic radio spectrum. *Nature* 168:350.

Edmondson, F. K. 1951. The nature of the universe. *Sky and Telescope* 10:273.

Fermi, E. 1934. Versuch einer Theorie der beta–strahlen. *Zeitschrift für Physik* 88:161.

Ferris, T. 1979. *The Red Limit*. London: Corgi Books.

Fowler, W. A., and F. Hoyle. 1964. Neutrino processes and pair formation in massive stars and supernovae. *Astrophysical Journal Supplement Series* 9:201.

Gamow, G. 1942. Concerning the origin of the chemical elements. *Journal of the Washington Academy of Sciences* 32:353.

Gamow, G. 1946. Expanding universe and the origin of elements. *Physical Review* 70:572.

Gamow, G. 1952. *Creation of the Universe*. New York: Viking.

Gamow, G. 1965. *Mr. Tompkins in Paperback*. Cambridge, UK: Cambridge University Press.

Gascoigne, S. C., K. M. Proust, and M. O. Robins. 1990. *The Creation of the Anglo-Australian Observatory*. Cambridge, UK: Cambridge University Press.

Gold, T. 1968. Rotating neutron stars as the origin of the pulsating radio sources. *Nature* 218:731.

Gold, T. 1993. The nature of pulsars. *Current Contents*, February 22, p. 8.

Guth, A. H. 1981. Inflationary universe: A possible solution to the horizon and flatness problems. *Physical Review D* 23:247.

Haselgrove, C. B., and F. Hoyle. 1956. A mathematical discussion of the problem of stellar evolution, with reference to the use of an automatic digital computer. *Monthly Notices of the Royal Astronomical Society* 116:515.

Haselgrove, C. B., and F. Hoyle. 1959. Main sequence stars. *Monthly Notices of the Royal Astronomical Society* 119:112.

Hawkins, G. S. 1963. Stonehenge decoded. *Nature*, 200:306.

Hawkins, G. S. 1964. Stonehenge as an astronomical observatory. *Nature* 202:1258.

Hawkins, G. S. 1965. *Stonehenge Decoded*. New York: Doubleday (London: Souvenir Press, 1966).

Hendry, J. 1984. In *Cambridge Physics in the Thirties*, J. Hendry, ed. Bristol: Adam Hilger.

Hewish, A., S. J. Bell, J. D. H. Pilkington, P. F. Scott, and R. A. Collins. 1968. Observation of a rapidly pulsating radio source. *Nature* 217:709.

Hey, J. S. 1971. *The Radio Universe*. Oxford: Pergamon Press.

Hill, E. R., and B. Y. Mills. 1962. Source corrections to the Sydney radio source survey catalogue. *Australian Journal of Physics* 15:437.

Howse, D. 1993. *Radar at Sea*. London: Macmillan Press.

Hoyle, F. 1937a. Capture of orbital electrons. *Nature* 140:235.

Hoyle, F. 1937b.The generalized Fermi interaction. *Proceedings of the Cambridge Philosophical Society* 33:277.

Hoyle, F. 1939a. Quantum electrodynamics I. *Proceedings of the Cambridge Philosophical Society* 35:419.

Hoyle, F. 1939b. Quantum electrodynamics II. *Proceedings of the Cambridge Philosophical Society* 35:438.

Hoyle, F. 1946. The synthesis of the elements from hydrogen. *Monthly Notices of the Royal Astronomical Society* 106:343.

Hoyle, F. 1948. A new model for the expanding universe. *Monthly Notices of the Royal Astronomical Society* 108:372.

Hoyle, F. 1949a. On the cosmological problem. *Monthly Notices of the Royal Astronomical Society* 109:365.

Hoyle, F. 1949b. *Some Recent Researches in Solar Physics*. Cambridge, UK: Cambridge University Press.

Hoyle, F. 1950. *The Nature of the Universe*. Oxford: Basil Blackwell.

Hoyle, F. 1952. Report to Commission. Transactions of the International Astronomical Union VIII. 28:397.

Hoyle, F. 1953. *A Decade of Decision*. London: Heinemann.

Hoyle, F. 1954. I. The synthesis of elements from carbon to nickel. *Astrophysical Journal Supplement Series* 1:121.

Hoyle, F. 1955. *Frontiers of Astronomy*. London: Heinemann.

Hoyle, F. 1957. *The Black Cloud*. London: Heinemann.

Hoyle, F. 1959. Machine calculations of main sequence stars. In *The Hertzsprung-Russell Diagram: Proceedings from IAU Symposium No. 10, Moscow, August 15–16, 1958*, J. L. Greenstein, ed. Published in *Annales d'Astrophysique Supplements*, Section 8. 1959.

Hoyle, F. 1962. *A for Andromeda*. New York: Harper & Brothers.

Hoyle, F. 1966. Stonehenge—an eclipse predictor. *Nature* 211(1966), p. 454.

Hoyle, F. 1968. *Encounter with the Future*. New York: Simon and Schuster.

Hoyle, F. 1972. *From Stonehenge to Modern Cosmology*. San Francisco: W. H. Freeman.

Hoyle, F. 1973a. *Nicolaus Copernicus*. London: Heinemann.

Hoyle, F. 1973b. Presidential Address: The origin of the universe. *Quarterly Journal of the Royal Astronomical Society* 14:278.

Hoyle, F. 1977. *On Stonehenge*. London: Heinemann.

Hoyle, F. 1981a. *Ice: The ultimate human catastrophe*. New York: Continuum International.

Hoyle, F. 1981b. *The Quasar Controversy Resolved*. Cardiff, Wales: University College Cardiff Press.

Hoyle, F. 1982a. The universe: Past and present reflections. *Annual Review of Astronomy and Astrophysics* 20:1–36.

Hoyle, F. 1982b. Two decades of collaboration with Willy Fowler. In *Essays in Nuclear Astrophysics*, C. A. Barnes, D. D. Clayton, and D. N. Schramm, eds. Cambridge, UK: Cambridge University Press.

Hoyle, F. 1982c. *The Anglo-Australian Telescope*. Cardiff, Wales: University College Cardiff Press.

Hoyle, F. 1986. *The Small World of Fred Hoyle: An Autobiography*. London: Michael Joseph.

Hoyle, F. 1994. *Home Is Where the Wind Blows: Chapters from a Cosmologist's Life*. Berkeley, Calif.: University Science Books.

Hoyle, F., and G. R. Burbidge. 1966. On the nature of quasi-stellar objects. *Astrophysical Journal* 144:534.

Hoyle, F., and J. Elliot. 1962. *A for Andromeda: A Novel for Tomorrow*. New York: Harper & Brothers.

Hoyle, F., and W. A. Fowler. 1960. Nucleosynthesis in supernovae. *Astrophysical Journal* 132:565.

Hoyle, F., and W. A. Fowler. 1963a. On the nature of strong radio sources. *Nature* 197:533.

Hoyle, F., and W. A. Fowler. 1963b. On the nature of strong radio sources. *Monthly Notices of the Royal Astronomical Society* 125:169.

Hoyle, F., and G. Hoyle. 1973. *The Inferno*. New York: Harper & Row.

Hoyle, F., and G. Hoyle. 1974. *Into Deepest Space*. New York: Harper & Row.

Hoyle, F., and G. Hoyle. 1977. *The Incandescent Ones*. New York: Harper & Row.

Hoyle, F., and G. Hoyle. 1978. *The Westminster Disaster*. New York: Harper & Row.

Hoyle, F., and G. Hoyle. 1982. *The Frozen Planet of Azuron*. London: Ladybird Books.

Hoyle, F., and R. A. Lyttleton. 1939a. The effect of interstellar matter on climate variation. *Proceedings of the Cambridge Philosophical Society* 35:405.

Hoyle, F., and R. A. Lyttleton. 1939b. The evolution of the stars. *Proceedings of the Cambridge Philosophical Society* 35:592.

Hoyle, F., and R. A. Lyttleton. 1940. On the physical aspects of accretion by stars. *Proceedings of the Cambridge Philosophical Society* 36:424.

Hoyle, F., and R. A. Lyttleton. 1941. On the accretion theory of stellar evolution. *Monthly Notices of the Royal Astronomical Society* 101:227.

Hoyle, F., and R. A. Lyttleton. 1942a. On the internal constitution of the stars. *Monthly Notices of the Royal Astronomical Society* 102:177.

Hoyle, F., and R. A. Lyttleton. 1942b. On the nature of red giant stars. *Monthly Notices of the Royal Astronomical Society* 102:218.

Hoyle, F., and R. A. Lyttleton. 1943. The theory of Cepheid variables and novae. *Monthly Notices of the Royal Astronomical Society* 103:21.

Hoyle, F., and J. V. Narlikar. 1962. On the counting of radio sources in the steady-state cosmology. *Monthly Notices of the Royal Astronomical Society* 125:13.

Hoyle, F., and J. Narlikar. 1980. *The Physics-Astronomy Frontier.* San Francisco: W. H. Freeman.

Hoyle, F., and M. Schwarzschild. 1955. On the evolution of type II stars. *Astrophysical Journal, Supplements Series* 2:1.

Hoyle, F., and N. C. Wickramasinghe. 1962. On graphite particles as interstellar grains. *Monthly Notices of the Royal Astronomical Society* 124:417.

Hoyle, F., and N. C. Wickramasinghe. 1978. *Lifecloud.* London: J. M. Dent.

Hoyle, F., and N. C. Wickramasinghe. 1979. *Diseases from Space.* London: J. M. Dent.

Hoyle, F., and N. C. Wickramasinghe. 1986. *Archaeopteryx, the Primordial Bird: A Case of Fossil Forgery.* London: Christopher Davies.

Hoyle, F., and N. C. Wickramasinghe. 2000. *Astronomical Origins of Life.* Dordrecht, The Netherlands: Kluwer Academic.

Hoyle, F., W. A. Fowler, G. R. Burbidge, and E. M. Burbidge. 1956. Origin of the elements in stars. *Astrophysical Journal* 124:611.

Hoyle, F., G. Burbidge, and W. L. W. Sargent. 1966. On the nature of quasi-stellar objects. *Nature* 209:751.

Hoyle., F., N. C. Wickramasinghe, and R. S. Watkins. 1985. Archaeopteryx. Problems arise—and a motive. *British Journal of Photography* 132:693.

Hoyle, F., G. Burbidge, and J. V. Narlikar. 1993. A quasi-steady state cosmological model with creation of matter. *Astrophysical Journal* 410:437.

Hoyle, F., G. Burbidge, and J. Narlikar. 2000. *A Different Approach to Cosmology—from a Static Universe Through the Big Bang Towards Reality.* Cambridge, UK: Cambridge University Press.

Jeans, J. 1928. *Astronomy and Cosmogony.* Cambridge, UK: Cambridge University Press, p. 352.

Jeffreys, H. 1929. *The Earth.* Cambridge, UK: Cambridge University Press.

Knox, K. C., and R. Noakes. 2003. *From Newton to Hawking.* Cambridge, UK: Cambridge University Press.

Kragh, H. S. 1990. *Dirac: A Scientific Biography.* Cambridge, UK: Cambridge University Press.

Kragh, H. 1996. *Cosmology and Controversy.* Princeton, N.J.: Princeton University Press.

Leedham-Green, E. 1996. *A Concise History of the University of Cambridge.* Cambridge, UK: Cambridge University Press.

Longair, M. S. 2004. A brief history of cosmology. In *Measuring and Modeling the Universe,* W. L. Freedman, ed. Cambridge, UK: Cambridge University Press.

Lovell, A. C. B. 1951. The new science of radio astronomy. *Nature* 167:96.

Lovell, B. 1985. Obituary of Martin Ryle. *Quarterly Journal of the Royal Astronomical Society* 26:358.

Lovell, B. 1991. The genesis of the Northern Hemisphere Observatory. *Quarterly Journal of the Royal Astronomical Society* 32:1.

Lovell, B. 2001. Professor Sir Fred Hoyle. *The Guardian,* August 23.

Lyttleton, R. A. 1936. On the origin of the solar system. *Monthly Notices of the Royal Astronomical Society* 96:559.

Lyttleton, R. A. 1953. *The Origin and Nature of Comets.* Cambridge, UK: Cambridge University Press.

Maddox, J. 2001. Obituary: Fred Hoyle (1915–2001). *Nature* 413:270.

McCrea, W. H. 1929. The hydrogen chromosphere. *Monthly Notices of the Royal Astronomical Society* 89:483.

McCrea, W. H. 1953. The rate of accretion of matter by stars. *Monthly Notices of the Royal Astronomical Society* 113:162.

Mendeleyev, D. 1891. *The Principles of Chemistry.* London: Longmans, Green.

Merrill, P. W. 1952. Technetium in the stars. *Science* 115:484.

Mestel, L. 1954. The influence of stellar radiation on the rate of accretion. *Monthly Notices of the Royal Astronomical Society* 114:437.

Mills, B. Y., and O. B. Slee. 1957. A preliminary survey of radio sources in a limited region of the sky at a wavelength of 3.5 m. *Australian Journal of Physics* 10:162.

Mitton, S. 1976. *Exploring the Galaxies.* New York: Scribner.

Mitton, S. 1979. *The Crab Nebula.* New York: Scribner.

Mitton, S., and J. Mitton. 1972. The emission line spectrum of Cygnus A. *Monthly Notices of the Royal Astronomical Society* 158:245.

O'Brien, M. C. 2003. *The Same Age as the State.* Dublin: O'Brien Press.

O'Connell, D. 1953. According to Hoyle. *Irish Astronomical Journal* 2:127.

Osterbrock, D. 2001. *Walter Baade: A Life in Astrophysics.* Princeton, N.J.: Princeton University Press.

Ostriker, J. P. 1968. A possible model for a rapidly pulsating radio source. *Nature* 217:1227.

Pawsey, J. 1957. Preliminary statistics of discrete sources obtained with the Mills Cross. In *Radio Astronomy,* H. C. van de Hulst, ed. Cambridge, UK: Cambridge University Press.

Payne-Gaposchkin, C. The nature of the universe. *Herald Tribune,* May 13, 1951, p. 22.

Peierls, R. 1984. Pp. 195–200 in *Cambridge Physics in the Thirties,* J. Hendry, ed. Bristol: Adam Hilger.

Peierls, R. 1985. *Bird of Passage.* Princeton, N.J.: Princeton University Press.

Penzias, A. A., and R. W. Wilson. 1965. A Measurement of Excess Antenna Temperature at 4080 Mc/s. *Astrophysical Journal* 133:355.

Perlmutter, S., et al. 1999. Measurements of omega and lambda from 42 high-redshift supernovae. *Astrophysical Journal* 517:565.

Pfahl, E., and S. Rappaport. 2001. Bondi-Hoyle-Lyttleton accretion model for low-luminosity x-ray sources in globular clusters. *Astrophysical Journal* 550:172.

Pilkington, J. D. H., A. Hewish, S. J. Bell, and T. W. Cole. 1968. Observations of some further pulsed radio sources. *Nature* 218:126.

Pontificiae Academia Scientiarvm (Pontifical Academy of Sciences, Vatican). 1958. Study week on the problem of stellar population, May 20–28, 1957. *Scripta Varia,* pp. lxvii–551.

Priestley, J. B. 1933. *English Journey.* London: Heinemann.

Rees, M. 2001. Fred Hoyle. *Physics Today* 45(November):75.

Roll, P. G., and D. T. Wilkinson. 1966. Cosmic background radiation at 3.2 cm—support for cosmic black-body radiation. *Physical Review Letters* 16:405.

Ryle, M. 1948. The generation of radio-frequency radiation in the Sun. *Proceedings of the Royal Society of London, Series* A 195:82.

Ryle, M. 1950. Radio astronomy. *Reports on Progress in Physics* 13:237.

Ryle, M. 1952. A new radio interferometer and its application to the observation of weak radio stars. *Proceedings of the Royal Society of London Series* A 211:351.

Ryle, M. 1955. Radio stars and their cosmological significance. *The Observatory* 75:137.

Ryle, M. 1958. The nature of cosmic radio sources. *Proceedings of the Royal Society of London Series* A 248:289.

Ryle, M., N. Kurti, and R. L. F. Boyd. 1971. *Search and Research.* London: Mullard.

Robertson, H. P. 1933. Relativistic cosmology. *Reviews of Modern Physics* 5:62.

Russell, H. N. 1929. On the composition of the Sun's atmosphere. *Astrophysical Journal* 70:11–82.

Rutherford, E. 1904. *Radioactivity and Radioactive Substances.* Cambridge, UK: Cambridge University Press.

Sadler, D. H. 1987. In *History of the Royal Astronomical Society*, vol. 2, R. J. Tayler, ed. Oxford: Blackwell Scientific.

Salpeter, E. E. 1952. Nuclear reactions in stars without hydrogen. *Astrophysical Journal* 115:326.

Schmidt, M. 1963. 3C 273: A star-like object with a large redshift. *Nature* 197:1040.

Schmidt, M. 1990. The discovery of quasars. In *Modern Cosmology in Retrospect*, B. Bertotti, R. Balbinot, S. Bergia, and A. Messina, eds. Cambridge, UK: Cambridge University Press.

Smith, F. G. 1987. Sir Martin Ryle. *Biographical Memoirs of Fellows of the Royal Society* 32:497.

Spear, R. 2002. The most important experiment ever performed by an Australian physicist. *The Physicist* 39:35.

Strömgren, B. 1983. Scientists I have known. *Annual Review of Astronomy and Astrophysics* 21:3.

Stubbings, F. 1995. *Bedders, Bulldogs, and Bedells: A Cambridge Glossary.* Cambridge, UK: Cambridge University Press.

Suess, H. E., and H. C. Urey. 1956. Abundances of the elements. *Reviews of Modern Physics* 28:53.

Sullivan, W. 2001. Fred Hoyle dies at 86; opposed 'big bang' but named it. *New York Times.* August 22, p. C15.

Tayler, R. J., ed. 1987. *History of the Royal Astronomical Society*, vol. 2. Oxford: Blackwell Scientific.

Thackeray, A. D. 1941. Interstellar molecules. *The Observatory* 64:57.

Unsöld, A. 1929. A spectroscopic determination of the pressure in the calcium chromosphere. *Astrophysical Journal* 69:209.

van de Hulst, H. C. 1945. *Nederlands Tijdschrift voor Natuurkunde* 11:210 (in Dutch).

van de Hulst, H. C., ed. 1957. *Radio Astronomy.* Cambridge, UK: Cambridge University Press.

White, M., and J. Gribbin. 1992. *Stephen Hawking.* London: Viking.

Wickramasinghe, C. 2001. Obituary: Professor Sir Fred Hoyle. *The Independent.* August 22.

Wickramasinghe, C., G. Burbidge, and J. Narlikar, eds. 2003. *Fred Hoyle's Universe*. Dordrecht, The Netherlands: Kluwer Academic.

Williamson, R. E. 1951. Fred Hoyle's universe. *Journal of the Royal Astronomical Society of Canada* 45:185.

Wilson, D. 1983. *Rutherford: Simple Genius*. London: Hodder and Stoughton.

Young, G. W. 1899. *The Roof Climbers Guide to Trinity*. Cambridge, UK: W. P. Spalding.

Acknowledgments

I am going to dispense with the normal convention that demands that the author's spouse or partner takes the final bow in the acknowledgments in a context of "last but not least." Therefore, I first thank Jacqueline Mitton, without whose constant support and encouragement the writing of this book would have spread over several years. We had a great many protracted discussions about my exciting findings in archives, my interviews with Hoyle's colleagues, and chance remarks picked up at scientific meetings. We made several visits together, most memorably to Fred's place of birth in West Yorkshire. We will always treasure our experiences of having worked in several of the institutions described in this book: Caltech and the Mount Wilson Observatory, the Royal Greenwich Observatory, the Cavendish Laboratory, the Cambridge Observatories, and the Institute of Theoretical Astronomy. We count as close colleagues numerous former associates and students of Hoyle's. And we have a joint interest in Hoyle's contributions to nuclear astrophysics. Together, we came to appreciate the places, the people, and the science that enrich this narrative. Furthermore, Jacqueline is a highly experienced author and editor who made invaluable suggestions for improving the drafts.

To write this book, I have been heavily dependent on expertise and knowledge of literally hundreds of scholars and scientists. On one occasion when I was working in the vastness of the Reading Room in University Library at Cambridge, I heard a guide offer the following explanation of the silent activity of the researchers: "These academics are reading the books we already own so they can write the books we will need in the future." Where I have drawn on the work of earlier writers, I have given the source and acknowledgment in the Notes and Bibliography.

I owe a debt of gratitude to the master and fellows of St. Edmund's College, Cambridge, for providing me with the facilities to conduct this research at the college, where I learned a little of the techniques of the academic historian from Michael Robson and Christopher Catherwood. The staff of Cambridge University Library were unfailingly helpful in locating source material. The facilities of the library itself are unmatched anywhere in the world for a project of this kind because most of the books I consulted are on open access. I would like to express my deep appreciation for the support of Jonathan Harrison who is the special collections librarian at St. John's College. Hoyle's personal papers are archived by the college, and Harrison extracted all the documents I needed to study. I thank the master and fellows of St. John's College for permission to quote from the papers. I would also like to thank the Archives Centre at Churchill College, Cambridge; the BBC Written Archives Centre, Caversham; Peter Hingley of the Royal Astronomical Society Library; the Oral History Archives of the Neils Bohr Library at the American Institute of Physics; and the archives center of the California Institute of Technology. I made extensive online use of the NASA Astrophysics Data Service, hosted by the Harvard-Smithsonian Center for Astrophysics. I would like to thank Guenther Eichhorn and his staff for the high quality of this invaluable tool for bibliographic research.

Several individuals contributed completely new material for this project, either by allowing me to record interviews or by sharing with me their autobiographical accounts of working with Fred Hoyle. For much invaluable help of this kind, I thank Leigh Bailey, Jocelyn Bell Burnell, Hermann Bondi, E. Margaret Burbidge, Geoffrey Burbidge, Al

Cameron, Donald Clayton, John Dougherty, Peter Eggleton, Bruce Elsmore, John Faulkner, Francis Graham-Smith, Geoffrey Hoyle, Joan Hoyle, Malcolm Longair, Bernard Lovell, Donald Lynden-Bell, Leon Mestel, Jayant Narlikar, Brian Pippard, Ed Salpeter, Ray Spear, Nigel Weiss, Ward Whaling, and Joyce Wheeler. This is also the place to place on record the passion and enthusiasm many of these individuals have for Fred Hoyle, which meant that they wanted to help me provide an accurate portrait. As Ward Whaling put it, "After you've gone to all this trouble you might as well let me help you get it right," which beautifully expresses the sentiments of many of my advisors.

Draft chapters were read and corrected by several colleagues. I am particularly grateful to Leon Mestel for improving my version of Hoyle's contributions to stellar evolution; to John Faulkner for helping me to get the nuance right in my descriptions of life at the Institute of Theoretical Astronomy; to Bernard Lovell for his assistance on Hoyle's achievements as an academic administrator; to Ward Whaling who improved my account of a crucial nuclear physics experiment at Caltech; to the Burbidges for helpful guidance on my chapter on the origin of the chemical elements; and to Michael Hoskin who spotted a glaring blunder. Donald Lynden-Bell and Jeremiah Ostriker read the final draft in its entirety, providing numerous corrections and improvements.

My literary agent Sara Menguc made important contributions during the genesis of the project through her critique of the first outline. I give my thanks to Chris Clare who transcribed several of the taped interviews and handled correspondence and administrative details. Jeffrey Robbins of the Joseph Henry Press gave wonderful support when the project started to run late for reasons outside my control. The final text has benefited hugely from his meticulous editing. I do feel very privileged to have had this opportunity of working with Jeffrey. The copy editor, Julie Phillips, did a magnificent job in reorganizing the notes and producing the bibliography. I also thank the production editor, Heather Schofield, for seeing the typescript through the press.

Another literary convention requires me to say that all the remaining mistakes are my own; clearly they are. My account is not complete; it would be impossible to recount every one of Hoyle's achievements as well as many colorful details from his everyday life. My references are not intended to be an exhaustive bibliography. I have refrained from giving a complete list of Hoyle's publications because that information can be retrieved from the Royal Society Library or online data services.

Index